Electronic Circuits
and Systems

Electronic Circuits and Systems

Robert King
Imperial College of
Science and Technology

A Halsted Press Book

John Wiley & Sons
New York

To
Gillian and Moti

Published in the USA
by Halsted Press, a Division
of John Wiley & Sons Inc, New York

First published in Great Britain by
Thomas Nelson and Sons Ltd, 1975
Copyright © R. A. King, 1975

Library of Congress Cataloging in Publication Data

King, Robert Ashford. Electronic circuits and systems.
'A Halsted Press book.'
Includes bibliographies
1. Electronic circuits. 2. Semiconductors.
3. Electronic apparatus and appliances.
I. Title.

TK7867.K56   1975   621.3815′3   74–31175
ISBN 0–470–47779–2

Printed in Great Britain

Contents

Preface

During the second half of the twentieth century we have, in the industrialized part of the world, become accustomed to the widespread application of electronics. This application ranges over a very wide field, from large scale uses in industry to simple domestic apparatus.

Probably the earliest application of electronics was in communication and it is here that we still see it making its first impact on the developing nations. From these beginnings a number of other important fields have developed, notably in control engineering and in the application to the construction of computers.

The initial discovery which allowed electrical engineers to start along the path from which all these systems have developed was the discovery of a device which could amplify the power in the fluctuating signal conveying a message. This device was the thermionic triode. Until that time all communication was restricted to relatively short distances; the initial transmitted power in the signal needed to be so large that, despite attenuation in the communication medium, sufficient power was received to render the message comprehensible.

The discovery of the triode valve permitted the distances between transmitter and receiver to be greatly increased since the received signal could be amplified many times at the receiver to render it audible. Alternatively amplifiers could be spaced at intervals along the communication channel to raise the power level as soon as it had fallen below an acceptable value.

These general principles could be applied to the transmission of speech signals or telegraphy signals along a telephone line or to the transmission of radio signals through space. Developments from the experience gained in the transmission of speech led to the design of systems which could transmit visual images, both stationary and moving, and still more recently to the transmission of colour pictures. Although the form of the message to be transmitted varies widely from colour television signals to computer data, the underlying principles are common to all.

In further studying the properties of amplifiers, it was soon realized that signals could not be amplified indefinitely. If the power was allowed to fall to too low a level before amplification, it was observed that the signal message had disappeared completely into a background 'fuzz' of random voltage variations known as noise; these were shown to be a fundamental property of nature and hence imposed a severe constraint on message transmission.

It was also appreciated that although speech signals having frequencies ranging from about 300 Hz to 5 kHz could satisfactorily be transmitted over short distances, over much larger distances the attenuation became excessive and it was found desirable to shift the frequency band to a higher position in the spectrum. This also permitted the transmission of many more messages simultaneously over the same channel. The same device, the triode, as well as the crystal discovered earlier, was capable of being used to change the frequency of a signal; in this application the triode was no longer used in a linear manner (amplifying the input signal without distortion).

During the 1930s, study was carried out on the effect on the performance of a system of 'feeding back' some of the output to the input. A number of improvements in performance were observed and also a number of ancillary effects were noticed. This important side issue developed into the complete science of control engineering. Frequently one requires to maintain the level of some variable at a fixed value despite environmental fluctuations. This may be achieved by comparing the output value of the variable with the desired value and using the error, after suitable amplification, to adjust the output closer to the desired value. In a great many such cases the power which it was desired to control was very large and thus the power capacity of valves was exploited. The ability of valves to handle large powers had already been established by the construction of high power radio transmitters to communicate over large areas.

One other property of the thermionic valves was their ability to act as a switch, changing from an open circuit to an approximate short circuit by the application of an external voltage. This could be used to control large electrical powers, but more importantly it provided the basis for a two stage device, one in which the output is either in a high state or a low state. This ultimately led to the development of the earliest digital computer.

Until about 1950 the thermionic valve was the principal device used in electronic circuits, although many modifications had been made to the original triode valve. At about that time the transistor became commercially available, in the first instance, as a point contact device, quickly to be superseded by the junction transistor. Since then other semiconductor devices have been produced, some of which have become firmly established as useful units and others which have been obsolescent almost as soon as they reached the point of commercial exploitation. The advent of semiconductor devices has not changed the overall principles of electronic circuits although the detailed circuit design has of necessity been modified. The small physical size has given the circuit designer considerably more scope in the fabrication of large and complex circuits. This is particularly noticeable in the construction of modern computers; the vast numbers of electronic devices used would have rendered these physically unrealizable using thermionic valves.

A further development from the transistor is the integrated circuit which embodies a large number of individual transistors and other circuit components constructed on a single slice of semiconductor material. The principles of operation do not represent a radical departure from those of discrete transistors but they again open a new dimension to the circuit designer in that the density of electronic devices is again an order of magnitude greater and also that his design problems move one stage further from the problems associated with the construction of circuits using discrete transistors to those relating to the interconnection of a number of integrated circuits.

In order therefore to undertake a first study of electronic circuits, it is necessary initially to consider the principal types of device which are available to the circuit designer, and then to study the application of these devices in various circuits and their use in some simple systems. The devices which are of interest at present are principally bipolar and field-effect transistors and their use in integrated circuits, although thermionic valves are also used in a limited field. As we are primarily interested in circuits we shall restrict our study of the construction and mechanism of devices to the minimum which will enable a satisfactory model to be constructed for the device and will allow us to appreciate under what conditions the chosen model will act as a valid representation of the device.

The circuits which we shall study will be selected from the vast number which are available in order to illustrate some of the fundamental principles of electronic devices. In a large class of applications, as we have seen, an amplifier is required which gives an output which is a faithful reproduction of the input; this is the class of linear amplifiers. In certain other applications we desire a device with a known or controllable degree of nonlinearity, so that the signal may be distorted in some predictable manner. Finally the device may be used as a switch between an *on* and an *off* state.

The above types of circuits reflect very closely the needs of the systems which we discussed earlier; a communication network principally employs linear amplifiers, although for modulators and frequency changes nonlinear devices are required. A control system is, in its simple form, essentially a linear system in which feedback is a fundamental factor; it may also be used as an example of a relatively high power application of semiconductor devices. An extreme use of semiconductor devices as switches (or digital circuits) is provided by a digital computer; this usually incorporates composite packaged digital units as integrated circuits in order to increase the capacity of a computer without affecting its physical size.

Although the above mentioned applications are now relevant, any study of electronic circuits must be sufficiently open-ended to permit any new device to be readily incorporated in the general theory and to allow the circuits studied to be extended to provide for the requirements of any new system which may be developed.

The plan of this book has been to outline the minimum acceptable theory of semiconductor properties which will enable satisfactory models for the circuit behaviour of the more important devices. A fairly extensive study of the properties of linear amplifiers has then been undertaken including the effect of feedback and the various modifications in design relevant to the construction of linear integrated circuits.

The final chapters deal with the simple properties of both bipolar and field-effect transistors as switches and their incorporation in integrated circuit construction. Some of the more basic applications of digital circuits are considered to illustrate the types of circuits used in digital computers.

The book is intended as a text to cover the electronics course relevant to the second year of an electrical engineering degree course, and as such it has been the basis for a course given at Imperial College over the past two years. It is also thought that it would be relevant to students studying other engineering disciplines who wish to become familiar with the principles of electronic circuits.

A comprehensive bibliography is included with each chapter in order to allow students to study any subject at greater depth than is possible in a book of this size. A number of problems are included with each chapter; these are, in the main, graded in complexity. The earlier problems should enable the student to confirm his understanding of the material of the text; later problems will allow him to apply this understanding to other allied problems not directly analysed in the text. Many of these problems have already been tried by students and modified in the light of their comments.

I should like to thank all my colleagues at Imperial College and also to students of successive second year electronics courses for their constructive criticism of the manuscript of this book. I should finally like to thank Christine Lourdan and Alison Elliott for their assistance in typing the manuscript.

Imperial College
February 1975

ROBERT KING

1
Conduction in Semiconductors and Diodes

1.0 Introduction

The majority of electronic devices are fabricated from semiconductor material, principally silicon, and thus a study of the movement of charges in semiconductors is a necessary preliminary to a study of the devices. Although the current flow is frequently confined to one type of semiconductor only, some devices incorporate the junction between two dissimilar types of material as an integral item in the construction; the current flow across such a junction, although obeying the same physical laws, results in an external behaviour which is considerably different from that of single material and therefore merits study on its own. Such a barrier is the basis of a semiconductor diode; the properties of such a diode and the application of the diode in a circuit are studied.

1.1 Charge availability in semiconductors

The majority of solids which are capable of conducting electricity are formed with a fairly regular crystal lattice in which the atoms are arranged in a regular pattern. In good conductors, such as copper, aluminium or silver, all the atoms are fully ionized, one electron becoming detached from each nucleus in the lattice; the detached electrons form a cloud which is free to move within the crystal lattice under the influence of an applied field; since there is a large number of free electrons available, a small electric field will cause a large movement of charge and hence a large current will flow; this accounts for the high conductivity of copper. The ionized atoms which remain fixed in the crystal lattice can take no part in this conduction process as their charges are immobile.

In a semiconductor such as silicon or germanium, a somewhat different situation exists. The energy required to ionize an atom of pure silicon is 1·2 eV. Since the thermal energies of electrons in a specimen are randomly distributed with an average energy at room temperature of 0·025 eV, only a minute fraction of the total number of atoms will be ionized. Furthermore the ionization of an atom causes the joint creation of a free electron and a positively charged ion which is deficient of one electron. The free electrons will be capable of movement through the crystal lattice as considered above. In addition to this motion of free electrons in the crystal lattice, bound electrons may move from an unionized atom to fill the deficiency in an ionized atom. We see therefore that the positions of ionized atoms may change; this situation may be modelled by postulating a positively charged particle, a hole, representing the deficiency of an electron in an atom, which is capable of 'moving' through a semiconductor. Thus in a pure semiconductor we have two types of charge carrier, holes and electrons, both relatively free and both capable of acting as charge carriers, their motion giving rise to a current. In a pure semiconductor the number of holes is exactly equal to the number of electrons. Despite the duplication in the number of charge carriers, the conduction of a piece of pure (or intrinsic) silicon is very low compared with copper since so few charge carriers are available. The resistivity of pure silicon is about 5×10^3 Ωm and is highly temperature dependent compared with that of pure copper, which has a value of $1 \cdot 6 \times 10^{-6}$ Ωm.

Pure silicon is of very little use in the manufacture of semiconductor devices such as diodes and transistors, although it is occasionally used in parts of the construction of more complex devices. However it is found that the addition of small quantities (of the order of one part in 10^7) of some impurity can improve the properties considerably. The impurities which are most commonly used are the trivalent elements boron and phosphorus, or the pentavalent elements arsenic and antimony.

Consider first the addition of a small quantity of pentavalent element, say antimony. Since silicon is a tetravalent element we note that the impurity atom has one electron in excess of each of the silicon atoms. When placed in a crystal lattice which is predominantly silicon this extra electron is very easily detached from the parent atom; in fact at room temperature an energy of the order of 0·01 eV is all that is needed to remove it. Thus in a doped piece of semiconductor containing a pentavalent impurity, known as a donor impurity, we may assume as a first approximation that all the donor atoms are completely ionized. We have therefore a considerable number of free electrons, considerably greater than in the intrinsic silicon. The conductivity of doped silicon at room temperature will consequently be considerably higher than that of pure silicon. Although a small number of holes and electrons will have been generated from the silicon atoms directly by thermal agitation, their number will be relatively small compared with the number of electrons which have been created from the donor impurity atoms. Silicon containing donor atoms is termed n-type silicon since free electrons (negative charge carriers) are available and its conductivity is predominantly by electron conduction. (In n-type silicon the electrons are termed majority carriers.)

On the other hand if a small quantity of a trivalent element such as phosphorus is introduced, the impurity atoms will each be deficient of one electron; in other words they will each contribute a hole; the hole can travel through the crystal lattice in the same way that holes in intrinsic silicon do. Again the energy required to fill a hole in one of these trivalent atoms, known as acceptor atoms, is also about 0·01 eV, and hence we may assume that at room temperature all the acceptor ions are fully ionized. Semiconductor material containing a small fraction of trivalent element is therefore termed a p-type semiconductor since conduction will take place predominantly by the movement of holes (positive charge carriers) within the crystal lattice. In p-type material, the holes are majority carriers.

In either n-type or p-type semiconductor the density of free charge carriers will closely approximate to the density of the impurity atoms introduced. Despite this it should be appreciated that both types of charge carrier will exist in both types of semiconductor; in equilibrium conditions the majority carriers will predominate and the density of minority carriers will be negligible. In situations where equilibrium is not established the density of minority carriers may be considerably greater than their equilibrium value but, generally, the minority carrier density may be always assumed to be small compared with the majority carrier density.

1.2 Conduction in a semiconductor

In a semiconductor at constant temperature there are two mechanisms which may give rise to a flow of charge carriers. The first is drift conduction which is the movement of the free charge carriers under the influence of an applied electric field. The second is diffusion conduction; this is the result of any non-uniformity in the distribution of charge carriers in the material. As a result the charges will tend to diffuse from a region of high charge density to one of lower charge density and this movement will constitute a current.

1.2.1 Conduction as a result of a potential gradient

The mechanism of conduction in any material due to an applied field, i.e., a potential gradient, is well known. In a conductor we are accustomed to write the current density J in terms of the applied field, $\mathscr{E}$, as

$$J = \sigma \mathscr{E} \tag{1.1}$$

where σ is the conductivity. This is usually applied to materials in which only one type of carrier exists, although it is still valid when both carrier types are present. We may develop equation 1.1 directly from a consideration of the motion of charge carriers in a sample. Consider first a material with electrons as

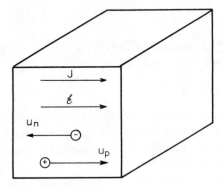

Fig. 1.1 Motion of holes and electrons drifting in a semiconductor

the only type of charge carrier. We should expect the current to be proportional to the density, n, of these carriers; also the current will be directly proportional to the electronic charge, e. Introducing a constant of proportionality, the electron mobility, μ_n, we may write the current density, $J_{n(drift)}$, due to the drift of electrons in the material under the influence of the field $\mathscr{E}$, as

$$J_{n(drift)} = ne\mu_n\mathscr{E} \tag{1.2a}$$

where the directions are shown in Fig. 1.1. The mobility is a constant which defines the ease with which electrons may move through the crystal lattice when under the influence of an imposed electric field. The mobility may be assumed constant for small fields, but it tends to rise in the presence of large fields. The mobility may also be shown to be the ratio of the drift velocity of the electrons u_n to the applied electric field

$$u_n = -\mu_n\mathscr{E} \tag{1.3}$$

Similarly a drift of holes in the field will give rise to a hole current density

$$J_{p(drift)} = pe\mu_p\mathscr{E} \tag{1.2b}$$

where p is the hole density in the material and μ_p is the hole mobility in the material. At room temperature the values for silicon are $\mu_n = 0{\cdot}16\,\mathrm{m^2/Vs}$, $\mu_p = 0{\cdot}04\,\mathrm{m^2/Vs}$.

The total current flow will now be made up of that due to both electron and hole movement. Although the holes and electrons move in opposite directions, their charges are also opposite and so the current density caused by each is in the same direction. The total current J is given by

$$J = ne\mu_n\mathscr{E} + pe\mu_p\mathscr{E} \tag{1.4}$$

Comparing this with equation 1.1 which is true for any material, we may write the conductivity in terms of the hole densities and mobilities as

$$\sigma = (n\mu_n + p\mu_p)e \tag{1.5}$$

For a material in which only one type of charge carrier exists or in which one type of charge predominates, the above expression reduces to only one term.

1.2.2 Conduction due to charge density gradient

The preceding discussion has assumed that the charge carriers are uniformly distributed throughout the conductor. In many situations this is not valid and the charge carrier density may be greater in one place than another. The tendency under these circumstances is for the charge carriers to diffuse within the conductor, in a way very similar to gaseous diffusion, in an attempt to establish a uniform charge density. There will thus be a flow of charge, constituting a current, proportional to the concentration gradient at any point. If we assume a one-dimensional system in which the charge density $n(x)$ only varies in the x-direction, an electron diffusion current density will flow as shown in Fig. 1.2; its value is given by

$$J_{n(\text{diff})} = eD_n \frac{dn(x)}{dx} \qquad (1.6a)$$

where D_n is the diffusion constant for electrons and is closely analogous to the diffusion constant for the motion of molecules in a gas.

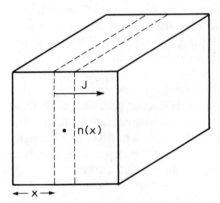

Fig. 1.2 Motion of charge carriers as a result of diffusion

Similarly, a concentration gradient of holes will cause a hole diffusion current density

$$J_{p(\text{diff})} = -eD_p \frac{dp(x)}{dx} \qquad (1.6b)$$

where D_p is the diffusion constant for holes. The diffusion constants of electrons and holes in silicon are 4100 and 1030 m^2/s.

1.2.3 Total current flow in a semiconductor

The total current will be caused by both the electric field and the charge concentration gradient. The current densities for electron flow will be given by the sum of equations 1.3a and 1.6a and for holes by equations 1.3b and 1.6b:

$$J_n = J_{n(\text{drift})} + J_{n(\text{diff})} = e\left(\mu_n n \mathscr{E} + D_n \frac{dn}{dx} \right) \qquad (1.7a)$$

$$J_p = J_{p(\text{drift})} + J_{p(\text{diff})} = e\left(\mu_p p \mathscr{E} - D_p \frac{dp}{dx} \right) \qquad (1.7b)$$

The total current density J will be given by

$$J = J_n + J_p \qquad (1.8)$$

It may be shown that mobility and diffusion constants are connected by the Einstein relationship

$$\frac{\mu_p}{D_p} = \frac{\mu_n}{D_n} = \frac{e}{kT} \qquad (1.9)$$

1.3 Junction with no applied voltage

We shall now consider the simplest device fabricated from semiconductor material—the junction diode. The essence of this is a junction between two types of semiconductor; one p-type and the other n-type. This junction is fabricated from a single chip of silicon, say, p-type, into the surface of which a donor impurity is diffused; this cancels the existing p nature and transforms a thin surface layer into n-type material. A junction with a rather gentle transition from p- to n-type now exists some small distance below the surface.

Consider initially a semiconductor junction in equilibrium, that is, with its terminals left unconnected. It is obvious that the total current across the junction is zero. Since the left side is p-type it will have an excess of holes and the right hand side being n-type will have an excess of electrons. There will, of course, be a few electrons on the left, and holes on the right—the minority carriers. Considering the electrons only, there is a concentration gradient across the junction which would tend to cause electrons to diffuse to the left. As this diffusion occurs, the right side near the junction is left deficient of electrons (depleted) and thus positively charged. An electric field is therefore set up across the junction as shown in Fig. 1.3a. This electric field will tend to set up a drift current causing electrons to flow to the right. Since no current can flow these two components must, in equilibrium, be equal and opposite. This is shown in equation 1.10a derived from equation 1.7 by setting $J_n = 0$. A similar argument will hold for the flow of holes leading to equation 1.10b.

$$0 = \mu_n n \mathscr{E} + D_n \frac{dn}{dx} \qquad (1.10a)$$

$$0 = \mu_p p \mathscr{E} - D_p \frac{dp}{dx} \qquad (1.10b)$$

where the positive directions of x and $\mathscr{E}$ are shown in Fig. 1.3a. Since the behaviour of holes in a semiconductor is independent of the behaviour of the free electrons, we are able to set the hole current density J_p and the electron current density J_n separately to zero in equations 1.10.

The above readjustment of charges across the junction takes place principally in a very small zone either side of the junction. This is known as the depletion region, transition region or space-charge region. At points more remote from the junction, charge neutrality is maintained and the hole and electron densities remain at their equilibrium values, p_{po} and n_{po} for holes and electrons in the

p material and p_{no} and n_{no} in the n material. Thus to a first approximation the net charge density remote from the junction must be zero.

$$p_{po} = N_A + n_{po} \simeq N_A \qquad (1.11a)$$

and $\qquad n_{no} = N_D + p_{no} \simeq N_D \qquad (1.11b)$

where N_A and N_D are the acceptor and donor concentrations in the p- and n-regions respectively and are shown in Fig. 1.3b.

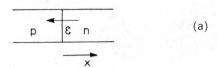

(a)

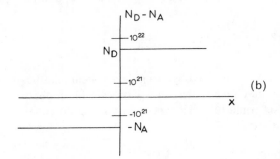

(b)

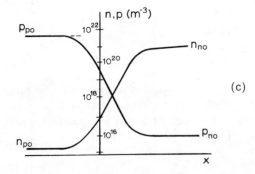

(c)

Fig. 1.3 Charge distributions in bulk regions of semiconductor (a) semiconductor junction (b) impurity densities (c) mobile charge densities

The electron and hole density profiles between these two regions must satisfy equations 1.10. Recalling that the electric field $\mathscr{E}(x)$ may be defined as the negative gradient of a potential $V(x)$ as

$$\mathscr{E}(x) = -\frac{dV(x)}{dx}. \qquad (1.12)$$

we may rewrite equation 1.10a as

$$\mu_n n(x) \frac{dV(x)}{dx} = D_n \frac{dn(x)}{dx} \qquad (1.13)$$

If this equation is now integrated from $x = -\infty$, corresponding to a region remote from the junction to the left where $n(x) = n_{po}$ and the potential $V(x) = V_P$, say, to $x = \infty$ at a remote region on the right where $n(x) = n_{no}$ and $V(x) = V_N$, we obtain

$$\ln \frac{n_{no}}{n_{po}} = \frac{\mu_n}{D_n} (V_N - V_P) \qquad (1.14a)$$

Similarly solving equation 1.10b with 1.12 between the same limits we obtain

$$\ln \frac{p_{po}}{p_{no}} = -\frac{\mu_p}{D_p} (V_P - V_N) \qquad (1.14b)$$

These two expressions may be further simplified by using the Einstein relationship, equation 1.9. Moreover, we may note that both equations 1.14 are functions of $V_N - V_P$ which we may set equal to ϕ, known as the contact potential. Equations 1.14 may now be rewritten

$$n_{po} = n_{no} \exp (-e\phi/kT) \qquad (1.15a)$$

$$p_{no} = p_{po} \exp (-e\phi/kT) \qquad (1.15b)$$

If we eliminate ϕ between these two equations we obtain

$$n_{no} p_{no} = n_{po} p_{po} \qquad (1.16)$$

This constant product of hole and electron densities in a semiconductor holds under all conditions at a given temperature. For intrinsic material $n_i = p_i$ and hence the hole and electron densities in a piece of doped semiconductor may be related to the intrinsic electron density by

$$np = n_i^2 \qquad (1.17)$$

1.3.1 Potential distribution across a junction in equilibrium

The total charge density at any point across the semiconductor junction may be obtained by summing all charge densities shown in Fig. 1.3b and c. The net charge density at a point distance x from the junction is

$$N_D - N_A + p(x) - n(x)$$

This is shown by the solid line in Fig. 1.4 (the vertical scale has now been made linear compared with the logarithmic scale of Fig. 1.3b and c). This total charge density may be approximated for convenience by the rectangular profile shown by the broken line. This transition between p- and n-type semiconductors has resulted in a dipole of charge across the junction, the charge being produced by the difference between the fixed charges due to the impurity atoms in the lattice and the free charges, the holes and electrons. This zone in the vicinity of the junction in which there is a reduction of charge is termed the depletion or transition region of the device (also known as the space charge region).

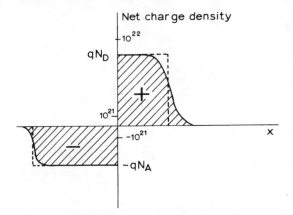

Fig. 1.4 Net charge density near semiconductor junction

We have also seen that the existence of this dipole of charge has set up an electric field across the junction and that this field will be the gradient of a potential distribution V shown in Fig. 1.5. The difference in potential across the junction $V_N - V_P$ is the contact potential defined in equations 1.15.

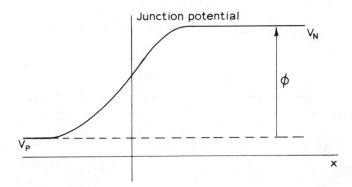

Fig. 1.5 Potential variation in vicinity of junction

1.4 Junction with applied voltage

Consider now the junction of Fig. 1.6a to which a voltage v_D is applied externally making the p-region more positive than the n-region. The total current density flowing across the junction is now given by equations 1.7 and 1.8. These equations may be solved to give the minority carrier densities n_p of electrons in the p-region and p_n of holes in the n-region just at the edges of the depletion layer as functions of the applied diode voltage v_D. The results which may be obtained by some simplifying approximations are given by equations 1.18.

$$n_p = n_{po} \exp\left(ev_D/kT\right) \tag{1.18a}$$

$$p_n = p_{no} \exp\left(ev_D/kT\right) \tag{1.18b}$$

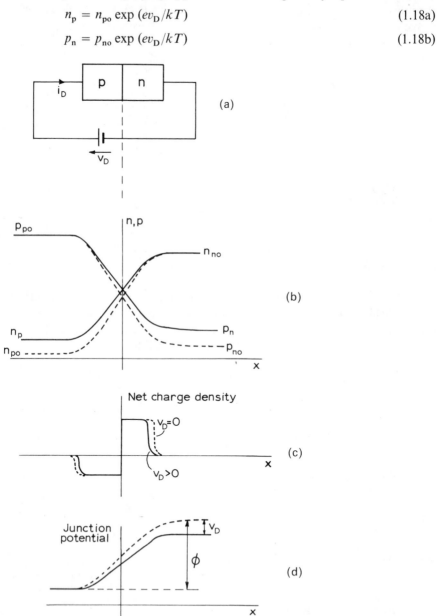

Fig. 1.6 Effect of forward bias at junction (a) biasing circuit (b) mobile charge distribution (c) net charge distribution (d) potential distribution

This charge distribution varies through the depletion region in the manner shown in Fig. 1.6b, in which the charge distributions for zero applied voltage are indicated by the broken line. It may be seen that the effect of applying a forward bias voltage is to increase the density of electrons injected into the p-region from n_{po} to n_p and similarly to increase the density of holes injected into the n-region from p_{no} to p_n.

If we also plot in Fig. 1.6c the total charge density at any point through the junction in a similar manner to Fig. 1.4, we observe that the total depletion charge on either side of the junction has decreased, causing the width of the depletion layer to become less. This corresponds with the potential distribution shown in Fig. 1.6d in which the junction potential has been reduced from ϕ to $\phi - v_D$. The application of a voltage which reduces the effective junction voltage is termed a forward bias.

A similar situation exists if v_D is made negative, reverse biasing the junction. The total potential across the junction is now increased to $\phi - v_D$ (v_D is numerically negative), the space charge layer now becomes wider and the minority electron and hole densities in the p- and n-regions at the edge of the space-charge layer are reduced; for negative values of v_D where $|v_D| \gg kT/e$, the minority carrier density tends to zero.

The calculations we have conducted apply only within the depletion layer and only allow us to determine the minority carrier densities n_p and p_n at each edge of the depletion layer. The current cannot be determined from the behaviour in the depletion layer.

1.5 Charges outside the depletion layer

We shall now concentrate our attention on the behaviour of the minority carriers outside the depletion region. In particular we shall study the behaviour of holes in the n-region; the behaviour of electrons in the p-region will be very similar.

Figure 1.7 shows the minority charge distribution in the n-region for forward

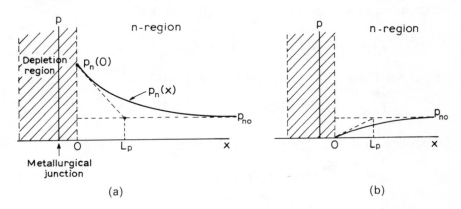

(a)   (b)

Fig. 1.7 Charge density beyond depletion layer in n-region (a) forward biased (b) reverse biased

and reverse biasing. (Note that the origin of x has now been shifted to the outside edge of the depletion layer.) At distances very remote from the depletion layer, equilibrium conditions will exist under all conditions and the hole density will be p_{no}. If no bias voltage is applied across the diode, the injection charge density (that is, the charge density just at the edge of the depletion layer) is, from equation 1.15b, also p_{no}. Thus the charge distribution throughout the n-region is constant at p_{no}, as shown by the broken line in Fig. 1.7a.

If a forward bias voltage, v_D, is applied across the junction, the injection density at the edge of the depletion layer rises to $p_n(0)$, the value of p_n at $x = 0$, given by equation 1.18b as

$$p_n(0) = p_{no} \exp (ev_D/kT) \tag{1.19}$$

The charge density $p_n(x)$ at a distance x from the junction must fall to a value p_{no} for large values of x by the mechanism of recombination. The holes are minority carriers in a region in which the electron density is many orders of magnitude greater than the hole density. There is thus a high probability that holes will collide with electrons, resulting in an annihilation of both. A reasonable assumption would be that the likelihood of a hole and electron colliding is directly proportional to the excess minority carrier density above the equilibrium value. The result will be that as the holes diffuse* through the n-type material they recombine with the majority carriers and the charge density falls exponentially as shown in Fig. 1.7a from $p_n(0)$ at the edge of the space-charge layer to p_{no} at a large distance. It is convenient to define a diffusion length L_p for holes in terms of the following expression:

$$\frac{p_n(0)-p_{no}}{L_p} = \text{slope of charge distribution curve at } x = 0$$

We are now in a position to determine the hole current flowing at the edge of the depletion layer. Since the field, $\mathscr{E}(x)$, is negligible outside the space-charge region we may condense equation 1.7b to read

$$J_p(x) = -eD_p \frac{dp_n(x)}{dx} \tag{1.20a}$$

At $x = 0$ this becomes

$$J_p(0) = eD_p \frac{p_n(0)-p_{no}}{L_p} \tag{1.20b}$$

Introducing equation 1.19 we obtain for the current carried by holes

$$i_p = \frac{eD_pAp_{no}}{L_p} [\exp (ev_D/kT)-1] \tag{1.21a}$$

* The charge motion is almost exclusively diffusion as the electric field here is negligibly small.

where A is the cross-sectional area of the junction. A similar current i_n will be carried by electrons. The total current will be given by

$$i_D = i_n + i_p$$
$$= I_0[\exp(ev_D/kT) - 1] \qquad (1.22)$$

where $\qquad I_0 = eA\left(\dfrac{D_n n_{po}}{L_n} + \dfrac{D_p p_{no}}{L_p}\right)$

This is the characteristic equation relating the current through a semiconductor junction to the voltage across it.

If we make v_D negative and large such that $|v_D| \gg kT/e$ corresponding to a reverse bias, the charge density at the edge of the depletion layer is $p_n(0) \simeq 0$ rising in an exponential manner towards p_{no} as shown in Fig. 1.7b.

The total reverse current is given when $v_D < 0$ as

$$i_{D(rev)} = -I_0$$

We may thus identify the current I_0 in equation 1.22 as the magnitude of the reverse current, flowing for large reverse bias voltages. It is known as the reverse saturation current.

One significant feature of the analysis of the junction which has been outlined above is that the magnitude of the current flow in the device has been determined principally by the properties of the field-free regions remote from the junction where the charge carriers move almost exclusively by diffusion. The boundary values of the charges in these regions are, however, determined by the behaviour at the junction where both the electric fields and the concentration gradient of charge carriers are very high. We shall later use this same subdivision of the device to simplify understanding of the operation of a transistor.

1.6 The semiconductor diode

We have seen in the foregoing analysis that a semiconductor junction has a current/voltage relationship which is given by equation 1.22. The current in the reverse direction is limited to I_0 and in the forward direction it may increase exponentially with forward voltage. The diode is just such a device, consisting of a junction between two dissimilar semiconductors. It has a characteristic which may ideally be represented by equation 1.22. The circuit symbol for a diode with the assumed reference direction of current and voltage is shown in Fig. 1.8.

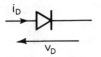

Fig. 1.8 Junction diode conventional symbol

The characteristics of a typical semiconductor diode for two ranges of current are shown in Fig. 1.9. These may be seen to have the shape which we might expect from equation 1.22, the reverse current being small and constant for reverse voltages greater than about $5kT/e$ and rising exponentially in the forward direction. A numerical comparison of equation 1.22 with the curves of Fig. 1.9 shows small discrepancies. We may account for these by noting that in

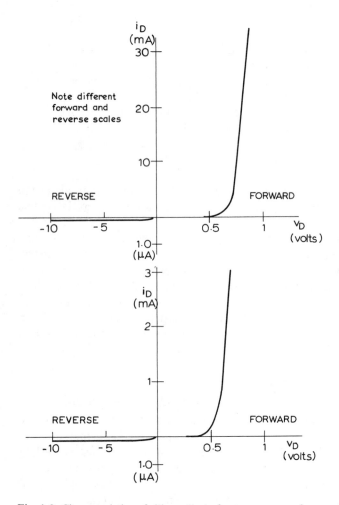

Fig. 1.9 Characteristics of silicon diode for two ranges of current

a physical diode the junction needs to be supported by some bulk semiconductor and the current flowing through this causes a small potential drop. The applied voltage is therefore slightly greater than the actual junction voltage and we should not expect exact agreement by inserting the applied voltage into equation 1.22.

1.7 Models of the static behaviour of a diode

The most obvious representation of the properties of a diode is to use the characteristics relating current and voltage. In situations where the diode is to be used in a circuit composed of batteries and resistors, this is the most convenient form of characterization. The graph presents one relationship between v_D and i_D, and the external circuit represents a second constraint relating v_D and i_D imposed by the external linear circuit components. This latter relationship may be written down analytically using Kirchhoff's laws. The diode current and voltage may be determined by solving these two relationships, one of which is analytical, the other graphical. This is most easily achieved by plotting a 'load line', whose equation is given by the analytic relationship superimposed on the diode characteristics and then obtaining the point of intersection. This may be illustrated by considering a diode, whose characteristics are given by Fig. 1.9, in series with a resistor of 60 Ω connected to a battery of 1·2 V. The diode voltage v_D is related to i_D by the equation $1·2 = v_D + 60\, i_D$. If this is plotted on the diode characteristic as shown in Fig. 1.10, an operating point is obtained at $v_D = 0·7$ V, $i_D = 7·5$ mA.

It is sometimes more convenient to use some form of approximation to the characteristic of the diode, obtained either graphically or analytically.

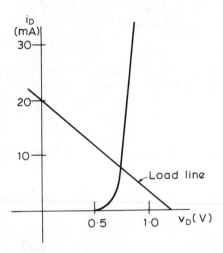

Fig. 1.10 Diode characteristic with load line superimposed

1.7.1 Piecewise linear approximation

One simple model for voltages ranging from positive to negative values is to represent the characteristic of Fig. 1.9 by the piecewise linear approximation of Fig. 1.11, consisting of two straight lines having a break point at $v_D = v_O$, $i_D = 0$. The value of R_r, the reverse resistance is large, of the order of 10^7 Ω for a silicon diode, and may frequently be assumed infinite. In the forward direction the value of both R_f, the forward resistance, and the break point (threshold) voltage, V_0, will depend on the magnitude of the current. The piecewise linear approximation shown will be valid when working with peak currents of the order of I_1 as shown in Fig. 1.11; however, for lower currents such as I_2 a smaller value of V_0 will be appropriate for the approximation, associated with a larger value of R_f.

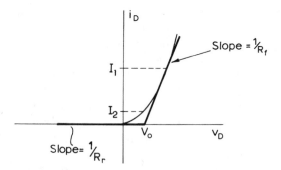

Fig. 1.11 Piecewise linear approximation to diode characteristic

For the diode characteristics shown in Fig. 1.9, at a current of about 20 mA, a satisfactory piecewise approximation would be to choose $V_0 = 0.75$ V and $R_f = 4.5\,\Omega$. At a much lower current of about 2 mA a suitable choice would be $V_0 = 0.6$ V, $R_f = 33\,\Omega$. This representation gives rise to the circuit model of Fig. 1.12 in which the ideal diode is a device which passes zero current when $v_D < 0$, and for which $v_D = 0$ when $i_D > 0$, in other words, it represents an open circuit when reverse biased and a perfect short circuit when forward biased.

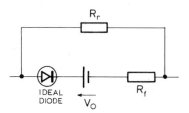

Fig. 1.12 Static model of junction diode

The characteristics of germanium diodes are similar in form. However, the break point occurs at about 0.2 to 0.4 V with the forward resistance from 50 Ω upwards. The reverse resistance of a germanium diode although still large is customarily of the order of 1 MΩ depending upon applied voltage.

This piecewise linear model may be used to represent a diode in a circuit, allowing us to perform a simple linear analysis to determine the currents and voltages. It naturally lacks the accuracy of the analysis obtained by using the diode characteristic with a load line, but it is much simpler where circuits involving several diodes are concerned.

1.7.2 Power series model

An alternative mathematical model for a diode is the expansion of the exponential characteristic given by equation 1.22b as a power series in v_D as

$$i_D = a_1v_D + a_2v_D^2 + a_3v_D^3 + \cdots. \tag{1.23a}$$

This may be done directly from equation 1.22b, obtaining the coefficients a_0, a_1, etc. from a Taylor series expansion. However, it is preferable when using this approach to obtain a direct approximation to some measured diode characteristic by a simple curve fitting procedure, since we have commented that real diodes do not accurately fit the exponential relationships of equation 1.22b. Equation 1.23a gives a power series expansion in terms of total voltages and currents in the diode. Sometimes a diode is biased away from the origin at a point where voltage and current are V_D and I_D as shown in Fig. 1.13. Using a Taylor series expansion about the operating point V_D, I_D we obtain

$$i_D - I_D = a_1'(v_D - V_D) + a_2'(v_D - V_D)^2 + \cdots$$
$$i_d = a_1'v_d + a_2'v_d^2 + \cdots \tag{1.23b}$$

where $v_d = v_D - V_D$ is the deviation of the voltage from the operating point. This power series model is useful where comparatively large voltages are applied and the distortion of the current introduced by the second and higher powers of voltage are desired in the output.

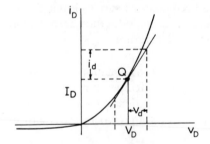

Fig. 1.13 Diode characteristic illustrating incremental conductance

1.7.3 Incremental model

A special case of the above approximation is useful in situations in which the device is used in a circuit where small variations of voltage are superimposed on a fixed bias voltage. This is illustrated in Fig. 1.13; a direct voltage V_D is applied with a small variation v_d superimposed on it. The variation in output current may be determined approximately from

$$\frac{i_d}{v_d} = \frac{di_D}{dv_D}$$

$$= g_d, \text{ the incremental conductance.}$$

From equation 1.22b

$$g_d = I_0(e/kT) \exp(ev_D/kT) \tag{1.24a}$$

$$= (e/kT)(i_D + I_0) \tag{1.24b}$$

$$\simeq eI_D/kT \tag{1.24c}$$

since the incremental conductance is only of practical significance when $i_D \gg I_0$.

At room temperature, (293 K), $e/kT = 40 \text{ V}^{-1}$. Thus at a current of 1 mA, the incremental conductance is 0·04 S, corresponding to a resistance of 25 Ω.

1.8 Effect of temperature on diode characteristic

Since semiconductor devices are likely to be used in a wide variety of environments, it is of interest to study the behaviour of a diode over a range of temperatures.

From equation 1.22 it is obvious that temperature will influence the diode current in an exponential manner; this will only be of significance in the forward biased direction since the exponential term is negligible for reverse voltages. In the reverse direction the saturation current I_0 dominates and as this is strongly dependent on the equilibrium minority charge density, variation of temperature will have a large effect on I_0.

The effect of temperature on saturation current may be determined theoretically from the simple physical model. However, these figures deviate considerably from measured values since no account has been taken of surface leakage current. Measurements show that at room temperature the saturation current of a silicon diode will double for every 10°C rise in temperature.

In the forward direction not only does the temperature affect I_0 but it also has a direct influence through the exponential term of equation 1.22. These two effects act in opposition, the overall magnitude of temperature dependence is less than that in the reverse direction; it is also a function of the forward bias point. The variation of forward diode characteristic with temperature is shown in Fig. 1.14 for a typical silicon diode.

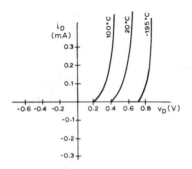

Fig. 1.14 Effect of temperature on silicon diode characteristic

At room temperature it is found that a typical value of the temperature variation of v_D is $-2\,\mathrm{mV/K}$. This may be confirmed from Fig. 1.14.

It is of interest to note the effect of temperature on the circuit models of a diode discussed in Section 1.6.

A comparison of Fig. 1.14 with the piecewise approximation of Fig. 1.11 shows that for a fixed magnitude of applied voltage, R_f decreases somewhat with temperature rise; the main effect, however, is a reduction in threshold voltage, V_O, by about $-2\,\mathrm{mV/K}$ for a silicon transistor at room temperature.

1.9 Dynamic characteristics of diode

So far we have assumed that the changes in applied voltage have been sufficiently slow that the time taken to alter the stored charges in the diode may be neglected. At higher frequencies this time delay must be included in the analysis and the effects of it on external behaviour must be determined.

There are two physical processes in a diode which are associated with a change of charge distribution consequent on a change of applied voltage. One of these is the distribution of charge in the depletion layer of the diode and the second is the variation of charge in the neutral regions of semiconductor more remote from the junction.

1.9.1 Charge stored in depletion layer

Consider first a diode in which the applied voltage is changed from $-V_1$ to a value $-V_2$ (namely from V_1 to V_2 volts in a reverse biased sense), where $V_1 < V_2$. When the voltage across the diode is $-V_1$, there is a dipole of charge across the metallurgical junction shown in Fig. 1.15 by the solid line. When the voltage, v_D, across the diode is changed to $-V_2$, the charge distribution has the profile given by the broken line.

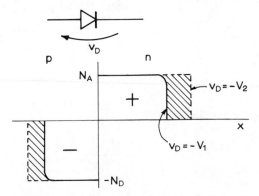

Fig. 1.15 Variation of net charge in depletion layer

It is seen, then, that a change of potential across the diode will cause a change of charge. Thus

$$Q_1 - Q_2 = \Delta Q = \mathrm{f}(-V_1 + V_2)$$
$$= \mathrm{f}(\Delta V)$$

where Q_1 and Q_2 are charges stored in either side of the junction when applied voltages are $-V_1$ and $-V_2$ respectively. This relationship is, in general, non-linear, the functional form of which depends on the manner in which the p-n junction has been fabricated. Although we have only considered the variation of depletion layer charge with variation of reverse bias voltage, it is obvious that this phenomenon exists for forward bias as well. However, in the forward direction the magnitude of the effect is usually swamped by the effect of changes of charge in the neutral regions (to be discussed in Section 1.9.2); thus for practical considerations depletion layer charge is of most significance in the reverse biased condition.

If we now consider the variations of voltage to be small we may define a depletion layer capacitance at the junction

$$C_j = \Delta Q / \Delta V$$

The magnitude of C_j will however be a function of the operating bias voltage.

1.9.2 Charge stored in the neutral region

We will now consider the effect of the charges stored in the bulk of the p- and n-regions of the diode. In Section 1.5 we showed that when a forward bias voltage is applied across the junction, the minority carrier density (of holes) in the n-region just outside the depletion layer is increased above the equilibrium value; this is illustrated in Fig. 1.7.

When a forward bias voltage $v_D = V_1$ is applied to the diode the charge density profile is shown by the solid line in Fig. 1.16. If the applied forward voltage is increased to V_2 the charge density profile is given by the broken line. There is thus an additional charge ΔQ represented by the shaded area which is a result of the change of voltage ΔV from V_1 to V_2.

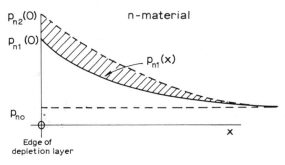

Fig. 1.16 Variation of charge in bulk region of diode

Again for small changes of voltage we may define an incremental capacitance which may represent the effect of the change of charge stored in the neutral region consequent on a change of applied voltage. The magnitude of this capacitance is also dependent on the applied voltage and only exists in the forward bias direction; the charge stored in the bulk semiconductor in the reverse direction is negligible.

The total capacitance across the diode is, therefore, the sum of the depletion layer capacitance (alternatively called the transition capacitance or space-charge capacitance) and the charge storage capacitance (sometimes termed the diffusion capacitance).

The variation of total capacitance, C_D, with applied voltage is shown for an appropriate diode in Fig. 1.17. This diode has been designed as a variable

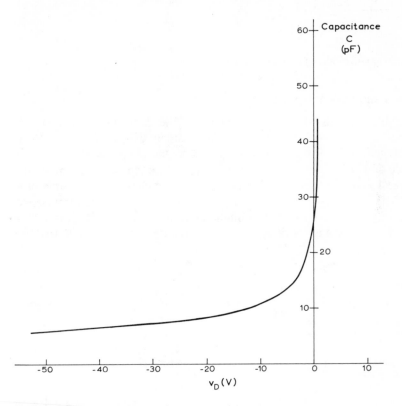

Fig. 1.17 Variation of diode capacitance with voltage

capacitance device. It is almost always used in the reverse biased direction, since the incremental resistance is then very high and thus will not introduce an appreciable loss across the capacitance.

1.10 Dynamic model of diode

The phenomena relating to the variation of stored charges in the depletion layers and in the bulk of the semiconductor may be represented by including a nonlinear capacitance in the model for the diode as shown in Fig. 1.18. The nonlinear capacitance will have a value depending upon the applied voltage such as that given in Fig. 1.17. Such a model is applicable in both the forward and reverse directions.

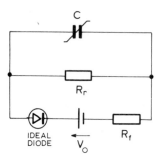

Fig. 1.18 Dynamic model of a junction diode

1.11 Reverse bias breakdown

The simplified analysis which we have conducted leads to the conclusion that the reverse current is small and effectively independent of applied voltage. Such a model is only valid for applied voltages below the breakdown voltage, the magnitude of which is dependent on the fabrication of the diode and may vary from several volts up to several hundred volts. The characteristic curve of a diode if extended beyond its normal operating range will have the form shown in Fig. 1.19 where the current rises suddenly at an applied voltage of about -170 V. If the current is not limited by a series resistance this could destroy the diode.

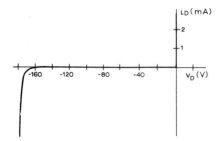

Fig. 1.19 Diode characteristic showing breakdown

There are two mechanisms which may give rise to breakdown phenomena; avalanche multiplication and Zener breakdown. As the reverse bias voltage is increased, the electric field in the depletion layer increases. The charge carriers passing through the depletion layer are thus accelerated by this field and many may acquire sufficient energy for each to generate a hole–electron pair when it collides with an atom in the crystal lattice; the generated hole and electron will move off in opposite directions adding to the current flow through the diode. At sufficiently high values of voltage, so many collisions will occur, the secondary holes and electrons themselves creating further pairs of charges, that a self-generating 'avalanche' of charge carriers results causing a sudden rise in current. Avalanche multiplication is likely to occur in silicon diodes whose material has a relatively low impurity density and consequently fairly wide depletion layer. The breakdown voltage is usually rather high.

The second mechanism which can cause breakdown is the Zener effect. This will be discussed in Section 6.1. It is not usually a limiting factor in normal power or signal diodes.

These phenomena will place an upper limit on the reverse voltage which may be applied across a junction diode if the current in the reverse direction should not reach a large value; this limitation is of relevance in applications such as rectification or detection of alternating voltages.

References

BECK, A. H. W. and AHMED, H., *An Introduction to Physical Electronics*, Arnold, 1968, Chapter 4.
ALLISON, J., *Electronic Engineering Materials and Devices*, McGraw-Hill, 1971, Chapter 9.
SPARKES, J. J., *Junction Transistors*, Pergamon, 1966, Chapters 1–3.
GRAY, P. E., *et al.*, *Physical Electronics and Circuit Models of Transistors*, Wiley, 1964, Chapters 1–6.
ADLER, R. B., *et al.*, *Introduction to Semiconductor Physics*, Wiley, 1964.
ANDERSON, J. C., LEAVER, K. D., ALEXANDER, J. M. and RAWLINGS, R., *Materials Science*, 2nd edn, Nelson, 1974, Chapter 13.
GIBBONS, J. F., *Semiconductor Electronics*, McGraw-Hill, 1966, Chapters 1–8.

Problems

1.1 The reverse saturation current of a silicon diode is 20 nA. At what voltage at room temperature is

  (a) the reverse current equal to 18 nA
  (b) the forward current equal to 20 nA ?

1.2 Two diodes are connected in series across a battery of voltage V. The reverse saturation current of diode A is 100 nA and of diode B is 150 nA. The breakdown voltage of both diodes is 150 V.
  Determine the current flowing and the voltage across each diode when

  (a) $V = 10$ V
  (b) $V = 200$ V

connected, in both cases, so that both diodes are reverse biased.

1.3 The two diodes in the circuit of Fig. P1.1 have the characteristics given in Fig. 1.9. When $R = 0$, determine the supply current flowing when the applied voltage is 1 V. What is the voltage drop across each diode ?

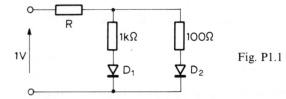

Fig. P1.1

1.4 Determine the supply current flowing in the circuit of Fig. P1.1 when $R = 20\ \Omega$.

1.5 A diode has a characteristic which may be represented by $i = 6 \times 10^{-3}\, v^2$ in the forward direction and $i = 0$ when reverse biased. Sketch the characteristic; by taking a suitable piecewise linear approximation to this curve, sketch and dimension the current flowing through a resistor of 200 Ω, when the resistor and diode are connected in series across a sinusoidal voltage of amplitude 1 V.

1.6 A diode has a reverse saturation current of 15 nA. Obtain a power series expansion valid at room temperature for the current flowing as the voltage is varied by small amounts (a) about 0 V and (b) about 0·45 V.

1.7 A voltage 0·45 + 0·02 cos 100 t is applied across the diode of Problem 1.6. Find the r.m.s. value of the fundamental, first, second, and third harmonic components of the current flowing.

1.8 A voltage 0·45 + V cos 100 t is applied across the diode of Problem 1.6. Determine the steady component of current flowing when (a) $V = 0$, (b) $V = 0·02$.

1.9 The pulse of voltage shown in Fig. P1.2 is applied to the input of the network of Fig. P1.3a. Sketch the output voltage v_o. What is the voltage if the network b is substituted for a?

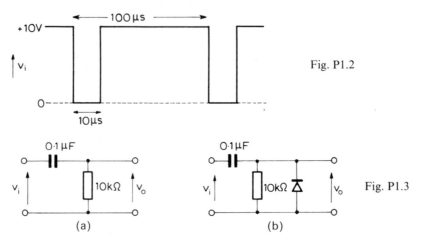

Fig. P1.2

Fig. P1.3

(a)   (b)

1.10 The circuit of Fig. P1.4 may be used as an attenuator of alternating voltages. The diode may be assumed to have an ideal exponential characteristic with reverse saturation current of 50 nA. If the input voltage, v_i, is 0·1V r.m.s. at a frequency of 50 Hz, determine the ratio v_o/v_i as the resistor R is varied over its full range.

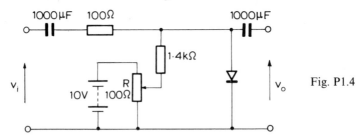

Fig. P1.4

1.11 The diode in the circuit of Fig. P1.4 has a capacitance when forward biased which varies linearly from 25 pF at zero current to 50 pF at a forward current of 7 mA.

Determine the frequency of the input signal at which the attenuation deviates from its low frequency value by more than 30% at (a) minimum and (b) maximum attenuation.

1.12 The diode in the circuit of Fig. P1.5 has a capacitance which varies with applied voltage as given in Fig. 1.14. It is used to vary the resonant frequency of the tuned circuit. Over what range of frequency may the circuit be tuned by variation of resistance R? How does the bandwidth of the tuned circuit vary with applied voltage? What is the purpose of the 1 MΩ resistor?

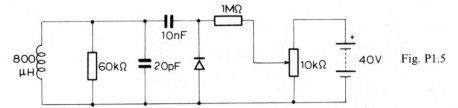

Fig. P1.5

1.13 A diode has a relationship between current, i, in milliamps and voltage, v, given by $i = 0.475v + 0.011v^2$. It is connected in series with a resistance R across a voltage $v = V_1 \cos 2\pi f_1 t + V_2 \cos 2\pi f_2 t$, where $V_1 = 2.0$ mV, $V_2 = 15.0$ μV, $f_1 = 1.05$ GHz, $f_2 = 1.00$ GHz. If $R = 100$ Ω, determine the voltage across the resistor. How could this be used as a microwave mixer and what would be the intermediate frequency?

What would be the effect of increasing R to 2 kΩ?

1.14 The circuit of Fig. P1.6 is used to generate a sine wave, v_o, at very low frequencies from a triangular wave, v_i. Study the circuit to see that you understand its operation. Draw a sine wave of amplitude 10 V. Approximate one half cycle of this from $-\pi/2$ to $\pi/2$ by five straight line segments. Choose the values of resistors R_1, R_2, and R_3 to give the correct voltages for each diode to start conducting; make their values small compared with 100 kΩ. Choose values of R_A and R_B to give the correct currents when the supply is fed from a triangular wave of appropriate amplitude. What amplitude of triangular wave, v_i, is necessary?

Why will a practical circuit give a better sine wave output than your circuit predicts?

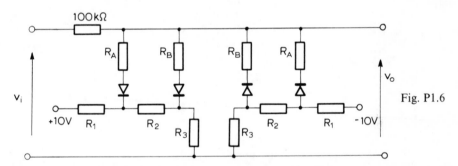

Fig. P1.6

1.15 Copper at room temperature has an electron concentration of 1×10^{29} m^{-3}. Its conductivity is approximately 0.58×10^8 S/m. Estimate the mobility of free electrons in copper.

If 100 m of copper wire is connected across a 1·5-volt battery, what is the drift velocity of electrons?

1.16 The mobilities of holes and electrons in silicon are given in Appendix B.

(a) Determine the conductivity of intrinsic silicon in which hole and electron densities are 1.5×10^{16} m^{-3}.

(b) What is the electron and hole concentration in silicon with donor impurity density of 5×10^{20} m^{-3}?

(c) What is the conductivity of the silicon of part (b)?

1.17 A circuit generates a square wave by successive clipping and amplification of a sine wave. A sine wave of r.m.s. value 10 V and frequency 50 Hz is clipped by the circuit of Fig. P1.7 using diodes whose break-point voltages are 0·7 V. The output voltage from the clipper is passed through an amplifier of voltage gain 12, clipped again in a similar circuit and further amplified by a factor of 12. What is the peak to peak magnitude of the output voltage and the rise time of the square wave?

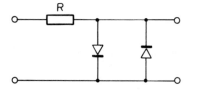

Fig. P1.7

2
Fets—Principles of Operation

2.0 Introduction

Of the many three-terminal semiconductor devices which have been invented, one of the simplest and at the same time one having wide applications is the field-effect transistor, sometimes known as unipolar since only one type of charge carrier is involved. Although the principle of operation of the junction field-effect transistor has been known since 1930, fabrication problems were such that reliable commercial devices were not available until about 1952. More recently a variant of the junction fet, the metal-oxide-semiconductor field-effect transistor (abbreviated MOSFET or MOST) has been developed. An alternative name is the IGFET, insulated gate field-effect transistor. The construction of this permits relatively simple fabrication in integrated circuit form. Although the mosfet was a later development, it is the more common device and so we shall study its construction and operation in some detail and look at the junction fet subsequently.

2.1 Construction of a mosfet

Since two types of semiconductor, p or n, are available, two types of mosfet may be constructed, depending on the type of silicon used for the substrate. In order to study the operation of the device we shall restrict ourselves to one type, namely that fabricated on n-type silicon; this choice will lead to a p-channel transistor. The complementary type formed on a p-type substrate will be considered later.

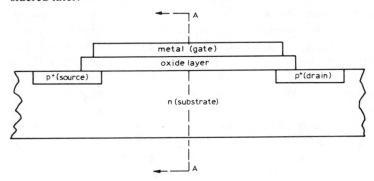

Fig. 2.1 Construction of a mosfet

A cross-section through the transistor wafer of a complete p-channel mosfet is shown in Fig. 2.1. It is formed by diffusing two highly doped p-type regions (p+regions) into the substrate; these are the source and drain electrodes. The region between the source and drain will form the channel along which conduction will occur. Over the surface of the substrate between source and drain an oxide layer is formed and on top of this a metal layer, frequently of aluminium, which will become the gate of the device. External leads are attached to source, gate, and drain to form a three-terminal device. This is shown in Fig. 2.2, together with reference directions of terminal currents and voltages. We shall now show that under certain conditions conduction can occur along the channel between source and drain and that the conductance of the channel may be varied by changes in the potential applied to the gate.

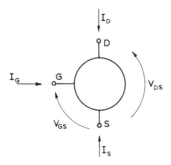

Fig. 2.2 Voltage and current convention for fet

2.2 Formation of conducting channel

In order to understand the manner in which charges flow through the channel of a mosfet and the way in which gate potential can influence this, we should look at the charge distribution in a cross-section of the transistor such as A–A in Fig. 2.1. We shall assume that, for the first analysis, no potentials are applied between source and drain and that the gate also is at zero potential with respect to the source. In order to develop the mechanism of conduction adequately we must accept a simplified explanation of a phenomenon occurring at the interface between the semiconductor and the oxide. As a result of the discontinuity of the crystal lattice at this interface, a number of fixed positive charges exist along this surface; these charges are immobile and may be pictured as fixed just inside the surface of the oxide layer. These fixed charges are shown as Q_{SS} in Fig. 2.3a. In order that the complete device remains electrically neutral, an equal negative charge must be provided as close as possible to the positive fixed charge to minimize the stored energy. This occurs by an increase in the concentration of majority carriers, electrons, immediately under the oxide layer. These mobile carriers are shown as Q_A in Fig. 2.3a and this is known as the accumulation charge; its magnitude is equal to Q_{SS}, but it is of opposite sign.

If a negative potential is now applied to the gate a charge Q_G of electrons will flow on to the gate. To maintain charge neutrality the charge Q_A must therefore be reduced, and as Q_G is progressively increased in magnitude, Q_A will ultimately vanish. Further increases of Q_G negatively will cause some of the free electrons normally present in n-type silicon to be repelled from under the oxide layer; this will leave a depletion layer having a net positive charge Q_D set up by the fixed positive ions in the semiconductor crystal (Fig. 2.3b). Still further increase of Q_G negatively will cause the depletion layer to continue to increase until a point is reached where it is no longer able to supply the charge necessary to maintain charge neutrality. Beyond this point the only way in which charge neutrality can be maintained is by free holes (minority carriers) being drawn in

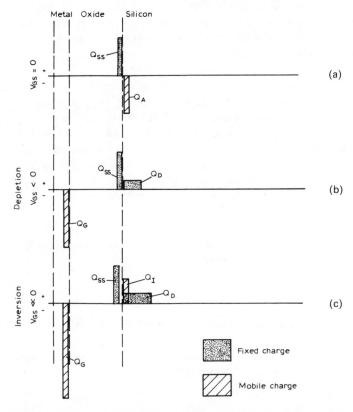

Fig. 2.3 Charge distribution between gate and substrate in mosfet

from the bulk of the semiconductor to form a layer of positive free charge Q_I consisting of holes (Fig. 2.3c). In this thin layer, immediately under the gate, the nature of the semiconductor has been changed from having a preponderance of electrons to an excess of holes; it has taken on the nature of a p-type semiconductor despite the fact that it is still composed of silicon containing a donor impurity. Such a region is called an inversion layer.

Consider now the application of a small negative potential on the drain with respect to the source. So long as an inversion layer exists, such a potential will cause holes to enter this layer from the source, pass along it to the drain and out through the drain electrode. Conduction between source and drain is thus seen to be possible when an inversion layer exists. The presence of the p-regions for source and drain shown in Fig. 2.1 may now be justified since this permits the complete conduction path through the transistor to occur in one type of semiconductor. This conduction layer under the oxide is known as the channel.

2.3 Conduction at small drain voltages

We have seen that the only state in which the right conditions prevail for conduction is when an inversion layer exists. The conduction charges in the case we are considering are holes and thus this is termed a p-channel mosfet. Since the inversion layer does not exist until a minimum magnitude of negative gate–source voltage exists, we note that no current will flow until a voltage V_{th}, the threshold voltage, has been exceeded in magnitude. For a p-channel mosfet, V_{th} is negative. As the gate voltage is further increased negatively, charge neutrality can only be maintained by an increase in Q_1. If we assume that a small negative potential is applied to the drain with respect to the source, V_{DS}, (say about -0.5 V) then, as soon as the threshold voltage has been exceeded, current will flow along the channel from source to drain due to drift under the influence of the potential gradient. The magnitude of the density of this current in the channel will be given by a modification of equation 1.2b as

$$J = -n_1 e \mu_p V_{DS}/l$$

where l is the channel length and n_1 is the density of charge carriers in the inversion layer. Thus the total drain current is given by

$$I_D = -JA = n_1 e A \mu_p V_{DS}/l \qquad (2.1)$$

using the convention shown in Fig. 2.2 that all currents are assumed entering the transistor.

Now the density of inversion charge carriers n_1 is a function of the difference between gate voltage and V_{th}, and so we may write

$$I_D = \frac{e A \mu_p V_{DS}}{l} \, f(V_{GS} - V_{th}) \qquad (2.2)$$

We see from equation 2.2 that the gate voltage may be used to control the drain current; an increase of V_{GS} negatively will cause an increase in I_D. A transistor in which an increase in magnitude of gate voltage causes an increase in the magnitude of drain current is termed an enhancement-type fet. A simplified analysis will show that, for values of $|V_{DS}| \leqslant |V_{GS} - V_{th}|$, the relationship between I_D and $V_{GS} - V_{th}$ is linear. The relationship is plotted in Fig. 2.4 for a commercial p-channel transistor, and the range of linearity is seen to be restricted to a very small range from $V_{GS} = -3.5$ V to -4 V. Nevertheless the assumption of linearity will enable us to obtain a model for the behaviour of the mosfet which, at least, is not too far removed from a real representation of the device behaviour.

Since the voltage drop between source and drain along the channel is assumed small compared with V_{GS}, we may make the approximation of assuming that the voltage between gate and channel is constant along the length of the channel

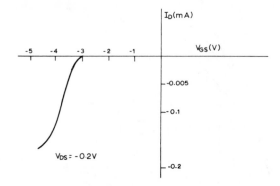

Fig. 2.4 Variation of drain current with gate voltage for low drain voltages

and equal to V_{GS}. This allows us to state that the total charge Q_I is effectively the charge induced on a capacitor whose plates consist of the metal gate and the channel separated by the uniform oxide layer and having a total capacitance C_{ox}. Then we may write, for the charge induced in the inversion layer,

$$Q_I = n_I e A l = -C_{ox}(V_{GS} - V_{th}) \qquad (2.3)$$

Thus equation 2.1 becomes

$$I_D = -\mu_p C_{ox}(V_{GS} - V_{th})V_{DS}/l^2 \qquad (2.4)$$

By definition C_{ox} is the capacitance measured between channel and gate when the drain current is zero but when V_{GS} is large enough to form an inversion layer.

Equation 2.4 may be rewritten in terms of the drain conductance G_D for small drain voltages as

$$G_D = I_D/V_{GS} = -\mu_p C_{ox}(V_{GS} - V_{th})/l^2 \qquad (2.4a)$$

Since $V_{GS} - V_{th}$ is negative, G_D is a positive linear conductance whose value may be varied by changing V_{GS}; in other words the fet acts as a voltage-controlled resistor. It should be noted that as a field-effect transistor is, in principle, a symmetrical device, there is no reason for nominating one electrode as source and the other as drain; however in practice certain transistors are made non-symmetrical and thus they will not give identical characteristics when inverted. The transistor can thus be used as a voltage-controlled resistor for small values of V_{DS} of either polarity,. This is illustrated in Fig. 2.5 which shows that symmetry about the origin, and linearity is better for larger values of V_{GS}.

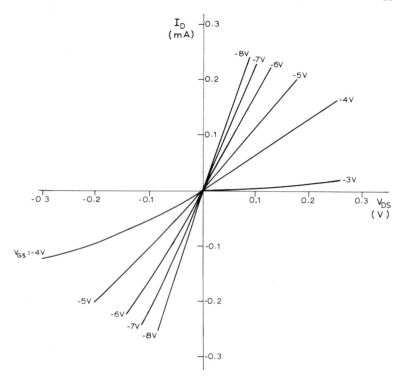

Fig. 2.5 Drain current versus drain voltage for small values of drain voltage

2.4 Pinch-off

It may be seen from equation 2.4 and also intuitively that when $V_{GS} = V_{th}$, no drain current will flow. This condition is termed pinch off, and the voltage at which it occurs is termed the pinch-off or threshold voltage. It may, of course, be alternatively considered as the gate voltage at which the inversion layer vanishes. The value of threshold voltage is not often stated by manufacturers, but its value may be inferred from other measurements as will be shown later. The terms pinch off and threshold are indiscriminately used to describe the same phenomenon, although the former term is usually restricted to junction fets.

2.5 Conduction below saturation

Up to this point we have assumed that the drain–source voltage is very small and hence the potential in the channel is constant from source to drain. If a larger negative voltage is applied between drain and source there will be a change in potential along the length of the channel and this will entail a variation in voltage across the oxide layer at different points along the length of the channel. Assume that, as shown in Fig. 2.6, there is a potential $V(x)$ with respect to source at some point in the channel at a distance x from the source, and that the density of charge carriers in the channel at that point is $n_1(x)$.

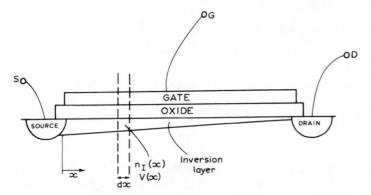

Fig. 2.6 Charge distribution along channel for large drain voltages, in triode region

Consider now the small cross-section of the transistor of width dx as shown. The potential difference between gate and channel is $V_{GS} - V(x)$. Since no inversion layer exists until this potential exceeds V_{th} we may rewrite equation 2.3 for this section as

$$n_1(x) = -\frac{C_{ox}}{eAl}[V_{GS} - V(x) - V_{th}] \tag{2.5}$$

Using equation 1.2b to determine the current density $J(x)$ at a distance x from the source we obtain

$$J(x) = -n_1(x)e\mu_p\frac{dV(x)}{dx} \tag{2.6}$$

and also that

$$I_D = -J(x)A \tag{2.7}$$

Solving these equations and integrating between $x = 0$ at which $V(x) = 0$ and $x = l$ where $V(x) = V_{DS}$ we obtain

$$I_D = \beta\left[(V_{GS} - V_{th})V_{DS} - \frac{V_{DS}^2}{2}\right] \tag{2.8}$$

where β is a constant having, for a p-channel transistor, a value

$$\beta = -\mu_p C_{ox}/l^2 \tag{2.8a}$$

We should note that an inversion layer will exist at all points along the channel only if

$$|V_{GS} - V_{th}| \geq |V_{DS}| \tag{2.9}$$

2.6 Conduction in the saturation region

We have noted from inequality 2.9 that the drain current will obey equation 2.8 only for values of $|V_{GS} - V_{th}| \geq |V_{DS}|$.

When $|V_{GS} - V_{th}| = |V_{DS}|$, the drain current is still given by equation 2.8 as

$$I_D = \frac{1}{2} \beta (V_{GS} - V_{th})^2 \tag{2.10}$$

Now if we increase the magnitude of V_{DS} we observe that at the drain end of the channel the inversion layer will have vanished completely, and only the depletion layer will remain; the charge distribution, therefore, in a small region close to the drain (shown as the depleted region in Fig. 2.7) will be as in Fig.

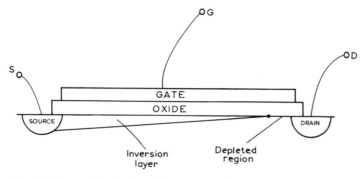

Fig. 2.7 Charge distribution along channel in saturated region

2.3b. This region will be completely depleted of charge carriers. There will, however, still be a field in this region along the direction of the channel caused by the negative voltage on the drain. Part of this voltage drop will occur along the major length of the channel at the source end, and part will occur along this depletion layer. The holes travelling along the channel and reaching the point where the inversion layer vanishes will be swept across this depleted part of the channel to the drain in a manner similar to the flow of charge carriers across the depletion layer in a diode. The voltage in the channel at the point where the inversion layer ceases is $V_{GS} - V_{th}$ and hence the voltage drop across the depleted part of the channel is $V_{DS} - (V_{GS} - V_{th})$. The depleted part of the channel is usually a very small fraction of the total channel length; as a consequence the electric field is high in that region and the charge carriers are swept through by the high drift velocities which they acquire.

Now the current flowing in this saturation region is controlled by the voltage which exists along the part of the channel in which an inversion layer exists, namely by $V_{GS} - V_{th}$ and this potential exists at a point in the channel slightly short of the drain. Thus the effective length of the channel is slightly less than the physical length l. This effective length will be progressively reduced as the magnitude of V_{DS} is increased. We should thus expect the drain current to be approximately constant in this region controlled by $V_{GS} - V_{th}$, rising slightly with increase of V_{DS} due to the reduction in effective length of channel. This may be observed on the drain characteristics shown in Fig. 2.8.

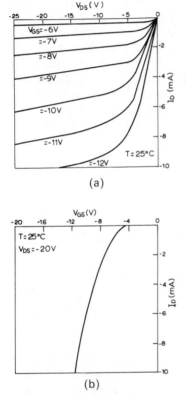

(a)

(b)

Fig. 2.8 Characteristics of a p-channel enhancement type mosfet (a) drain characteristics (b) transfer characteristics

This modulation of channel length with V_{DS} may frequently be ignored and equation 2.10 used to predict the current in the saturation region when $|V_{DS}| \geqslant |V_{GS} - V_{th}|$.

It may be further noted from equation 2.10 that the saturation current is a square law function of the gate–source voltage. This may be seen on the characteristics of the device, shown in Fig. 2.8b. It is interesting to compare the measured transfer characteristic relating I_D with V_{GS} with the analytic expression, and we find that over a large range of voltage, the transfer characteristic is very close to the square law relationship predicted.

2.7 Characteristics of p-channel mosfet

The operation of the device which we have developed in the preceding sections may be compared with the measured characteristics of a p-channel mosfet shown in Figs. 2.8a and b.

In Fig. 2.8a are shown the drain characteristics. For a given gate–source voltage we can see that the current increases almost linearly with V_{DS} at first as predicted by equation 2.4 and then flattens out towards saturation value at the knee of the curve as shown by equation 2.8. Beyond the knee the current remains approximately constant in agreement with equation 2.10 with a slight rise for increased values of V_{DS}.

In Fig. 2.8b the transfer characteristic shows how the current increases above the threshold value as an almost square law function of $V_{GS} - V_{th}$.

To complete the specification of the transistor we need information on the input current–voltage relationship. Since the gate is electrically isolated from the channel, the current flowing in the gate is almost entirely due to current leakage across the surface, and this is rarely greater than about 1·0 nA at room temperature, a typical value being 10 pA. Thus the gate characteristic of the device is usually considered as an open circuit.

2.8 n-channel mosfets

Although similar operating conditions exist in an n-channel mosfet constructed on a lightly doped p-type substrate, there are one or two differences in detail.

It should firstly be noted that surface-state charges on an oxide passivated silicon surface, such as that occurring in the majority of mosfets, is equivalent to a fixed positive surface charge density irrespective of the type of the substrate, whether p or n. The effect of this will be that, when a p-type substrate is used at zero gate bias voltage, a depletion layer will certainly be formed and, depending on the relative magnitudes of surface-state charge and substrate acceptor density, an inversion layer may even be formed. Thus, even at zero bias, conduction charges are available.

It is thus possible to construct an n-channel mosfet which conducts for zero V_{GS}. (In fact this is the most common form of n-channel mosfet.) If the gate voltage is now increased (positively) the drain current will increase since this will cause an increase in the inversion layer. This mode of operation, in which an increase in the magnitude of gate voltage results in an increase in the magnitude of drain current, is termed enhancement. If the gate voltage is increased in magnitude negatively from zero, the drain current will be reduced; this condition is termed a depletion mode. Thus an n-channel mosfet is, in principle, capable of both enhancement and depletion modes of operation, although the enhancement mode is frequently restricted by power dissipation limitations.

The drain characteristics of a typical n-channel mosfet are shown in Fig. 2.9a and the transfer characteristics in Fig. 2.9b.

In the case of depletion mode mosfets, a significant current flows at zero gate voltage. This current I_{DSS} is often specified for such transistors. For the transistor shown in Fig. 2.9 it has a value of about 14 mA.

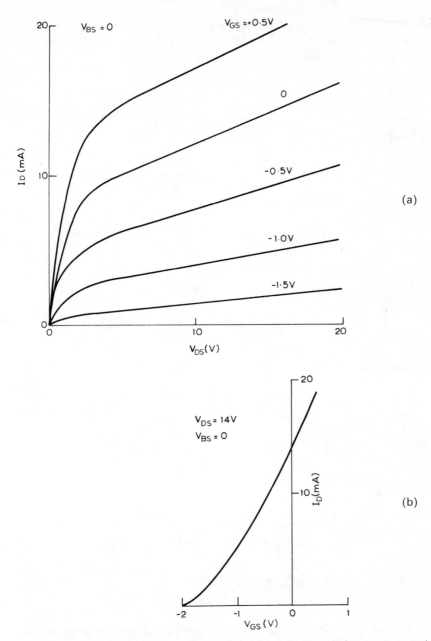

Fig. 2.9 Characteristics of a n-channel depletion type mosfet (a) drain characteristics
(b) transfer characteristic

2.9 Symbols and characteristics of mosfets

A variety of symbols is in use for mosfets. Figure 2.10 shows the British Standard symbols adopted for various types of mosfet. In the symbols the L-shaped
gate is so drawn that the horizontal arm is placed at the source end of the fet.

The substrate is shown separated from the gate as an indication of the presence of the insulating oxide layer. The direction of the arrow on the substrate lead indicates the type of conductivity of the substrate.

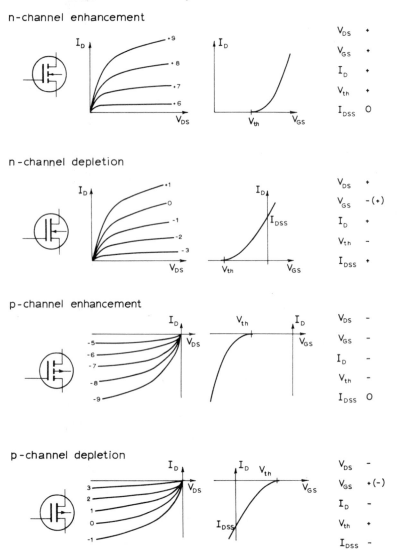

Fig. 2.10 Conventional symbols, transfer characteristics, and normal voltage and current polarities for different types of mosfets

Enhancement-type devices do not have current flow from source to drain for zero bias and are thus shown with broken lines. Depletion-type devices are drawn with a continuous line; since current will flow for zero gate bias voltage. Note that, p-channel depletion-mode devices are not commercially manufactured; the reason for this is that there is no need for such a device if the other types are available. For single gate devices the substrate (sometimes termed the backgate) is internally connected to the source.

Also shown in Fig. 2.10 are typical drain and transfer characteristics of each type of device and the polarities of applied voltage and current for normal operation as an amplifier. The convention for current which has been adopted for all types of mosfets is that all currents are assumed to be flowing into the transistor as shown in Fig. 2.2.

2.10 Dynamic properties of mosfets

The relationships given by equations 2.8 and 2.10 are adequate to represent the transistor when the applied voltages are constant. However, for changing voltages we must take note of the dynamic effects due to variation in magnitude of charge.

The most significant charges exist jointly in the channel and on the gate; these are both caused by the gate–source potential. The relationship between channel charge and gate-source potential may be readily determined. Referring to Fig. 2.6 we see that the total charge contained in the cross-section of channel of length δx is $\delta q_G = n_i e A \delta x$.

Using equation 2.5 and 2.6 (with lower case letters substituted for capitals, since we are now considering time-varying quantities), we obtain

$$\delta q_G = -\frac{C_{ox}^2 \mu_p}{l^2 I_D} [v_{GS} - v(x) - V_{th}]^2 \, \delta V(x) \tag{2.11}$$

Integrating over the length of the channel from $v(0) = 0$ to $v(l) = V_{DS} = v_{GS} - V_{th}$ at saturation gives

$$q_G = -\tfrac{2}{3} C_{ox}(v_{GS} - V_{th}) \tag{2.12}$$

This shows that the gate–source acts like a linear capacitance of value $\tfrac{2}{3} C_{ox}$ subjected to a voltage $v_{GS} - V_{th}$. Note that equation 2.12 is valid for operation in the saturation region only.

For small values of drain source voltage, the potential along the channel is uniform and hence the gate charge q_G is given by

$$q_G = -C_{ox}(v_{GS} - V_{th}) \tag{2.13}$$

Thus for saturation values of v_{DS} the effective gate–source capacitance

$$C_{GS} = \tfrac{2}{3} C_{ox} \tag{2.14}$$

and for low values of v_{DS}

$$C_{GS} = C_{ox}. \tag{2.15}$$

2.11 Circuit model of a mosfet

A simple circuit model of a mosfet is shown in Fig. 2.11, where the charge on the capacitance is given by equation 2.12 for the saturation region or equation 2.13 at low drain voltages, and where the magnitude of the capacitance depends upon the drain current, varying from a value of C_{ox} at low drain currents to $\frac{2}{3}C_{ox}$ at full saturation current.

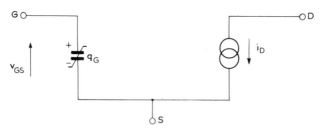

Fig. 2.11 Simple dynamic model of mosfet

2.12 Junction field-effect transistor

An alternative type, the junction fet, may be constructed by alloying an n-type layer, the gate, on the surface of a p-type substrate to form the structure shown diagrammatically in Fig. 2.12. The main flow of current between source and drain is parallel to the n–p junction forming the gate and channel.

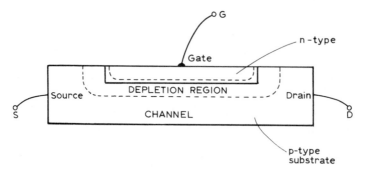

Fig. 2.12 Construction of junction fet and charge distribution at small drain voltages

If the gate is connected to the source and a very small negative voltage applied to the drain, a depletion region is set up at the p–n junction which will have a boundary as shown by the broken line. Within this region there are no mobile carriers and hence the only part of the substrate which is available for conduction between source and drain, the channel, is below the depletion layer.

If the gate is made slightly positive with respect to the source thereby reverse biasing the junction, the depletion layer will get wider and the channel width will be reduced; thus for a given small drain potential the drain current will be smaller. It may be seen therefore that variation of reverse bias on the gate will effectively control the flow of current in the channel. This control will be maintained up to the point (pinch-off) where the gate voltage is such that the depletion region extends right across the substrate; the channel has now vanished and the drain current is reduced to zero. Thus the terminal behaviour of the device is very similar to the mosfet.

In practice appreciable voltages are applied to the drain (in this case negative with respect to the source) and hence the depletion layer will not be of uniform width along the channel as in Fig. 2.12 but will increase in width towards the drain end as in Fig. 2.13. In this case increase in the magnitude of the gate voltage will ultimately cause the depletion layer to extend right across the substrate such that over a short length of the channel no conduction layer exists

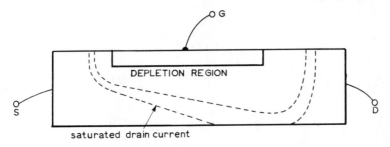

Fig. 2.13 Charge distribution in junction fet at large drain voltages

at all. This behaviour is similar to that observed in mosfets where the channel vanishes near the drain for large gate voltages due to the disappearance of the inversion layer. The current through the device under these saturation conditions is determined, as in the mosfet, by the voltage along the undepleted portion of the channel; the profile of the depletion layer is shown by the lower dotted line in Fig. 2.13. Thus the general shape of the drain characteristics of a junction fet are very similar to those of a mosfet. It should be noted that for zero gate voltage some drain current will flow and that, as the magnitude of the gate voltage is increased, the drain current is reduced and hence the device is a depletion type fet.

The complementary type of junction fet may be made by starting with n-type substrate and diffusing a p-type layer on its surface to form an n-channel fet.

The drain and transfer characteristics of a typical junction fet are shown in Fig. 2.14 in which one may note the typical linear region for low values of V_{DS}, the triode region as V_{DS} is increased up to the ultimate saturation region for large values of V_{DS} where the current settles to a constant value.

An analysis of the current flow may be made if prior assumptions are made about the nature of the junction between the gate and substrate. The simplest hypothesis is that of an abrupt junction between n-and, p-regions. Such analysis leads to expressions which are rather too cumbersome to be of much value in circuit design. However, from an experimental determination, the drain current of a junction fet may be seen to be an almost exact square law function of the gate–source voltage, and hence a useful empirical relation for a junction fet in saturation is

$$I_D = I_{DSS}(1 - V_{GS}/V_{th})^2 \qquad (2.16)$$

where I_{DSS} is the drain current at zero gate–source voltage. V_{th} is the pinch-off voltage, namely the value of gate voltage needed to reduce the drain current to zero. (Frequently the symbol V_p is used.)

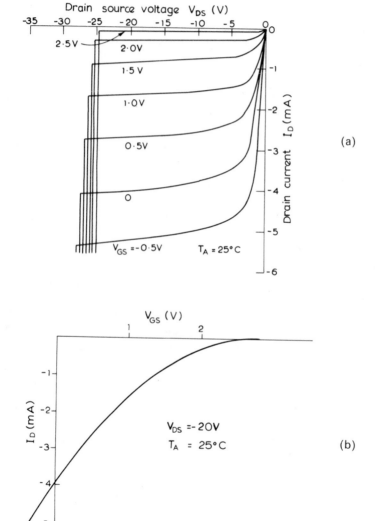

Fig. 2.14 Characteristics of p-channel junction fet (a) drain characteristics (b) transfer characteristic

Equation 2.16 is valuable since it is of the same form as equation 2.10 for a mosfet and allows us to use comparable arguments for both types of transistor.

The input conductance of the device, namely the ratio of gate current to gate–source voltage is very high and is frequently approximated by an open circuit. This has less justification than in the case of the mosfet; however, in the junction fet, so long as the gate is reverse biased the current flowing will only be the reverse saturation current of a silicon diode, which is usually of the order of 100 nA, and so its neglect is not usually serious.

The symbols used for junction fets of the two types are shown in Fig. 2.15, which also shows typical drain and gate characteristics and also polarity of normal operating voltages and currents.

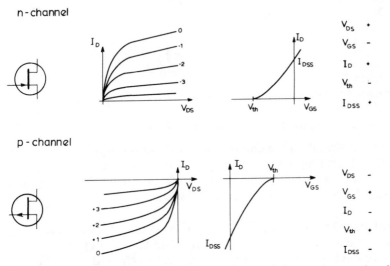

Fig. 2.15 Conventional symbols, transfer characteristics, and normal voltages and currents for various types of junction fet

The dynamic behaviour of the junction fet may again be represented by a capacitor between gate and source. In this case the value of the capacitor is a function of the applied gate–source voltage, the nature of the functional relationship being determined by the form of the semiconductor junction between gate and channel. Rather than attempt to obtain an analytic expression for this, it is preferable to use measured values at known values of drain voltage.

2.13 Effect of temperature on mosfets

The effect of variation of temperature on the properties of mosfets is seen in Fig. 2.16, which shows the transfer characteristic of a mosfet type MTO1 at three different temperatures. It will be noted that the principal effect is a variation in the saturation drain current for a given gate–source voltage. As the temperature increases the drain current falls. The cause of this may be seen by reference to equation 2.10 where it is seen that for a mosfet in the saturation region at a given value of V_{GS}, I_D is proportional to the mobility which is highly temperature dependent. Measurements show that drain current is approximately proportional to T^{-2}.

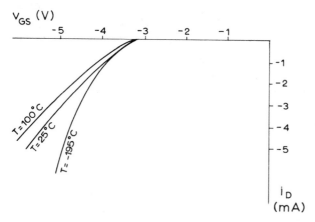

Fig. 2.16 Effect of temperature on transfer characteristic of mosfet

2.14 Effect of temperature on junction fets

For junction fets the saturation drain current is proportional to I_{DSS} and it may be shown that this is proportional to μV_{th}. Now both mobility, μ, and V_{th} vary with temperature. The effect of these on the transfer characteristic for a junction fet shows that I_{DSS} is strongly dependent on temperature but V_{th} is more weakly so.

An additional effect noticeable in a junction fet is that the gate current will increase as the temperature is raised since the gate–source is identical with a reverse biased diode in which the saturation current has been shown in Section 1.7 to increase with temperature.

2.15 Voltage limitations in fets

Figure 2.14a shows that in the junction fet considered a drain–source voltage in excess of about 25 V will cause a large increase in drain current. This is the same phenomenon as the breakdown which occurs in semiconductor diodes and has been discussed in Section 1.11.

Since breakdown occurs across the junction between gate and channel, it is the maximum value of gate–channel voltage which limits operation. The maximum value of this will occur at the drain and is equal to V_{GD}. This is confirmed by Fig. 2.14a from which it may be seen that as V_{GS} becomes more positive, breakdown occurs at slightly lower values of V_{DS}. Since the mechanism of breakdown in a junction transistor is an avalanche process, the transistor will not be damaged if the breakdown voltage is exceeded so long as the current is limited.

In a mosfet a similar rise in drain current may be observed if the voltage between drain and gate is too large. In this case, however, the breakdown is completely destructive since it is caused by the rupture of the insulating oxide layer between channel and gate. It is essential, therefore, to ensure that no excessive voltages exist between gate and source during operation or assembly of a circuit. Since the gate–channel resistance is very high, any charges which may be induced on the gate may take hours to leak away. Such accumulated charges on the gate, induced perhaps by friction during handling, may cause destructive voltages to appear across the oxide layer. Most mosfets are, therefore, supplied fitted with a ring which shorts all electrodes together and this should only be removed immediately before connection in a circuit. Other mosfets are constructed with protective diodes connected between gate and source to remove any excess charge before any large voltage develops.

References

ALLISON, J., *Electronic Engineering Materials and Devices*, McGraw-Hill, London, 1971, Chapter 12.

SEVIN, L. J., *Field-effect Transistors*, McGraw-Hill, New York, 1965, Chapters 1 and 2.

WALLMARK, J. T. and JOHNSON, H., *Field Effect Transistors*, Prentice Hall, 1966, Chapters 5 and 8.

CRAWFORD, R. H., *Mosfet in Circuit Design*, McGraw-Hill, 1967, Chapters 1, 2 and 3.

COBBOLD, R., *Theory and Application of Field-effect Transistors*, Wiley-Interscience, 1970, Chapters 3, 4 and 6.

TODD, C. D., *Junction Field-effect Transistors*, Wiley, 1968, Chapters 1 and 2.

Problems

2.1 The characteristics of a p-channel mosfet are given in Fig. 2.8 and of an n-channel mosfet in Fig. 2.9. In each case sketch the characteristic of the one-port nonlinear element obtained by connecting (a) gate to source, (b) gate to drain.

2.2 Sketch the voltage transfer characteristic relating V_i to V_o for the network of Fig. P2.1 using a transistor whose characteristics are given in Fig. 2.8.

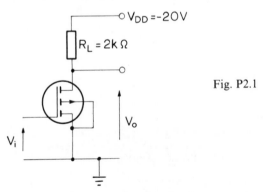

Fig. P2.1

2.3 A one-port nonlinear element is used as the load resistance in a fet amplifier. Which of the devices investigated in Problem 2.1 would be suitable.

2.4 A one-port network element is constructed by using a transistor whose characteristics are given in Fig. 2.8 with the gate connected to its drain. This device is used as a load resistor in an amplifier using an identical transistor with a supply voltage of -20 V. Compare with the results obtained in Problem 2.2 for the voltage transfer characteristic.

2.5 A one-port network element is constructed by using a transistor whose characteristics are given in Fig. 2.9. It is used as a load resistor for an identical transistor supplied from 20 V. Sketch the voltage transfer characteristic.

2.6 The two mosfets in Fig. P2.2 form an exactly balanced complementary pair. The p-channel device has characteristics given in Fig. 2.8. Sketch the voltage transfer characteristic relating V_o to V_i. Also plot I_D as a function of V_i.

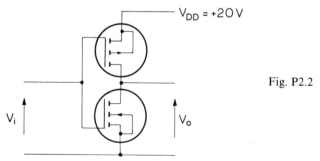

Fig. P2.2

2.7 (a) In the amplifier of Problem 2.2, choose a value for V_{GS} to give maximum range of voltage excursion about that point. What is the maximum deviation of input voltage possible to ensure that the output is approximately linearly related to input? (This involves a purely subjective judgement.) What would be the voltage gain in such an amplifier? Compare this with the voltage gain of the amplifiers of Problems 2.4 and 2.5 for the same excursion of input voltage.

(b) Reverting to the amplifier of Problem 2.2, the drain current needs to be increased by 25%; what value of V_{GS} is needed? How has this affected the input voltage excursion?

2.8 The drain current of a mosfet measured at $V_{DS} = -20$ V, $V_{GS} = -5$ V is $I_D = -0.8$ mA. It is used in the circuit of Fig. P2.3. Determine the quiescent values of V_{GS}, V_{DS}, I_D in the circuit (a) when A is connected to C, (b) when A is connected to B. $V_{th} = -3$ V.

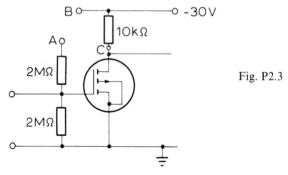

Fig. P2.3

2.9 The characteristics of a p-channel mosfet are given in Fig. 2.5. It is used in the circuit of Fig. P2.4. Assume that the maximum permitted value of V_{GS} is (a) -8 V, (b) -12 V. Determine the range of attenuation (v_o/v_i) to small variations of input voltage, which is possible.

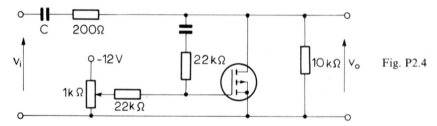

Fig. P2.4

2.10 The mosfet in the circuit of Fig. P2.5 has $V_{th} = 2$ V. When $V_{GS} = 7$ V, $V_{DS} = 5$ V the transistor is in its saturated region and $I_D = 4$ mA. The gate to source capacitance in the absence of drain current is 4 pF. The voltage v_i is suddenly raised from zero to 7 V. Discuss the manner in which the gate–source voltage changes with time and determine its steady value. Discuss also how the drain current changes with time (analytic solutions to these problems are of little practical value).

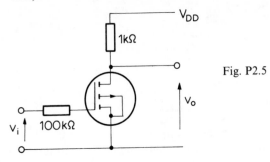

Fig. P2.5

3
Fets—Small Signal Models

3.0 Introduction

One of the main fields of application of both unipolar and bipolar transistors is in the amplification of small fluctuations of voltage and current. In order to use the device in the most efficient manner these small fluctuations are usually superimposed on a steady voltage or current. The performance of the device as an amplifier of small signals is governed by the parameters of the device in the vicinity of the operating point which has been established by the steady voltages in the circuit. In Section 1.7.3 we examined the incremental conductance of a diode which gives a measure of its effect on a small signal voltage superimposed on a steady voltage. In this chapter we shall apply the same techniques to a study of the small signal parameters of field-effect transistors relating the terminal currents and voltages when these are small in magnitude and superimposed at a fixed operating point.

In the field-effect transistor we may note that the largest change of drain current for a given change of gate–source voltage will occur when the transistor is on the saturated part of its characteristic and so we shall always select our operating point in that region. As we have shown that the characteristics of both mos and junction fets are almost identical, we shall develop this model for mosfets only; the only difference will be in the magnitude of certain components.

We may break down the elements of a device model into several groups; firstly the intrinsic elements, which are used to represent the fundamental mode of operation of the device; secondly the extrinsic elements, which are inevitably present but which are undesirable in an idealized device and which the device designer attempts to minimize; thirdly the parasitic elements which result from mounting and packaging the device in a convenient form. This subdivision is merely one of convenience since it is not possible to eliminate any of the elements. We shall look first at the intrinsic transistor model and then at the extrinsic and parasitic elements.

3.1 Intrinsic small signal model

The basic equations which govern the operation of the device in the saturation region are 2.10 and 2.12. As they are now relevant to time varying voltages they are rewritten here using lower case letters for the total voltages and currents.

$$i_D = \frac{1}{2}\beta(v_{GS} - V_{th})^2 \tag{3.1}$$

$$q_G = -\frac{2}{3}C_{ox}(v_{GS} - V_{th}) \tag{3.2}$$

We now select an operating point defined by the voltages V_{GS}, V_{DS} at which drain current I_D flows and gate charge Q_G exists. The variations of current i_d and charge q_g due to variations v_{gs} and v_{ds} may be determined by differentiation of equations 3.1 and 3.2.

$$\delta i_D = i_d = \beta(v_{GS} - V_{th})\delta v_{GS}$$
$$= \beta(V_{GS} - V_{th})v_{gs} \tag{3.3}$$

since $v_{gs} \ll V_{GS}$ and thus $v_{GS} \simeq V_{GS}$; also

$$\delta q_G = q_g = -\frac{2}{3}C_{ox}\delta v_{GS}$$

$$q_g = -\frac{2}{3}C_{ox}v_{gs} \tag{3.4}$$

These two equations may be modelled by the linear circuit of Fig. 3.1 where

$$C_{gs} = \frac{2}{3}C_{ox} \tag{3.5}$$

and

$$g_m = \beta(V_{GS} - V_{th}) \tag{3.6}$$

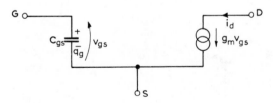

Fig. 3.1 Simple small signal model of mosfet

It should be noted that g_m is a linear function of excess gate voltage and therefore must not be considered as a fixed constant of the device. Combining equations 3.6 and 3.1 gives

$$g_m = (2\beta I_D)^{1/2} \tag{3.6a}$$

Observations show that this relationship is closely realized for both mosfets and junction fets.

Equation 3.5 shows that the gate–source capacitance of a mosfet is independent of V_{GS}. However as noted in Section 2.12 the capacitance in a junction fet is often a strong function of V_{GS} as evidenced by Fig. 3.2 for a typical junction fet.

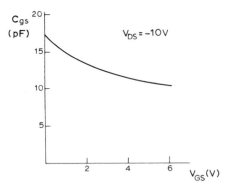

Fig. 3.2 Variation of gate–source capacitance with drain current

3.2 Extrinsic elements of small signal model

The elements we shall now consider are those which, although not essential to the operation of the device, are nevertheless still associated with the semiconductor material and therefore may be reduced or modified by careful design without impairing performance, but which cannot be eliminated.

3.2.1 Drain conductance

As stated in Section 2.6 increase of drain voltage will cause a slight increase in drain current. Small fluctuations of drain–source voltage will therefore give rise to small fluctuations of drain current, and these may be assumed to be linearly related for small variations. This may be modelled by including a small

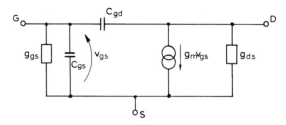

Fig. 3.3 Small signal model of mosfet

signal conductance, g_{ds}, $(= 1/r_{ds})$ the drain–source conductance, connected between drain and source in the model. This is shown in Fig. 3.3. It may be directly determined from the slope of the drain characteristic in the saturation region. We may note, from Figs. 2.8a, 2.9a and 2.14a, that the slope of these characteristics is dependent on the magnitude of drain current and this is shown for a junction fet in Fig. 3.4.

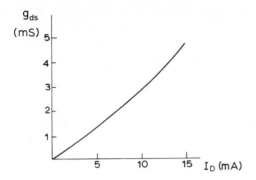

Fig. 3.4 Variation of drain conductance with drain current

3.2.2 Gate–drain capacitance

From the simplified diagram of the fet of Fig. 2.1, we observe that the gate electrode overlaps the drain and hence there will be a small capacitance C_{gd} directly between gate and drain. This is also included in Fig. 3.3. This will exist both in a mosfet and a junction fet. The magnitude of this capacitance will normally be very small but nevertheless it proves to be one of the more undesirable elements in the transistor when used as a high frequency amplifier. Attempts may be made to reduce it by careful fabrication so that the gate electrode does not overlap the drain. Since the transistors are mass produced a certain tolerance must be allowed in fabrication, and it may be shown that it is possible to fabricate depletion type mosfets such that the gate does not overlap the drain, in fact it may fall short of it, but that this is not possible in enhancement mosfets.

3.2.3 Gate conductance

We have seen in Sections 2.6 and 2.12 that the direct current flowing in the gate electrode will be negligibly small for a mosfet. As a consequence the incremental resistance between the gate and source will be large. In the case of a junction fet, the resistance will still be large, although several orders of magnitude less than for a mosfet. This element of the circuit model of conductance, g_{gs}, is included in Fig. 3.3, although in most cases it may be approximated by an open circuit.

3.2.4 Source and drain conductances

One further phenomenon which needs modelling is the flow of holes from the source end of the channel to the source electrode and from the drain end of the channel to the drain electrode. Since this flow consists entirely of majority carriers through the bulk semiconductor, its effect can be modelled in these two regions by a resistance, $r_{ss'}$, between source electrode S and source end of channel S' and by a resistance, $r_{dd'}$, between drain electrode D and drain end of channel D'.

3.3 Parasitic elements

In addition to the elements considered in Section 3.1 and 3.2 which are inherent in the wafer of the transistor, a number of other phenomena relate to the packaging of the transistor, in particular, capacitance between the leads on the 'header' on which the transistor is mounted, and the self inductance of these leads. However, at frequencies less than about 100 MHz these may either be ignored or combined with existing capacitances such as C_{gs}, C_{gd}. Sometimes an additional capacitance C_{ds} between drain and source as shown in Fig. 3.5 is included to account for strays. At even higher frequencies it may be necessary to include an inductance in series with each of the three leads to account for the inductance of the finite length of the leads; such refinements are rarely required below about 0·5 GHz.

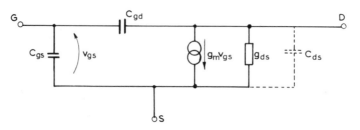

Fig. 3.5 Practical small signal model of fet

3.4 Values of small signal parameters

The model for a field-effect transistor which we shall adopt is given in Fig. 3.5. For most low- and medium-frequency work this will be adequate. Occasionally it may be necessary to include C_{ds}.

Typical values of elements for a mosfet are

$$
\begin{aligned}
C_{gs} &= 5\cdot5 \text{ pF} \\
C_{gd} &= 0\cdot12 \text{ pF} \\
C_{ds} &= 1\cdot4 \text{ pF} \\
g_{m} &= 10\cdot0 \text{ mS} \\
g_{ds} &= 1\cdot0 \text{ mS}
\end{aligned}
$$

Values for a junction fet are

$$
\begin{aligned}
C_{gs} &= 11\cdot0 \text{ pF} \\
C_{gd} &= 5\cdot0 \text{ pF} \\
C_{ds} &= 1\cdot5 \text{ pF} \\
g_{m} &= 4\cdot0 \text{ mS} \\
g_{ds} &= 3\cdot0 \text{ mS}
\end{aligned}
$$

Values of $r_{ss'}$ and $r_{dd'}$ are typically of the order of 50 Ω.

3.5 Temperature variation of parameters

The effect of temperature on the small signal characteristics of fets is seen most in the variation of g_m. In Section 2.13 and 2.14 it is noted that the transfer characteristics depend significantly on temperature. Since g_m is defined as the slope of the transfer characteristic we should expect this also to be related to temperature. Typical variation of g_m for a junction fet is shown in Fig. 3.6. It is seen that even over normal ranges of ambient temperature considerable variations in g_m are evident.

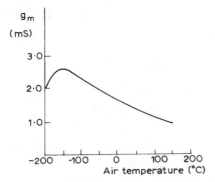

Fig. 3.6 Variation of transconductance with temperature

3.6 Low frequency small signal model

At relatively low frequencies the model of the transistor shown in Fig. 3.5 may be considerably simplified. At frequencies where the reactances of the capacitances is large compared to the values of the resistors the former may be replaced by open circuits and we obtain the very simple low frequency small signal model shown in Fig. 3.7.

3.7 Representation by y parameters

The model which we have developed in Fig. 3.7 is based on a linearized representation of the various physical phenomena which we can identify in the operation of a fet. Such a model may be progressively refined by the addition of further elements to account for other observed phenomena. However, a point is reached where the circuit model becomes so complex that its utility is reduced.

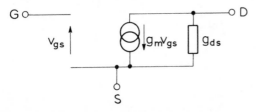

Fig. 3.7 Low frequency small signal model of fet

An alternative form of representation may be obtained if the values of the currents and voltages at the terminals of the transistor are measured in magnitude and phase at a number of discrete frequencies. If this is done we may define a set of admittance parameters for the transistor with one of its terminals, say the source, common between input and output circuits, as outlined in appendix A. We may therefore specify the transistor by quoting the real and imaginary parts, the conductance and susceptance, of the four admittance parameters measured in common source,

$$y_{is}(= g_{is}+jb_{is}); \quad y_{fs}(= g_{fs}+jb_{fs});$$
$$y_{rs}(= g_{rs}+jb_{rs}); \quad y_{os}(= g_{os}+jb_{os})$$

Thus in common source configuration

$$i_g = y_{is}v_{gs}+y_{rs}v_{ds}$$
$$i_d = y_{fs}v_{gs}+y_{os}v_{ds} \tag{3.7}$$

where the four y parameters are dependent on frequency.

A typical set of y parameters for a junction fet is given in Fig. 3.8.

The merit of this representation is that the parameters are known precisely at a given frequency, operating point, temperature, etc., and no reliance is made on the accuracy of the model assumed for some physical process. The disadvantage of the two-port parameter representation also lies in the fact that it is so precisely specified. No inferences about these parameters may be made under any other conditions than those at which measurements were made since their functional dependence on biasing voltage or frequency is unknown, or, if known, too complex for utility. The two-port parameters are useful, however, when a computer is used to assist in circuit design since it is relatively easy, although maybe tedious, to store large amounts of information giving the properties over a wide range of external conditions.

3.7.1 Comparison of y parameters and physical model

The relationship between the y parameters and the parameters of the small signal hybrid π model may be investigated by analysing the network of Fig. 3.5, including $r_{ss'}$ and $r_{dd'}$, to obtain its two-port parameters. For frequencies where $\omega \ll 1/C_{gd}r_{dd'}$ and $\omega \ll 1/C_{gs}r_{ss'}$ and assuming $1/g_m \gg r_{ss'}$ and $r_{dd'}$, we obtain:

$$\left.\begin{aligned}
y_{is} &= \omega^2(C_{gd}^2 r_{dd'}+C_{gs}^2 r_{ss'})+j\omega(C_{gd}+C_{gs}) \\[2mm]
y_{fs} &= g_m/(1+g_m r_{ss'})-j\omega C_{gd} \\[2mm]
y_{rs} &= -\omega^2 C_{gd}^2 r_{dd'}-j\omega C_{gd} \\[2mm]
y_{os} &= g_{ds}+\omega^2 C_{gd}^2 r_{dd'}+j\omega(C_{gd}+C_{ds})
\end{aligned}\right\} \tag{3.8}$$

These expressions may be plotted for a given transistor and they show that reasonable agreement is obtained with the measured two-port parameters over a limited range of frequency.

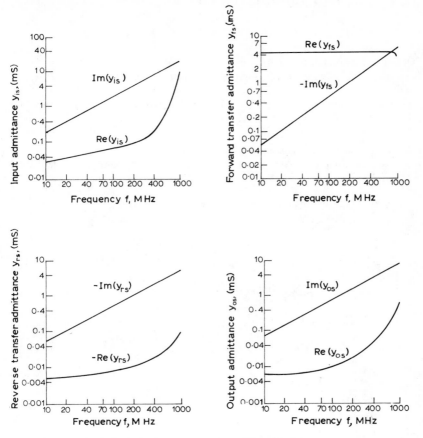

Fig. 3.8 Variation of admittance parameters of fet with frequency

References

Sevin, L. J., *Field-effect Transistors*, McGraw-Hill, 1965, Chapter 3.

Crawford, R. H., *Mosfet in Circuit Design*, McGraw-Hill, 1967, Chapter 4.

Wallmark, J. T. and Johnson, H., *Field Effect Transistors*, Prentice Hall, 1966, Chapters 10 and 11.

Cobbold, R., *Theory and Application of Field-effect Transistors*, Wiley-Interscience, 1970, Chapter 5.

Todd, C. D., *Junction Field-effect Transistors*, Wiley, 1968, Chapter 3.

Problems

3.1 The transconductance of an n-channel junction fet at $V_{DS} = 20$ V, $V_{GS} = 0$, $I_D = 10$ mA is 6 mS. Estimate the value of the threshold voltage. Determine the value of transconductance and the gate–source voltage at $V_{DS} = 20$ V, $I_D = 5$ mA. What assumptions are made?

3.2 Determine as many parameters as possible of the small signal model of the fet whose characteristics are given in Fig. 2.8 at $V_{GS} = -9$ V, $V_{DS} = -20$ V.

3.3 The parameters of a certain mosfet whose threshold voltage, V_{th}, is -5 V, measured at $V_{DS} = 15$ V, $V_{GS} = -10$ V at a frequency of 1 kHz are

$$y_{fs} = 4000 \ \mu S$$
$$C_{iss} = 11 \ pF$$
$$C_{rss} = 1 \cdot 6 \ pF$$

(a) Obtain an incremental model for the fet.

(b) Find new values of g_m and C_{gs} appropriate to an operating point of $V_{DS} = -30$ V, $V_{GS} = -20$ V.

(c) If the output admittance measured at a frequency of 1 MHz is $(32 + j \ 17 \cdot 6) \ \mu S$ extend your simple circuit model of (a).

Note:
C_{iss} = input capacitance in common source with output short circuited.
C_{rss} = feedback capacitance between output and input in common source with output short circuited.

3.4 Prove the first of the four equations 3.8 by analysis of Fig. P3.1

$$y_{is} = \omega^2 (C_{gd}^2 r_{dd'} + C_{gs}^2 r_{ss'}) + j\omega(C_{gd} + C_{gs})$$

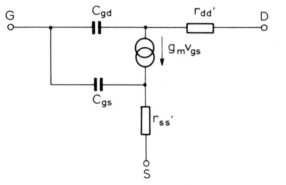

Fig. P3.1

3.5 Using the data curves of Fig. 3.8 estimate the parameters g_{gs}, C_{gs}, g_{ds}, C_{ds}, g_m, C_{gd}, $r_{ss'}r_{dd'}$ of a small signal physical model of the transistor.

4
Bipolar Transistor—Principles

4.0 Introduction

In Chapter 2 we studied the operation of one type of three-terminal electronic device. We shall now turn our attention to a second device, the more common bipolar transistor. Historically this was the first commercially available type of transistor, although its principle of operation was not discovered until many years after that of the field-effect transistor. Like the field-effect transistor it is capable of acting both as an amplifier or as a switch of large voltages. We shall look briefly at the construction and operation of the device without delving too deeply into the underlying physical mechanism responsible for its operation. Nevertheless a general understanding of the relevant physical processes is necessary for a clear appreciation of some of its properties and limitations, and this also will allow us to develop a model which can represent its external behaviour.

4.1 Basic operation of a transistor

A transistor essentially is constructed from three regions of semiconductor, the emitter, base, and collector regions, being either n, p, n respectively or p, n, p, giving rise to the two basic types of bipolar transistor. Nowadays they are almost invariably constructed using silicon as the basic material with appropriate added impurities to produce either n or p type as required. The conventional symbols for the two types of transistor are shown in Fig. 4.1, in addition

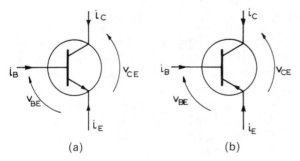

Fig. 4.1 Conventional symbol and current and voltage conventions in bipolar transistor, (a) n-p-n, (b) p-n-p

to the reference directions adopted for currents and voltages; the convention is that the positive sense of all currents is directed into the transistor and the polarity of voltages is indicated by the order of the suffixes (regardless of whether the transistor is p-n-p or n-p-n).

The device is seen to consist of two semiconductor junctions fabricated back to back. However, it is not possible to study the behaviour simply in terms of two isolated diodes as the base region common to the two diodes is so narrow that the two junctions interact.

We shall restrict our initial study to an n-p-n transistor operating in the forward active region with the emitter–base junction forward biased and the collector–base junction reverse biased. Since the emitter junction is forward biased, electrons will be injected into the base from the emitter exactly as they are in the diode. In the base these electrons become minority carriers and their density $n_B(0)$ in the base just at the inner edge of the emitter–base depletion layer is determined by the forward bias potential across the junction as we saw for the diode in Section 1.4. Since in a transistor the base width is very narrow little recombination can take place before the electrons have diffused (because of the concentration gradient) across the base and reached the collector junction. As the collector is reverse biased any electron arriving at the left side of the collector depletion layer will come under the influence of the high field and be swept through to the collector; this may be confirmed by reference to Fig. 1.7b from which we see that the minority charge density on either side of a reverse biased junction is effectively zero.

We see that in simple terms an n-p-n transistor operates by injecting electrons from the emitter into the base; the electrons then diffuse, with very little recombination occurring, through the narrow base, to be collected almost in their entirety by the collector. Thus the current flowing out of the collector is sensibly the same as that flowing in at the emitter; nevertheless since it is flowing from a relatively high voltage, considerably more power is available at the collector than is provided at the emitter, where the applied voltage is of the order of 0·6 V. In addition we may note that a very small base current will flow; this is essentially an imperfection in the device and in modern transistors is limited to a very small fraction of the collector current.

4.1.1 Charge distribution and current flow in the transistor base

Let us consider first the distribution of charge within the base region of the transistor. As we have seen in Section 1.5 in the case of the diode it is essentially the charge distribution in the neutral regions which controls the current flow and the charge in the depletion regions either side of the junction which determines the junction voltages. We shall first look at the charge distribution in the base region between the two inner edges of the emitter–base and collector–base depletion layers. We have noted in Section 1.5 that outside the depletion layer the electric field is small and that nearly all the minority charge carriers move by diffusion only. Thus for the n-p-n transistor we are considering, in which minority carriers are electrons, the current density flowing through the base,

J_B, may be developed from equation 1.7a by neglecting the drift term and changing suffixes as

$$J_B = eD_n \frac{dn_B(x)}{dx} \tag{4.1}$$

where n_B is the electron density in the base. For a first approximation we shall assume there is no recombination or generation, i.e. no charge carriers are created or destroyed between emitter and collector, and hence the current injected by the emitter i_E and the current received by the collector i_C will be equal and

$$i_C = -i_E = -AJ_B \tag{4.2}$$

where A is the effective cross-section of the base. Since J_B is constant we see from equation 4.1 that the gradient of the charge distribution is constant, hence the charge distribution in the base is linear with distance. We may also observe that in all transistors of this nature the current in the base is proportional to the gradient of the charge distribution.

Again, since the emitter–base junction is forward biased, electrons will be injected across the junction into the base in a manner almost identical with that which we developed for a diode; the electron density just inside the base is thus an exponential function of the base–emitter voltage and may be written, by suitable modification of the suffixes and charge carrier type, from equation 1.19

$$n_B(0) = n_{BO} \exp(ev_{BE}/kT) \tag{4.3}$$

where $n_B(0)$ is the electron density at the emitter side of the base and n_{BO} is the equilibrium density in the base. At the collector, the junction is reverse biased and hence the electron density in the base at the edge of the depletion layer around the collector junction will be forced to be approximately zero.

We have thus established that a finite charge distribution exists at the emitter side of the base, a zero distribution at the collector and a linear charge/distance

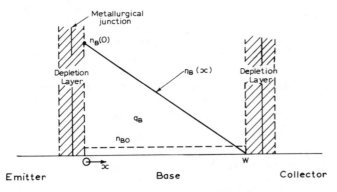

Fig. 4.2 Simplified charge distribution in base region of transistor

relationship between. The charge distribution profile is therefore a straight line from $n_B(0)$ at the emitter to zero at the collector as shown in Fig. 4.2. We may derive a circuit model for this simple analysis; since the slope of the charge distribution is linear, the collector current is equal to the emitter current. This gives a very simple model for the transistor, involving a single current generator in the collector circuit whose magnitude is equal to the emitter current.

From Fig. 4.2 we may observe that q_B, the total charge stored in the base in excess of the equilibrium charge, (i.e. the charge lying between $n_B(x)$ and n_{BO}), is given very closely by

$$q_B = -\frac{1}{2}eAW[n_B(0) - n_{BO}] \tag{4.4}$$

$$= Q_{BO}[\exp(ev_{BE}/kT) - 1] \tag{4.4a}$$

where $\qquad Q_{BO} = -\frac{1}{2}eAWn_{BO}$

For an n-p-n transistor q_B and Q_{BO} are numerically negative. We may now determine the collector current from equations 4.1 and 4.2. Since

$$\frac{dn_B(x)}{dx} = -\frac{n_B(0)}{W}$$

we obtain

$$i_C = \frac{eAD_n n_B(0)}{W} \tag{4.5}$$

Using equation 4.4 and noting that $n_B(0)$ is large compared with n_{BO} we obtain

$$i_C = -\frac{2D_n}{W^2}q_B \tag{4.6a}$$

$$= -q_B/\tau_F \tag{4.6b}$$

where $\tau_F \triangleq W^2/2D_n$ and is the transit time for the electrons to travel across the base.

By combining equations 4.6a, 4.4, 4.3, and 4.2, we obtain

$$i_E = -\frac{D_n eA}{W} n_{BO}[\exp(ev_{BE}/kT) - 1]$$

$$= -I_1[\exp(ev_{BE}/kT) - 1] \tag{4.7}$$

where $I_1 = D_n e A n_{BO}/W$, showing that the emitter current is an exponential function of the base–emitter voltage (for $v_{BE} > 5kT/e$ we may neglect the second term in brackets).

The model which will represent this simple explanation of transistor action is shown in Fig. 4.3, the diode being an ideal diode with reverse saturation current I_1 and the current generator forcing the collector and emitter currents to be equal.

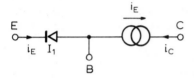

Fig. 4.3 Simple model for transistor under normal operating bias conditions

4.2 Causes of base currents

We now need to develop the model of the transistor further to include some of the additional physical effects. The electrons in the base which are causing transistor action are minority carriers and they are flowing through a layer of p-type silicon in which there is a high density of holes. It is therefore to be expected that the probability of collision between an electron and a hole is relatively high. When such an event occurs the electron will fill up the hole and one minority carrier which was available for conduction between emitter and collector will have been lost.

Thus we should expect that more electrons need to be injected into the base from the emitter than are collected by the collector; this results in the emitter current being greater than the collector current. There will consequently be a current flowing into the base which is a result of this recombination and is equal to the difference between the emitter and collector currents. Since the recombination occurs within the base itself the current entering the base through the base lead must consist of holes replacing those which have combined with electrons.

It is reasonable to assume that the probability of an electron combining with a hole in the base is proportional to the total excess charge q_B and thus the recombination current flowing into the base is proportional to q_B.

In addition to this we must recall that we have only considered the flow of electrons from emitter to base across the junction. By analogy with the diode we should also expect there to be a flow of holes across the emitter–base junction from base to emitter. In general this flow will be considerably smaller than that of the injected electrons as a result of correct choice of doping levels in the base and emitter. This hole current will only flow across the emitter–base junction and will not contribute to the collector current. It will therefore be an additional component of the base current and will add further to the emitter current. Again it may be shown that the magnitude of this current is directly proportional to the excess base charge q_B.

Adding these two components we obtain the total base current i_B and we note that this is proportional to the base charge q_B. Introducing a constant of proportionality we obtain

$$i_B = -q_B/\tau_{BF} \qquad (4.8)$$

where τ_{BF} is the base charging time, being the time required to change the base charge by q_B at a constant current through the base.

The minority charge distributions in the three regions of the transistor are shown in Fig. 4.4. The charge profile in the base is almost linear, the slight

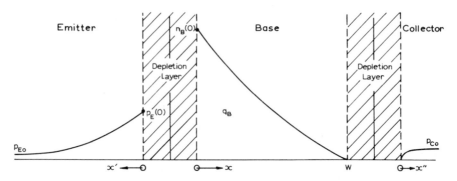

Fig. 4.4 Charge distribution in emitter, base, and collector when normally biased

curvature being due to recombination. As we expect, the gradient at the emitter is greater than that at the collector to account for the fact that the emitter current exceeds the collector current by the base recombination current. In the emitter the charge distribution falls exponentially to zero due to recombination, since the emitter is considerably wider than the base. The gradient of the charge density in the emitter is proportional to the hole current injected into the emitter from the base. (This may be compared with Fig. 1.7a.) In the collector the charge profile starts at zero, rising to the equilibrium value some distance from the junction. (This too may be compared with Fig. 1.7b for a reverse biased diode.) Again the charge profile has a finite gradient at the edge of the depletion layer which is completely independent of the collector current flowing across the base from the emitter. We see that this suggests that at the collector there is a flow of holes from right to left across the junction irrespective of any electrons which may be flowing in the opposite direction. This current from collector to base may be measured as the collector–base current I_{CBO} which will flow with the emitter on open circuit and there is no injected electron current. Thus the total collector current is composed of this reverse saturation current, in addition to the injected collector current given by equation 4.6b. Hence

$$i_C = -\frac{q_B}{\tau_F} + I_{CBO} \qquad (4.9)$$

4.3 Models of transistor in active region

We are now in a position to derive a model for the transistor under steady conditions which is an improvement on that of Fig. 4.3.

The relationship of equation 4.8 must be modified by I_{CBO} to

$$i_B = -\frac{q_B}{\tau_{BF}} - I_{CBO} \tag{4.10}$$

In addition, using equation 4.9 and noting that

$$i_B + i_C + i_E = 0 \tag{4.10a}$$

we obtain

$$i_E = \frac{q_B}{\tau_{BF}} + \frac{q_B}{\tau_F} \tag{4.11}$$

Also we have as before

$$i_E = -I_1[\exp(ev_{BE}/kT) - 1] \tag{4.12}$$

Since from equation 4.11 we see that i_E is directly proportional to q_B we may rewrite equations 4.9 and 4.10 as

$$i_C = -\alpha_F i_E + I_{CBO} \tag{4.13a}$$

$$i_B = -(1 - \alpha_F)i_E - I_{CBO} \tag{4.13b}$$

where $\alpha_F = \tau_{BF}/(\tau_{BF} + \tau_F)$ and is the short circuit current gain between emitter and collector with base terminal common. These equations are used to obtain

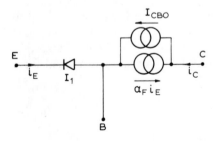

Fig. 4.5 Total model of transistor: common base

the circuit model of Fig. 4.5. α_F is less than, but very close to, unity, deviating from it by about 1% or even less.

An alternative method of connection of a transistor is to have the emitter terminal common between input and output. The relationship between base and collector currents in this case is obtained by eliminating i_E between equation 4.13 to give

$$i_C = \beta_F i_B + I_{CEO} \tag{4.14}$$

where
$$\beta_F = \frac{\alpha_F}{1 - \alpha_F} = \frac{\tau_{BF}}{\tau_F} \tag{4.15a}$$

$$I_{CEO} = I_{CBO}/(1 - \alpha_F) \tag{4.15b}$$

β_F is the large signal common emitter current gain of the transistor, being the ratio of collector to base current. The symbol h_{FE} is sometimes used. I_{CEO} is the common emitter, collector saturation current measured between collector and emitter with the base on open circuit.

Equation 4.14 in conjunction with equation 4.12 leads to the circuit model of Fig. 4.6.

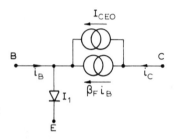

Fig. 4.6 Total models of transistor: common emitter

It should be stressed here that these models are only valid when the transistor is operating in its normal amplifying mode with the emitter forward biased and the collector reverse biased.

Similar equations may be derived for a p-n-p transistor; in this case it is simpler to retain the form of the equations we have already developed and to consider the reverse saturation currents I_{CBO} and I_{CEO} to have a negative magnitude; the base charge, being holes, will be positive. The only change that is necessary is in the exponent in equation 4.12 which must now read for a p-n-p transistor

$$i_E = -I_1[\exp(-ev_{BE}/kT) - 1] \tag{4.12a}$$

4.4 Dynamic behaviour of transistor

The equations and models we have derived so far are valid so long as the currents and voltages are steady or, at least, only changing very slowly. If the fluctuations are fast we must take into consideration the time taken to establish the various charge distributions in the transistor and the delays in response that this will incur.

We have seen in Section 1.9 that any change in the potentials across a diode will involve a change in the charges stored (a) in the space-charge (depletion) region and (b) in the neutral regions. The same situation exists in a transistor where the flow of current is determined predominantly by the charge q_B stored in the neutral region of the base. Change in current can only occur as a result of change in charge; although the charge at the emitter side of the base will change immediately as a result of the change of junction voltage, the total charge will only reach its equilibrium value after some delay. In addition there are two junction depletion layers in which space charges exist, and, as in the diode, any change of terminal voltage will cause a consequent change in depletion layer charge, giving rise to a transient component of terminal current. We shall look at these two phenomena in an n-p-n transistor.

4.4.1 Charge stored in the base region

Let us consider now a situation in which the base–emitter voltage v_{BE} is changed to a new value v'_{BE}. We must assume that this change is sufficiently slow that the charge profile at any instant of time may be represented by its static distribution as in Fig. 4.5; this restriction implies that redistribution of charge carriers within the base occurs almost instantaneously, whereas time is required to inject the carriers into the base. For simplicity of analysis we shall assume that $I_{CBO} = 0$.

Now at any instant of time t, the collector current is given by equation 4.6b as

$$i_C = -q_B/\tau_F \qquad (4.6b)$$

Further, by noting that the base current is essentially a flow of majority carriers to maintain neutrality in the base when the injected minority carriers are changed we may consider the injected base current as serving two purposes. The first is to provide for an increase in base charge given by dq_B/dt; the second is to provide for replacement of charge carriers lost by recombination according to equation 4.8. Thus the total base current is

$$i_B = -\frac{q_B}{\tau_{BF}} - \frac{dq_B}{dt} \qquad (4.16)$$

Whence using equation 4.10a we may write

$$i_E = \frac{q_B}{\tau_F} + \frac{q_B}{\tau_{BF}} + \frac{dq_B}{dt} \qquad (4.17)$$

This allows us to improve on the circuit model of Fig. 4.5 by including these dynamic effects. This improvement is shown in Fig. 4.7 for a transistor in common base configuration with the currents expressed in terms of the stored charge as a controlling parameter. The nonlinear capacitor, having charge

q_B, accounts for the last term in equation 4.17 since equation 4.4a shows that the base charge is not linearly related to the base–emitter voltage.

The emitter diode saturation current may be expressed in terms of Q_{BO} by combining equations 4.4a, 4.11 and 4.12 to give

$$I_1 = -Q_{BO}\left(\frac{1}{\tau_F}+\frac{1}{\tau_{BF}}\right) \tag{4.18}$$

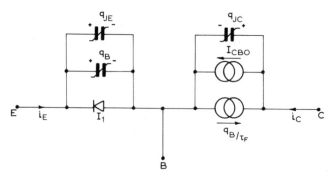

Fig. 4.7 Dynamic model of complete transistor when normally biased

4.4.2 Charge stored in depletion layers

In Section 1.9 we observed that change of applied junction voltage will cause a change in the charge stored in the depletion layer either side of the junction. This will cause an additional transient current to flow across both junctions. If we designate the charge on either side of the emitter–base depletion layer at some instant of time as q_{JE} then, in order to change this charge, we must supply an emitter current dq_{JE}/dt. Thus the total emitter current may be written by adding this term to equation 4.17

$$i_E = \frac{q_B}{\tau_F}+\frac{q_B}{\tau_{BF}}+\frac{dq_B}{dt}+\frac{dq_{JE}}{dt} \tag{4.19}$$

Similarly if q_{JC} is the charge stored in either half of the collector depletion layer, then equation 4.6b becomes

$$i_C = -\frac{q_B}{\tau_F}+\frac{dq_{JC}}{dt} \tag{4.20}$$

and utilizing equation 4.11 we obtain

$$i_B = -\frac{q_B}{\tau_{BF}}-\frac{dq_B}{dt}-\frac{dq_{JE}}{dt}-\frac{dq_{JC}}{dt} \tag{4.21}$$

These additional effects are shown on the circuit model of Fig. 4.7 by the two nonlinear capacitors with charges q_{JE} and q_{JC}.

4.5 Effect of temperature on characteristics

The static characteristics of a common emitter silicon transistor are shown in Fig. 4.8. The effect of variation of the temperature on the collector characteristics is seen in the three sets of curves (a), (b), (c) for three different temperatures. The only effect noticeable is the change in spacing of the members of the

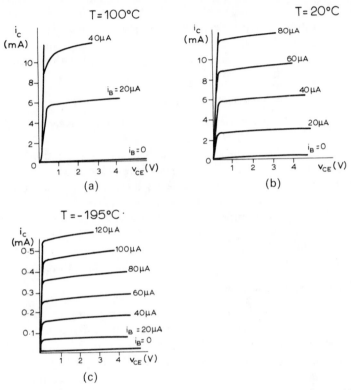

Fig. 4.8 Variation of collector characteristics of silicon transistor with temperature

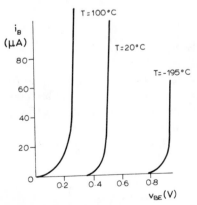

Fig. 4.9 Variation of base characteristics of silicon transistor with temperature

family of collector characteristics; this is a result of the variation of β_F with temperature. The base characteristics of the same silicon transistor are shown in Fig. 4.9 for the same three temperatures. This shows a variation of base–emitter voltage for a fixed base current.

4.5.1 Effect of temperature on β_F

The variation of β_F is one of the major effects in a silicon transistor operating under variable temperature. The physical mechanisms which give rise to this phenomenon are somewhat complex and no explanation will be attempted. It is sufficient to note that the current gain β_F increases, in general, with increase in temperature.

4.5.2 Variation of base characteristics with temperature

The variation of the base characteristic may be explained qualitatively as for a junction diode.

In Section 1.8 it is noted that the forward current across a junction will increase as an exponential function of the temperature. This applies almost without any modification to the base–emitter diode of a transistor and accounts for the variation of base characteristics of Fig. 4.9.

The base characteristic may be modelled in a piecewise linear way like the diode in Section 1.7.1. For silicon, the value of V_O, the break voltage, is about 0·3 V. As the temperature is increased, the value of V_O decreases as may be inferred from Fig. 4.9. It will fall by about 2 mV for a rise in temperature of 1 K.

4.5.3 Variation of I_{CBO} with temperature

Again as we discussed in Section 1.8 for a diode, the reverse saturation current across the collector–base junction will increase exponentially with temperature. For a silicon transistor it will approximately double for each 10 K temperature rise. This phenomenon is of negligible significance in silicon transistors since the value of I_{CBO} is usually so low at room temperature that even a tenfold rise is not usually appreciable. (It was of considerable significance in the obsolescent germanium transistors.)

4.6 Operation outside the active region

We must now develop a model for the transistor operating with either junction arbitrarily biased—either forward or reverse.

As a first approximation in this study we may assume that there is no re-combination in the base. Thus the charge distribution in the base will be deter-mined by equation 4.1 and since J_B is constant, across the base, the charge density will vary linearly with distance. The charge density at the emitter and collector depletion layer boundaries of the base are by similarity to equation 4.3,

$$n_B(0) = n_{BO} \exp\left(-ev_{EB}/kT\right) \tag{4.22}$$

and
$$n_B(W) = n_{BO} \exp(-ev_{CB}/kT) \tag{4.23}$$

The total charge distribution in the base is shown in Fig. 4.10. Also shown are the hole distributions in the emitter and collector region due to injection of holes from the base.

Since equation 4.1, the diffusion equation, is linear, we may study the relationship between current and charge by means of superposition. The charge distribution in Fig. 4.10 may be considered to be the sum of the two distributions given in Fig. 4.11a and b. In Fig. 4.11a the emitter–base junction is assumed

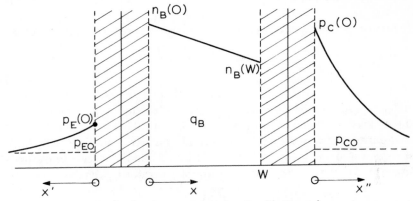

Fig. 4.10 Charge distribution in transistor when hard bottomed

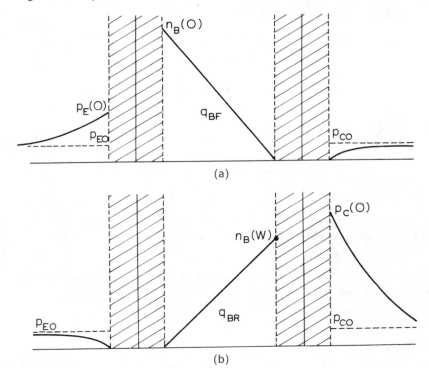

(a)

(b)

Fig. 4.11 Component charge distributions summing to Fig. 4.13 (a) forward component (b) reverse component

forward biased and the collector junction reverse biased; in Fig. 4.11b the collector–base junction is forward biased and the emitter–base junction reverse biased. This is equivalent to two transistors back to back, both in the active region, the forward transistor having stored base charge q_{BF} and the reverse transistor of q_{BR}.

The total base charge is given by

$$q_B = q_{BF} + q_{BR} \tag{4.24}$$

and the currents flowing are similarly linearly superposed according to equation 4.1 as

$$J_B = J_{BF} + J_{BR} \tag{4.25}$$

where J_{BF} and J_{BR} are the currents flowing separately due to q_{BF} and q_{BR}.

Now we may develop a circuit model to represent these two components of the total current flow, and these are shown in Figs. 4.12a and b. The forward transistor has a model identical to that of Fig. 4.5 and the reverse transistor a reciprocal model; the reverse saturation current generator I_{CBO} has been omitted.

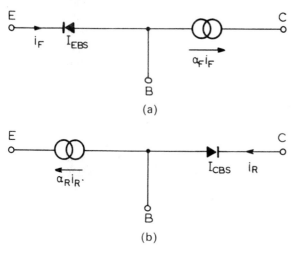

Fig. 4.12 Component models of transistor (a) forward model (b) reverse model

The total model is given in Fig. 4.13, in which the two component models are connected in parallel since the currents add linearly, as in equation 4.25.

From Fig. 4.13 we obtain

$$i_E = i_F - \alpha_R i_R \tag{4.26a}$$

$$i_C = -\alpha_F i_F + i_R \tag{4.26b}$$

if i_F and i_R are related, by direct analogy with equation 4.12, to the relevant junction voltages by

$$i_F = -I_{EBS}[\exp(-ev_{EB}/kT)-1] \tag{4.27a}$$

$$i_R = -I_{CBS}[\exp(-ev_{CB}/kT)-1] \tag{4.27b}$$

The terminal currents I_E and I_C may now be obtained from equations 4.26 and 4.27 as

$$i_E = -I_{EBS}[\exp(-ev_{EB}/kT)-1]+\alpha_R I_{CBS}[\exp(-ev_{CB}/kT)-1] \tag{4.28a}$$

$$i_E = \alpha_F I_{EBS}[\exp(-ev_{EB}/kT)-1]-I_{CBS}[\exp(-ev_{CB}/kT)-1] \tag{4.28b}$$

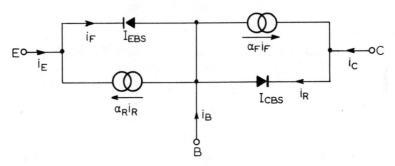

Fig. 4.13 Total model for transistor with arbitrary bias

These equations relate the terminal currents of a transistor with the voltages applied between the two pairs of terminals. They are usually known as the Ebers–Moll equations and, as written, are relevant to an n-p-n transistor. In addition, in order to determine the base current, we have

$$i_B = -i_E-i_C \tag{4.29}$$

The two currents I_{EBS} and I_{CBS} may be easily identified physically. If $v_{CB} = 0$, collector shorted to base, and $v_{EB} \gg kT/e$ then $i_E = I_{EBS}$. Thus I_{EBS} is the reverse saturation current flowing across the emitter–base junction with collector shorted to base. Similarly I_{CBS} is the reverse saturation current flowing across the collector junction with the emitter shorted to base. The term α_F is the forward common base current gain when emitter–base is forward biased and collector–base reverse biased and α_R is the similar current gain when the transistor is operated in the reverse manner.

It may be noted that if the emitter is open circuited and the collector reverse biased the emitter current is zero and the collector current equal to I_{CBO}. Solving for I_C under these stated conditions we obtain

$$I_{CBO} = (1-\alpha_F\alpha_R)I_{CBS}$$

It is interesting to note that a rigorous analysis of the transistor leading to equations 4.28 shows that

$$\alpha_F I_{EBS} = \alpha_R I_{CBS} \tag{4.30}$$

and thus only three parameters are necessary to define the properties of a transistor.

A similar analysis performed on a p-n-p transistor will give the following as the Ebers–Moll equations, the only difference being in the polarity of the voltages in the exponents.

$$i_E = -I_{EBS}[\exp{(ev_{EB}/kT)} - 1] + \alpha_R I_{CBS}[\exp{(ev_{CB}/kT)} - 1] \tag{4.31a}$$

$$i_C = \alpha_F I_{EBS}[\exp{(ev_{EB}/kT)} - 1] - I_{CBS}[\exp{(ev_{CB}/kT)} - 1] \tag{4.31b}$$

Being a p-n-p transistor and using our sign convention, both I_{EBS} and I_{CBS} will be numerically negative. The value of these equations lies in their ability to determine the terminal currents in a transistor in terms of the potentials between pairs of terminals.

4.7 Charge distribution in transistor base

In Section 4.6 we developed equations showing how the currents in a transistor in the active region of operation depend on the charge stored in the base and in the depletion regions at the junctions. Of these the most significant is the total charge stored in the base, and it is primarily this charge which influences the speed at which the transistor can respond to a change of applied voltage.

This becomes even more significant when the transistor is being switched over a very large range of voltage from cut-off, when the current is very small, to fully conducting. It is interesting therefore to study the distribution of charge in the base of a transistor in any of the four possible conditions of operation:

   (a) Normal mode: emitter forward biased, collector reverse biased.
   (b) Cut-off: emitter and collector reverse biased.
   (c) Bottomed (sometimes termed saturated): emitter and collector forward biased.
   (d) Inverse mode: emitted reverse biased, collector forward biased.

The currents flowing under these various conditions may be directly determined from the Ebers–Moll equations developed in the preceding section. Since the charge distribution in the base is almost exclusively governed by the diffusion of charges through the base, we may see from the diffusion equation 4.1 that the charge distribution is a linear function of distance across the base. The actual distribution is determined by the value of charge density at the boundaries with the emitter and collector space-charge layers. When a junction is forward biased a finite charge density will exist, given, for example at the emitter, by equation 4.3. When a junction is reverse biased the charge density there is effectively zero. The charge profiles in these four regions of operation are shown in Fig. 4.14.

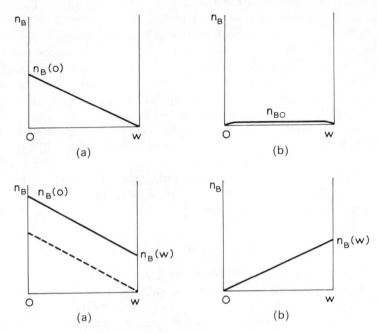

Fig. 4.14 Charge distribution in base
        (a) emitter forward biased, collector reverse biased
        (b) emitter and collector reverse biased
        (c) emitter and collector forward biased
        (d) emitter reverse biased, collector forward biased

These profiles may be related to the operating points on the typical transistor collector characteristics of Fig. 4.15. Point A, anywhere within the active region where the characteristics are almost horizontal, will correspond to a charge profile of Fig. 4.14a. When the transistor is cut off as at B the charge stored in

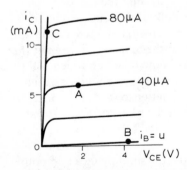

Fig. 4.15 Operating points on collector characteristics corresponding to active region at A, cut-off at B, bottomed at C

the base will be negligibly small as in Fig. 4.14b. At point C the collector current is fixed at about 12 mA and is independent of base current; the same collector current will flow for all base currents in excess of approximately 80 µA. This may be seen on the charge profile by noting that at the base current at which the point C is just at the knee of the characteristics (about 80 µA) the charge profile is given by the broken line on Fig. 4.14c. As the base current increases the operating point will still remain approximately at C but more charge will be stored in the base, the charge profile being given by the solid line. Since the gradient of this base charge profile does not change we should not expect any alteration in collector current and this is confirmed by reference to the collector characteristics. It is significant to note that this excess charge above the broken line in Fig. 4.14c must be removed from the transistor before the collector current can change; this is the cause of much of the time lag in transistor switching circuits. It is not possible to illustrate the operating point corresponding to profile 4.14d since the characteristics of Fig. 4.15 are only relevant to transistors operating in the normal mode, and Fig. 4.14d applies only to the reverse mode.

4.8 Transistors with non-uniform base

Although the above analysis gives a good conceptual idea of the mechanism of operation of a bipolar transistor, it is only numerically valid for some of the earlier types of transistor in which the base might be considered to be homogeneous across its width. In modern transistors the base is intentionally made nonhomogeneous in order to improve the high frequency performance of the device.

We assumed in Section 4.1 and subsequently that motion of charge carriers across the base was governed solely by diffusion and that, since the electric field is small, drift plays a negligible part. The result of this is that the motion of charge carriers across the base by diffusion is a relatively slow process and thus the response at the collector to a change of base current is comparatively slow. The transit time of charge carriers across the base may be reduced by establishing an electric field across the base and hence causing current to flow by drift as well as diffusion. Such an electric field will not be set up by any externally applied voltage, as such a voltage will only alter the junction potentials and not affect the potential gradient in the bulk of the semiconductor.

A potential gradient may be produced in the base region of a semiconductor by arranging for the impurity density $N_A(x)$ within the base to be greater at the emitter than at the collector. An electric field will now be set up which will be proportional to the gradient of the impurity density and, in the case considered of an n-p-n transistor, directed from the emitter towards the collector. This will accelerate the minority charge carriers in the base, the holes, towards the collector and thus reduce the transit time across the base. The overall effect will be to improve the frequency response.

We shall see that no longer are we able to make the simplifying assumptions that the hole density in the base is a linear function of the distance across the base, since this was founded on the assumption that holes moved within the base by diffusion only. However, although the charge profiles we have developed will no longer be valid, the general results and the model we have postulated will still give an accurate representation of the behaviour of the device in an external circuit.

The production of the graded base is relatively simple since the diffusion process which is used to construct the collector may be carried out relatively slowly allowing some of the donor ions to diffuse fairly deeply into the base neutralizing the existing acceptor atoms. This neutralization will be greater near the collector surface than deeper in the base towards the emitter and this will result in an impurity acceptor density within the base which falls from a relatively high value at the emitter to zero at the collector.

References

ALLISON, J., *Electronic Engineering Materials and Devices*, McGraw-Hill, 1971, Chapter 11.

SPARKES, J. J., *Junction Transistors*, Pergamon, 1966, Chapters 4, 5 and 6.

GRAY, P. E., *et al.*, *Physical Electronics and Circuit Models of Transistors*, Wiley, 1964, Chapters 7, 9 and 10.

Problems

4.1 The base of a p-n-p transistor is estimated to have an average width of $1\cdot5$ µm. Determine the value of the collector forward charge control time constant τ_F.

4.2 Measurements on the transistor of Problem 4.1 at a fixed collector voltage of -10 V show that when $I_B = 15$ nA, $I_C = -15$ nA and when $I_B = -15$ nA, $I_C = -1\cdot45$ µA. Determine the base forward charging time constant τ_{BF}.

4.3 Estimate the collector and emitter currents flowing in the transistor of Problem 4.1 when $I_B = 0$.

4.4 The transistor of Problem 4.1 is operated at a collector current of 5 mA. Estimate the charge stored in the base and the emitter and base currents.

4.5 The transistor whose characteristics are given in Figs 4.8 and 4.9 is operated at 20°C with its collector and base connected together. Sketch the current/voltage characteristic of the diode so formed.

4.6 A p-n-p transistor is operated with a collector–emitter voltage of -15 V. The transistor has $\tau_F = 4\cdot5$ ns and $\tau_{BF} = 0\cdot3$ µs. The base current is suddenly changed from zero to -50 µA. Sketch and dimension the collector current as a function of time.

4.7 The p-n-p transistor of Problem 4.6 is used in the circuit of Fig. P4.1. The input voltage is changed instantaneously from 0 to $-1\cdot5$ V. Sketch the collector current as a function of time when
  (a) $C = 5$ pF, (b) $C = 10$ pF, (c) $C = 20$ pF
The base–emitter voltage when the transistor is conducting may be assumed to be zero.

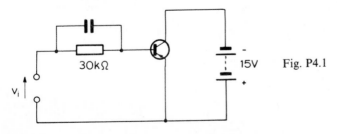

30kΩ   15V   Fig. P4.1

v_i

4.8 An n-p-n transistor has an emitter reverse saturation current with the collector open of 10 nA and a collector reverse saturation current with the emitter open of 40 nA. The forward current gain with collector open is 0·99. The device is connected as a diode in the following configurations. Determine the equivalent reverse saturation current of the diode under the following conditions:

(a) Collector and base connected together
(b) Base and emitter connected together
(c) Collector left disconnected
(d) Emitter left disconnected
(e) Collector and emitter connected together.

4.9 The transistor of Problem 4.8 is operated with both emitter and collector reverse biased. Determine the base current flowing.

4.10 The transistor of Problem 4.8 is operated with the collector connected through a resistor of 4 kΩ to a 15 V supply and the base is connected to the same supply (a) through a 2 MΩ resistor, (b) through a 50 kΩ resistor. Determine collector and base currents and emitter–base voltage.

4.11 The transistor of Problem 4.8 is left with its base on open circuit and 10 V is applied between collector and emitter. Determine the base–emitter voltage.

4.12 A transistor operated in the forward active mode has charge control parameters $\tau_F = 4\cdot5$ ns, $\tau_{BF} = 300$ ns; when operated in the reverse active mode the corresponding parameters are $\tau_R = 4\cdot5$ ns, $\tau_{BR} = 100$ ns. Determine the forward and reverse short circuit current gain between emitter and collector.

4.13 The transistor of Problem 4.12 is operated in the active region with a collector current of 10 mA. Determine the charge stored in the base and the base current.

4.14 The transistor of Problem 4.13 is operated in the bottomed region with a collector current of 10 mA and an emitter current of -20 mA. Determine the charge stored in the base.

5
Bipolar Transistor—
Small Signal Model

5.0 Introduction

The circuit model for a bipolar transistor evolved in Section 4.6 is useful for predicting the behaviour of the device in circuits in which large changes of voltage occur.

As discussed in Section 3.0 in relationship to the field-effect transistor, when we are interested in the application of a device as an amplifier of small signals the incremental parameters are of greater significance. We shall proceed to study the manner in which a bipolar transistor may be modelled when we are studying the interrelations which exist between small signal voltages and currents superimposed on a fixed operating point. The model will be broken down as in the fet into intrinsic, extrinsic, and parasitic elements.

5.1 The intrinsic transistor

The intrinsic performance of a transistor has been defined by the equations 4.6b and 4.16 with q_B given by equation 4.4a. We may now assume that small variations Δi_C, ΔI_B, Δq_B and Δv_{BE} occur around the steady values of I_C, I_B, Q_B and V_{BE}. Thus

$$i_C = I_C + i_c = I_C + \Delta i_C$$

$$i_B = I_B + i_b = I_B + \Delta i_B$$

$$q_B = Q_B + q_b = Q_B + \Delta q_B$$

$$v_{BE} = V_{BE} + v_{be} = V_{BE} + \Delta v_{BE}$$

Differentiating equations 4.6b, 4.16 and 4.4a and making these substitutions gives

$$i_c = -\frac{q_b}{\tau_F} \qquad (5.1)$$

$$i_b = -\frac{q_b}{\tau_{BF}} - \frac{dq_b}{dt} \qquad (5.2)$$

$$q_b = Q_{BO} \frac{ev_{be}}{kT} \exp\left(ev_{BE}/kT\right)$$

$$= (q_B + Q_{BO})ev_{be}/kT$$

$$\simeq q_B ev_{be}/kT$$

since $q_B \gg Q_{BO}$

$$\simeq Q_B ev_{be}/kT \tag{5.3}$$

since $q_B = Q_B + q_b$ and $q_b \ll Q_B$.

At the operating point we may obtain from equation 4.6b

$$I_C = -\frac{Q_B}{\tau_F} \tag{5.4}$$

From equations 5.1, 5.3 and 5.4 we obtain

$$i_c = \frac{e|I_C|}{kT} v_{be} \tag{5.5a}$$

and from 5.2, 5.3 and 5.4

$$i_b = \frac{\tau_F}{\tau_{BF}} \frac{e|I_C|}{kT} v_{be} + \tau_F \frac{e|I_C|}{kT} \frac{dv_{be}}{dt} \tag{5.5b}$$

Equations 5.5 are two linear relationships between the two terminal currents, i_c and i_b, and the emitter base voltage, v_{be}. Since they define the terminal behaviour of the transistor we may attempt to compare them with the network equations for a linear circuit. Let us rewrite equations 5.5 as

$$i_c = g_m v_{be} \tag{5.6a}$$

$$i_b = g_\pi v_{be} + C_b \frac{dv_{be}}{dt} \tag{5.6b}$$

where we define

$$g_m = \frac{e|I_C|}{kT} \tag{5.7a}$$

$$g_\pi = \frac{1}{r_\pi} = \frac{\tau_F}{\tau_{BF}} \frac{e|I_C|}{kT} \tag{5.7b}$$

$$C_b = \tau_F \frac{e|I_C|}{kT} \tag{5.7c}$$

The equations 5.6 give the relationship between terminal currents and voltages of the network of Fig. 5.1 and thus we may consider this as a satisfactory model of the intrinsic small signal properties of a transistor.

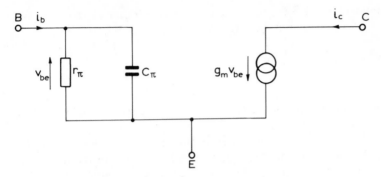

Fig. 5.1 Simple small signal model of bipolar transistor

We may now consider the three elements of the circuit model in more detail. The transconductance, g_m, is seen to be a linear function of the collector current. This is confirmed from measurements on a transistor which show that linearity is maintained over a large range of current. The transconductance is seen to be independent of transistor type, having the same value at a given collector current for all transistors. If the collector current is measured in amperes, the transconductance in siemens at room temperature of 293 K is given by

$$g_m = 40|I_C| \tag{5.8}$$

It should also be noted that g_m is independent of frequency up to very high frequencies.

From equations 5.6a and 5.6b for slowly varying voltages where dv_{be}/dt may be assumed small, the low frequency short circuit current ratio

$$\frac{i_c}{i_b} = \frac{g_m}{g_\pi}$$

and this will be given the symbol β_0.* Whence, from equations 5.7a and b,

$$g_m r_\pi = \beta_0 \tag{5.9}$$

For low currents β_0 and β_F will be identical but due to nonlinearity in the current ratio this will not be the case at large currents. It may be seen from

* The symbol h_{fe} is sometimes used in place of β_0. This is only truly valid at very low frequencies. At higher frequencies the value of h_{fe} is frequency dependent and not directly related to β_0.

equations 5.7 that β_0 should be independent of I_C. Measurements confirm this at large values of collector voltage over a limited range of collector current; at lower collector voltages β_0 falls from its optimum value at both high and low currents.

Finally from equations 5.7b and c, the base charge storage capacitance, C_b, may alternatively be written in the form

$$C_b = \frac{\tau_{BF}}{r_\pi} \tag{5.10}$$

from which it is seen that the time constant $C_b r_\pi$ in the circuit model is equal to the equivalent base time constant τ_{BF}.

We have thus obtained a circuit model for the small signal behaviour of the transistor as shown in Fig. 5.1. The three circuit elements in the model are specified in terms of I_C, τ_{BF} and τ_F. Since g_m is directly related to the collector bias current, it is not possible to divorce the design of an amplifier having a specified small signal performance from the design of the optimum bias conditions.

5.2 Extrinsic elements

We must now look at other phenomena occurring within the semiconductor wafer but outside the neutral base region, for example in the depletion layers at emitter and collector and in the bulk of the base region remote from the main flow of minority carriers between emitter and collector.

At both the emitter and collector junctions, whether forward or reverse biased, a depletion layer will exist. At the emitter, as discussed in Section 1.9, a change of emitter–base voltage will result in a change in the charge stored in the emitter depletion region. For small variations of voltage the change of charge may be assumed to be linearly related to the change of voltage and thus this process may be represented on a circuit model by a capacitor C_{je} between emitter and base.

Similarly the effect of small changes in collector–base voltage may be modelled by a capacitor C_{jc} between collector and base. These are shown in Fig. 5.2 added to the simple model of Fig. 5.1.

The two capacitors between emitter and base, C_b and C_{je}, may be replaced by a single capacitor C_π and the collector–base capacitor renamed C_μ. These

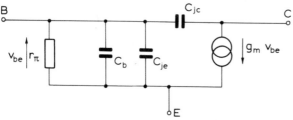

Fig. 5.2 Intrinsic transistor small signal model

modifications are shown on Fig. 5.3. Usually C_{je} is not more than about 10% of C_π and equation 5.10 is usually approximated by

$$C_\pi = \frac{\tau_{BF}}{r_\pi} \tag{5.11}$$

We may alternatively consider equation 5.11 as redefining τ_{BF} directly from circuit measurements, so that it is now no longer related exactly to the base charging time constant.

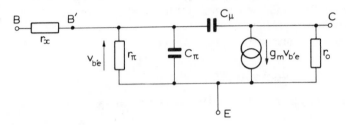

Fig. 5.3 Complete small signal model of transistor

In the simple model of Fig. 5.1, the base current providing for recombination of majority charge carriers flows to the emitter through r_π to represent the voltage drop across the emitter–base junction as a result of this base current. The points in the base at which recombination occurs are randomly distributed in that part of the base between the emitter and collector junctions. The base current flows to these recombination centres from the base lead connection through the bulk material of the semiconductor and in so doing creates a small voltage drop. The net effect of the current flow to these separate recombination centres may be approximated by assuming one equivalent point, B′, within the base at which recombination occurs and considering the base current to flow from the point of connection of the external lead, B, to this fictitious point B′ through a resistor of value r_x, simulating the bulk resistance of the base. This also is shown in Fig. 5.3.

In the same context it may be noted that the bulk resistance of the emitter, and particularly the collector, may also be modelled by resistors $r_{ee'}$ and $r_{cc'}$ in series with the emitter and collector leads. These are not shown in Fig. 5.3 since at the frequencies which we shall consider they will not be of great significance.

One further phenomenon which should be considered in our model is the increase in the collector current in response to an increase in collector–emitter voltage as may be seen in Fig. 4.8b. This suggests that the incremental collector current i_c is proportional to v_{ce} and we may represent this by a resistor r_o connected between collector and emitter as in Fig. 5.3. The physical mechanism causing this is, in part, leakage across the reverse biased collector–base junction, but a more predominant cause is the modulation of base width with variation of collector–base voltage.

5.3 Parasitic elements

The model of Fig. 5.3 may be accepted as a fairly accurate representation of a transistor between the three points at which the external leads are connected to the semiconductor chip. For final assembly the chip is mounted on the header by three short wires and finally assembled in the protective can. Associated with this packaging there will be stray capacitances between the three external leads and each of the connecting leads will itself possess a small lead inductance. At frequencies below about 100 MHz these may generally be ignored.

The circuit model of Fig. 5.3, known as the hybrid π model is thus a good representation of the small signal behaviour of transistors up to frequencies of the order of 10 MHz (the upper limit to the validity of the model depends upon the particular geometrical construction of the device).

5.4 Values of hybrid π elements

Values of the elements of the hybrid π model appropriate to a general purpose n-p-n transistor operating at a collector–emitter voltage of 10 V, and a collector current of 10 mA at a temperature of 25°C are given below

$$r_\pi = 200 \ \Omega$$
$$C_\pi = 80 \ \text{pF}$$
$$C_\mu = 3 \ \text{pF}$$
$$r_x = 25 \ \Omega$$
$$g_m = 0{\cdot}4 \ \text{S}$$
$$r_o = 50 \ \text{k}\Omega$$

As has been discussed above all these values are dependent on the operating point of the device. The hybrid π circuit model is of considerable value in circuit design since it relates very closely to the physical phenomena in the device; it is valid over a very wide range of frequency and the element values are usually relatively simple functions of bias voltage and current and temperature. Unfortunately manufacturers rarely quote these values in their data sheets.

It is however possible to determine a number of these parameters directly from data given by manufacturers. The small signal data for a transistor working at a given operating point are often specified in terms of β_0, the low frequency short-circuit common-emitter current gain, and f_T, the frequency at which the magnitude of the current gain falls to unity.

5.5 Low frequency model of bipolar transistor

The hybrid π model for the transistor developed in the preceding sections may be simplified for use at relatively low frequencies. If the susceptance of C_π at the frequency of interest is small compared with the conductance g_π, we are justified in omitting C_π from the equivalent circuit. Similarly we shall justify later that if $\omega g_m R_L C_\mu \ll g_\pi$ we may omit C_μ from the circuit model. We are now left with the relatively simple circuit model shown in Fig. 5.4.

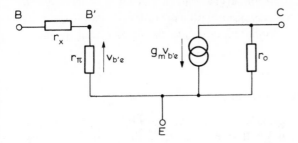

Fig. 5.4 Low frequency small signal model of transistor

It is often convenient to simplify this still further. If $r_x \ll r_\pi$ we may replace r_x by a short circuit and, if we assume that r_o is large compared with any load resistor connected between collector and base, we may omit r_o. This gives the even simpler low frequency model of Fig. 5.5a which may alternatively be drawn as in Fig. 5.5b where the current generator is now expressed as a function of the input base current rather than the base–emitter voltage. This last circuit model is very valuable in enabling us to obtain a quick, although perhaps crude, understanding of the mode of operation of a circuit and its basic properties.

There are, of course, a large number of alternative models which may be used

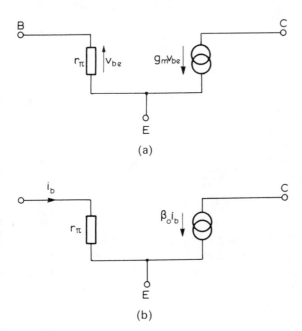

Fig. 5.5 Simplified small signal models of transistor

to represent the performance of a transistor, particularly at low frequencies. In general these low frequency models may all be related to the circuit of Fig. 5.4 and in certain situations some models have advantages over others. In order to prevent the confusion which may occur with a proliferation of low frequency models, all the analyses in this book will be conducted using one of the models developed in this chapter.

5.6 Frequency dependence of current gain

The common emitter current gain, β, of a transistor, as a function of frequency, may be defined from equations 5.6. If we assume that voltages and currents are sinusoidal at angular frequency ω these two equations may be written in phasor form as

$$I_c = g_m V_{be} \tag{5.12a}$$

$$I_b = g_\pi V_{be} + j\omega C_\pi V_{be} \tag{5.12b}$$

(Note that C_b has been replaced by its approximate equivalent of C_π, as discussed in Section 5.3.)

If β is defined as the small signal current gain at frequency ω we have

$$\beta = \frac{I_c}{I_b} = \frac{g_m}{g_\pi + j\omega C_\pi}$$

$$= \beta_0/(1 + j\omega C_\pi r_\pi) \tag{5.13}$$

It is thus possible to sketch the relationship between $|\beta|$ and ω given by

$$|\beta| = \frac{\beta_0}{(1 + \omega^2 C_\pi^2 r_\pi^2)^{1/2}} \tag{5.14}$$

This is shown in Fig. 5.6, using the parameters quoted in Section 5.5.

From equation 5.14 it is immediately seen that the magnitude of the current gain has fallen to $\beta_0/\sqrt{2}$ at an angular frequency

$$\omega_\beta = 2\pi f_\beta = 1/C_\pi r_\pi \tag{5.15}$$

f_β is usually termed the β cut-off frequency and is a parameter which may be directly measured. It is however even more convenient to measure the frequency f_T at which the magnitude of the common emitter current gain has fallen to unity. If it is assumed that $f_T \gg f_\beta$ we may determine its value from equation 5.14 by writing

$$|\beta| \simeq \frac{\beta_0}{\omega C_\pi r_\pi}$$

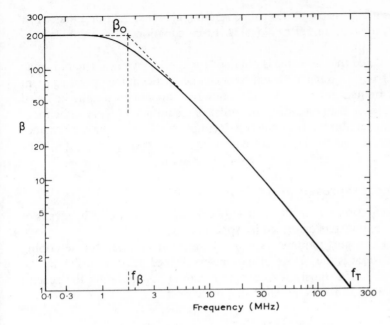

Fig. 5.6 Variation of current gain with frequency

and setting $|\beta| = 1$ at $\omega = \omega_T = 2\pi f_T$, we obtain

$$\omega_T = \frac{\beta_0}{C_\pi r_\pi} \tag{5.16a}$$

$$= \beta_0 \omega_\beta \tag{5.16b}$$

$$= g_m / C_\pi \tag{5.16c}$$

$$= 1/\tau_F \tag{5.16d}$$

Now the product of current gain and bandwidth may be determined as

$$\beta_0 \omega_\beta = g_m / C_\pi = \omega_T$$

and is a fixed parameter of the device. There is thus an upper bound placed on the current amplification available over any given band of frequencies. For the transistor whose parameters are given in Section 5.5, $f_T = 795$ MHz, $f_\beta = 10$ MHz.

5.7 Determination of hybrid π elements

A typical set of data for a bipolar transistor measured at $V_{CE} = 5$ V, $I_C = 5$ mA is $\beta_0 = 120$, $f_T = 200$ MHz, $C_{obo} = 4$ pF.*

* C_{obo} is the capacitance measured between collector and base with the emitter on open circuit, namely the output capacitance, in common base with remaining terminal open.

From this it is possible to determine g_m from equation 5.7a or 5.8 as $g_m = 200$ mS. From equation 5.9 $r_\pi = 600 \, \Omega$. From equation 5.16c $C_\pi = 265$ pF. $C_\mu \simeq C_{obo} = 4$ pF.

It may be noted that no value is obtained for r_o or r_x. An estimate of r_o may be obtained from the slope of the collector characteristics at the operating point since at low frequencies $r_o = \partial V_{CE}/\partial I_C$; however this cannot usually be relied upon to give more than an idea of the order of magnitude of r_o. A value for r_x is not usually available but fortunately it does not usually affect the performance of an amplifier except at very high frequencies; at lower frequencies it may be considered to add to the value of the source resistance.

5.8 Two-port representation of transistor

As we saw in Section 3.7, an alternative specification of a device is by means of the two-port parameters described in Appendix A1.

The set of two-port parameters which is most widely used for describing bipolar transistors is the 'h' set of parameters defined in Appendix A by the equations A2. For a transistor in common emitter configuration the general nomenclature used there is modified slightly to read

$$\begin{bmatrix} V_{be} \\ I_c \end{bmatrix} = \begin{bmatrix} h_{ie} & h_{re} \\ h_{fe} & h_{oe} \end{bmatrix} \begin{bmatrix} I_b \\ V_{ce} \end{bmatrix} \tag{5.17}$$

The first suffix of the h parameters relates to input, reverse, forward and output conditions, and the second suffix denotes that the measurements are made in common emitter configuration. Similar parameters h_{ib}, h_{rb}, h_{fb}, h_{ob} could be defined in common base configuration, relating V_{eb}, I_c with I_e, V_{cb}. The relationship between these two sets of parameters may be determined relatively easily, as shown in Appendix A.

More recently the use of short circuit admittance parameters has been adopted to specify transistors, particularly at high frequencies. These are defined in equations A4.

5.9 Two-port and hybrid π parameters

One advantage of the use of two-port parameters is in the analysis and optimization of amplifiers and other devices using a computer program. Since most network analysis programs are written in terms of nodal admittance matrices, it is more convenient to specify the device in this form also. As stated previously the specification of a transistor over a wide range of frequency, biasing conditions, etc. will entail the storage of a considerable bank of data in the computer; this will however usually offset the added complexity entailed when using the hybrid π model which will require the computation of a more complex network at each frequency.

It is thus seen that the two forms of a transistor modelling are complementary. The hybrid π, or similar network, throws considerable light on the physical process occurring in the device and lends itself readily to situations where the transistor may be operated over a considerable range of frequency or where the operating point or environment is liable to change.

5.10 Typical presentation of two-port data

The y parameters of a bipolar transistor measured in common emitter are shown in Fig. 5.7; the real parts, g, and imaginary parts, b, are shown separately.

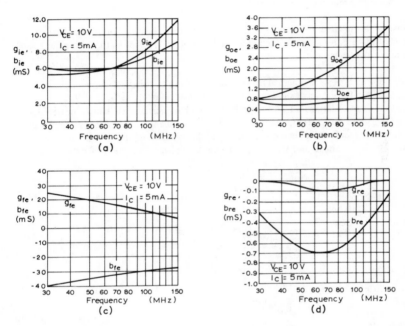

Fig. 5.7 Variation of admittance parameters with frequency

It is of interest to note that any attempt to predict these curves based on the simple hybrid π model we have proposed is unsuccessful even if we assume values for the unknown components. This suggests that the model of Fig. 5.5 is not accurate at frequencies of the order of 100 MHz. In order to obtain better agreement it would be necessary to include the effects of the capacity between the three leads as they pass through the header and also their self-inductance.

References

GRAY, P. E. , et al., Physical Electronics and Circuit Models of Transistors, Wiley, 1964, Chapter 8.
LINVILLE, J. G. and GIBBONS, J. F., Transistors and Active Circuits, McGraw-Hill, 1961, Chapters 6, 9.
SEARLE, C. L., et al., Elementary Circuit Properties of Transistors, Wiley, 1964, Chapters 3 and 4.

Problems

5.1 The transistor whose characteristics are given in Figs. 4.8 and 4.9 is operated at room temperature at a bias point of $V_{CE} = 3$ V, $I_B = 40$ μA. Determine as many parameters as possible of the hybrid π model.

5.2 A transistor is operated at a collector current of 5 mA. The two charge control parameters are $\tau_{BF} = 300$ ns, $\tau_F = 5$ ns. Determine a small signal model for the transistor.

5.3 A transistor operated at a collector current of, 2 mA has a low-frequency small-signal current gain β_o of 200. The frequency at which the magnitude of the current gain $|\beta|$ falls to units is 250 MHz. Obtain a small signal model of the transistor.

5.4 A transistor is operated at $V_{CE} = -5$ V, $I_C = -1.25$ mA. The common emitter current gain β_o measured at low frequency is 120, and at a frequency of 25 MHz $|\beta| = 10$. Determine a small signal model for the transistor.

5.5 The data of Fig. P5.1 were obtained by measurements on a transistor. Determine the values of g_m, r_π, C_π, C_μ, r_x, r_o for the small signal model of the transistor at $V_{CE} = -20$ V, $I_C = -10$ mA. Fig. P5.1(a) shows the relationship between collector and base currents, (b) shows the dependence of base charge on collector current, (c) shows the dependence of emitter depletion layer charge with emitter–base voltage and (d) the similar relationship at the collector. (e) shows base current dependence on base–emitter voltage and (f) shows collector current as a function of collector–emitter voltage.

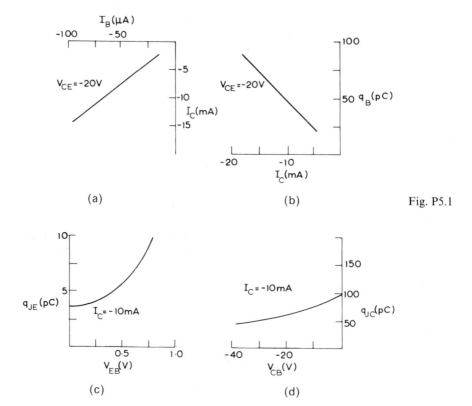

Fig. P5.1

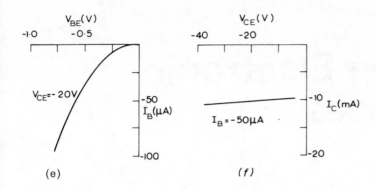

(e)   (f)

5.6 Using the circuit model of Fig. 5.3 but omitting r_x determine expressions for the small signal y parameters of the transistor.

5.7 Compute the values of the y parameters at 100 MHz and compare them with those given in Fig. 5.7. Take the following as appropriate values for the elements of the hybrid π model.

$$r_\pi = 600\ \Omega$$
$$C_\pi = 265\ \text{pF}$$
$$C_\mu = 4\ \text{pF}$$
$$r_o = 50\ \text{k}\Omega$$
$$g_m = 200\ \text{mS}$$

Large discrepancies will be found between your computed values and those of Fig. 5.7; can you account for them?

5.8 Determine expressions for the h parameters of the simple transistor model used in Problem 5.6.

5.9 Include r_x in your model and obtain expressions for the y parameters. Compute the values of the y parameters at 100 MHz, using the data of Problem 5.7 and assuming $r_x = 25\ \Omega$.

5.10 Write a computer program to determine the real and imaginary parts of the four y parameters at a range of frequencies and compare this data with the graphs of Fig. 5.7. To save considerable repetition just compute g_{fe} (the real part of y_{fe}). Alternatively you may use an existing network analysis program. Choose r_x as 25 Ω. Does this improve the fit to the graphical data?

6
Other Electronic Devices

6.0 Introduction

Since the advent of the bipolar transistor, many other semiconductor devices have been suggested. In this chapter we shall look at a few of the less specialized devices. In addition we shall consider the properties of some simple thermionic devices.

6.1 Zener diode

In Section 1.11 we discussed the breakdown which may occur in diodes subjected to large reverse voltages. In the context of the signal and power diodes we were considering there this phenomenon is undesirable. However, a useful practical application may be found in Zener diodes (otherwise known as reference diodes) in which the voltage across the diode is held constant over a wide range of current. There are two physical mechanisms which may give rise to such a diode characteristic; avalanche multiplication and Zener breakdown.

The first, avalanche multiplication, has been described in Section 1.11. The second mechanism is Zener breakdown. When a high electric field exists in a semiconductor, such as occurs in the depletion region of a reverse biased junction, new hole–electron pairs may be directly created by the disruption of the bonds between atoms in the crystal lattice. Once the electric field has risen to a value high enough to break the atomic bonds, large numbers of charge carriers become available for a very slight increase in applied voltage. Thus at the Zener voltage the current can rise to very high values.

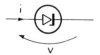

Fig. 6.1 Conventional symbol for Zener diode

Zener diodes are designed to have a characteristic in which very little change of voltage is necessary for very large changes of current. Such characteristics are shown in Fig. 6.2 for various Zener diodes, the circuit symbol for which is shown in Fig. 6.1. The mechanism of breakdown in a Zener diode may be either by the true Zener effect or by avalanche multiplication. For breakdown voltages greater than about 10 V, avalanche multiplication is predominant and for breakdown voltages lower than about 5 V, the mechanism responsible is Zener breakdown. In the intermediate range both mechanisms occur. The wide range of breakdown voltages displayed by Zener diodes, from 1·3 V to 75 V or above, are obtained by variation of the depletion layer width; this is achieved by control of the impurity density in the two sides of the p-n junction.

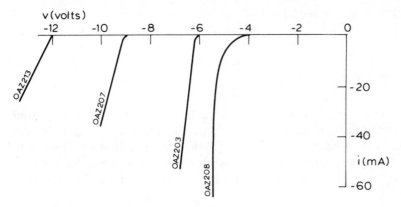

Fig. 6.2 Characteristics of typical Zener diode

The current range of Zener diodes is also large, extending from 1 mA up to 500 A. The main limitation to current capacity is imposed by the power handling capacity of the diode and this is limited by the heat which can be dissipated at the junction.

6.1.1 Modelling of a Zener diode

The Zener diode characteristic shown in Fig. 6.3 may be represented in the reverse biased direction by the two broken lines having a break point at the Zener voltage, $-V_z$. (The characteristic in the forward direction is similar to a normal diode, but we shall not consider this as it would be very unusual to operate a Zener diode with forward bias.) This piecewise linear approximation

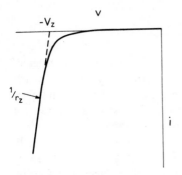

Fig. 6.3 Piecewise linear approximation to Zener diode characteristic

to the Zener diode characteristic may be modelled by the circuit of Fig. 6.4 in in which R_z represents the reciprocal of the slope of the characteristic in the region where the voltage is approximately constant, the battery has a value V_z and the diode represents an ideal diode with infinite reverse resistance and zero forward resistance.

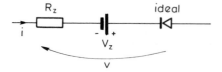

Fig. 6.4 Circuit model of Zener diode

Typical values of V_z range from 1·3 V to 75 V and values of R_z vary from 0·2 Ω to 50 Ω depending on Zener voltage and operating current. Since R_z is a function of operating current, increasing as the current falls, it is customary to stipulate a minimum current where the slope of the characteristic has fallen to an unacceptable value. This minimum current is often specified by the manufacturers.

6.1.2 Effect of temperature on Zener diode characteristics

The effect of temperature on the characteristics of Zener diodes depends upon whether the principal mechanism is avalanche or Zener. For high voltage diodes with Zener voltages greater than about 7 V, avalanche multiplication predominates and the reference voltage rises with increase in temperature. This is probably due to a reduction in the average distance between collisions as temperature is increased and thus the lower probability of an electron having sufficient energy to create a hole–electron pair when it does make a collision. Below about 7 V the breakdown voltage falls with increasing temperature. Since breakdown is primarily due to the Zener effect, this may be explained by realizing that the atoms in the crystal lattice will have higher energy at elevated temperature and thus will require less additional energy from the electric field to create new hole–electron pairs. Typical values of temperature coefficients vary from 0·05%/°C to 1%/°C.

6.2 Silicon controlled rectifier

Another device which is used principally in the control of large quantities of power is the silicon controlled rectifier (SCR), alternatively known as the thyristor. It is a device with three terminals, termed anode, cathode, and gate, as shown in Fig. 6.5a and it is constructed from four layers of semiconductor material alternately p- and n-type, as in Fig. 6.5b.

A sketch of the characteristics of the device relating anode current with anode–cathode voltage for a variety of gate–cathode voltages is shown in Fig. 6.6. We can explain the performance of the thyristor by considering the division of the device into two transistors, Tr1 being p-n-p and Tr2 being n-p-n, as

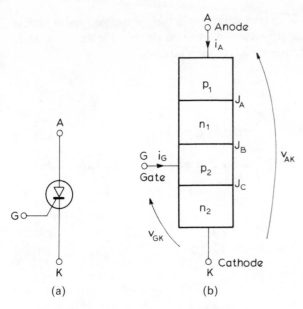

Fig. 6.5 Thyristor (a) conventional symbol (b) outline construction

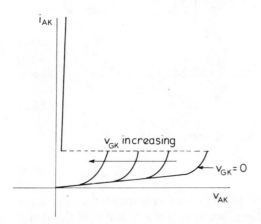

Fig. 6.6 Anode current characteristics of thyristor

shown in Fig. 6.7, interconnected between two of the p-regions and two of the n-regions to form the four layer thyristor. This is not an entirely academic exercise as a thyristor may be simulated in practice by two transistors connected as in Fig. 6.7.

Consider first the situation in which the gate is left on open circuit and hence $i_G = 0$. Assume that a positive voltage v_{AK} is applied between anode and cathode. We may note that, in the thyristor the centre junction is reverse biased and the other two junctions forward biased or, with reference to the two transistor model of Fig. 6.7, both collectors are reverse biased and both emitters forward biased, and thus both transistors are in their active regions of operation. We may now analyse the situation using equation 4.13a and writing I_1 and I_2 as the saturation currents I_{CBO} for each transistor

$$i_{C1} = -\alpha_{F1}i_A + I_1 \tag{6.1a}$$

$$i_{C2} = -\alpha_{F2}i_K + I_2$$

$$= \alpha_{F2}i_A + I_2 \tag{6.1b}$$

assuming that $i_K \simeq -i_A$ (this is exactly true when $i_G = 0$ and approximately true otherwise).

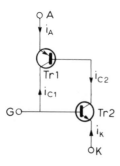

Fig. 6.7 Circuit model of thyristor

We may now solve equation 6.1 with the added constraint that

$$i_{C1} + i_A - i_{C2} = 0 \tag{6.2}$$

obtained by summing all currents entering transistor Tr1 to zero. Thus

$$i_A = I_0/(1 - \alpha_{F1} - \alpha_{F2}) \tag{6.3}$$

where $I_0 = I_2 - I_1$ (I_0 cannot be measured directly. It is the reverse saturation current which would flow across the junction between n_1 and p_2 if this diode were isolated from the remainder of the thyristor.)

Under normal conditions when the collector currents are small both α_{F1} and α_{F2} are small and hence the anode current is also small and of the order of magnitude of the reverse saturation current of a diode. This situation will exist for low values of v_{AK}. As the anode voltage is increased, the reverse voltage across the centre junction will also increase and a high electric field will be established in that region. This will allow avalanche multiplication to occur; the mechanism is the same as that responsible for avalanche breakdown but the field is not sufficiently great as to initiate a self-maintained current. The result is that the common-base current gain, α_F, of each transistor rises and the denominator in equation 6.3 falls towards zero. The anode current will thus increase suddenly to a very large value, limited entirely by external circuit resistance and the voltage between anode and cathode will fall to a low value of the order of 1 V.

Assuming that both transistors are identical, we may note that for either transistor, the base current is equal to the collector current and is sufficiently large to drive the transistor to the bottomed state. As we have seen in Section 4.9 the collector–base junction is forward biased in this condition and thus all three junctions are forward biased. The total voltage v_{AK} may be obtained by summing the junction voltages in this bottomed state. From Fig. 6.7 we see that $v_{AK} = v_{AC2} + v_{C2K}$ where v_{AC2} is the voltage between base and emitter of Tr1 when bottomed and of the order of 0·7 V and v_{C2K} is the collector bottoming voltage of Tr1 and is approximately 0·3 V. Hence v_A in the ON state is about 1 V as quoted above.

The thyristor is now stable in this condition since the collector and base currents of both transistors are equal. In this bottomed state equation 6.3 is no longer valid. Once the thyristor has been switched ON by an increase in anode voltage, it will remain ON so long as the current flowing through it exceeds a specified minimum holding current I_H. If the supply voltage falls to such a low value that this cannot be maintained, conduction will cease since the base currents are not large enough to keep either transistor bottomed.

We must now consider the effect of the introduction of a gate current i_G to the base of the thyristor, or in the two transistor model to the base of Tr2. This base current flowing through junction J_C on Fig. 6.5b (or the emitter–base junction of Tr2) will cause α_{F2} to increase and hence initiate the onset of avalanche multiplication at a lower value of anode–cathode potential. It is thus not only possible, but customary, to initiate firing of a thyristor by injecting a gate current into the thyristor at the instant when conduction is desired.

Although switch on by means of an injected gate current is possible, removal of this gate current will leave the thyristor in its conduction state. In order to switch it off it is necessary, not only to remove the gate current, but also to reduce the anode voltage so that the anode current falls below the holding current, I_H. It is possible to switch off a thyristor by a large reverse gate current but the required current is so large as to render this impracticable.

The thyristor thus lends itself to applications in which the anode is fed through the load from an alternating supply, which reverses once every cycle. It is therefore possible to switch the circuit off at any time by removal of the gate current; the anode current will fall to zero on the next occasion when the anode voltage reverses in sign. The gate supply for thyristor circuits may be provided either by a direct voltage, or an alternating voltage shifted in phase relative to the anode voltage, or by pulses of voltage applied to the grid at the appropriate time intervals. In this last case a minimum pulse length must be maintained in order that firing should be satisfactory. This may be seen by noting that to change the state of the thyristor the centre junction changes from reverse to forward bias and therefore considerable charge needs to be built up in the n_1 and p_2 regions. The two-transistor model in conjunction with the discussion in Section 4.5 will give a conceptual picture of the situation; a finite time is required to establish this bulk charge and the thyristor cannot be

switched ON until this has occurred. In order to ensure firing it is often convenient to apply a rapid sequence of pulses to the gate to ensure that, even with
inductive loads, the maintaining current has been exceeded.

6.3 Light sensitive elements

An entirely different application of semiconductor material makes use of the
variation of some of its properties under the influence of light. When light falls
on a semiconductor material, the energy of the incident radiation may be given
up to the crystal lattice. If the wavelength of the light lies within certain limits,
the energy given up may be sufficient to disrupt some of the crystal lattice bonds
and release an electron together with its associated hole. Thus luminous radiation may result in the generation of hole–electron pairs in a semiconductor.

Two basic techniques are available for the detection of these additional
photo-generated carriers. Their presence in a semiconductor will result in an
increase in the conductivity of the material. Alternatively, if the photo-genera-
tion occurs in one of the two regions of a semiconductor diode, the flow of
carriers across the junction may be detected and used as a measure of the
incident light.

6.3.1 Photovoltaic cell

One very important application of the photoelectric effect in a semiconductor
is in the generation of an e.m.f. across the open circuit terminals of a diode when
illuminated.

Under equilibrium conditions we have seen in Section 1.3 that a potential
difference is set up between the p- and n-regions of a junction diode. This occurs
as a result of the diffusion of majority carriers across the boundary to the
opposite region. The construction of a typical photodiode is shown in Fig. 6.8a
and the circuit symbol in Fig. 6.8b.

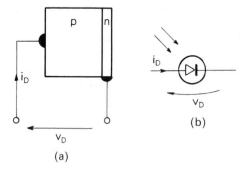

(a)

(b)

Fig. 6.8 Photodiode (a) construction (b) conventional symbol

If we now illuminate the n-region of the semiconductor, additional holes and
electrons are generated. The holes which are so generated will diffuse randomly
and those which reach the barrier will be swept into the p-region. This will tend
to reduce the negative charge in the p-region and thus the potential difference

across the diode. Thus a high impedance voltmeter connected across the diode would now show that the p side of the diode was positive with respect to the n side. The open circuit voltage for a photodiode as a function of luminous intensity is shown in Fig. 6.9.

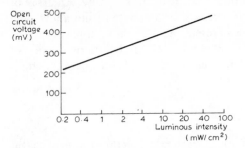

Fig. 6.9 Open circuit voltage characteristics of photodiode

If, however, a current is taken from the diode, the terminal voltage will fall. Since the current is composed of holes flowing from the n- to p-regions we see that this is in the same direction as the reverse saturation current in a diode. The variation of current with voltage for different degrees of illumination is shown for the same photodiode in Fig. 6.10. This shows that the characteristic

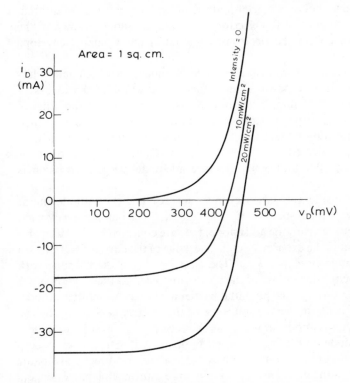

Fig. 6.10 Current versus voltage characteristic of photodiode

of a photodiode is almost identical with that of the diode under dark conditions but displaced vertically by a current which is directly proportional to the intensity of illumination.

Such photovoltaic diodes have two principal uses. Constructed in miniature form, they may be used to read computer punched cards or tape. An array of such cells positioned behind the punched cards, illuminated from above, will give a voltage when a hole in the card is present. An alternative application is as a primary generator of electricity. A bank of relatively large cells, up to areas of say 10 cm^2, may be used to charge batteries in remote places where the sun is probably the most convenient energy source.

6.4 Thermionic devices

All the electronic devices studied so far have been based on the motion of charge carriers in semiconducting materials. Another very large class of devices relies for its performance on the motion of electrons in a vacuum. Historically, vacuum tubes (or valves) preceded solid state devices by several decades; nowadays their use is limited, with one major exception, to specialized devices for high frequency work. The major exception is the use of simple triodes and pentodes (the basic thermionic devices) as amplifiers at frequencies in excess of about 30 MHz delivering power up to several hundred kilowatts.

A thermionic device consists of a cathode, from which electrons are emitted, one or more grids to control the electron flow, and an anode (known in the USA as a plate) to collect the electrons, enclosed in an evacuated envelope usually made of glass.

The cathode is usually cylindrical and made of a material which emits electrons when heated. In low power devices this is a mixture of the oxides of barium, strontium and calcium. When heated to a temperature of the order of 800 to 1200 K, large numbers of electrons are emitted from the surface. In high power valves the emitting surface is usually either a pure tungsten filament or a tungsten filament activated with thorium; this needs to be heated to a temperature of between 2000 and 3000 K before adequate electron emission is obtained.

The anode is a considerably larger cylinder surrounding the cathode and at a potential positive with respect to it. The emitted electrons are accelerated across the space between cathode and anode; they are collected by the latter and give rise to a current in the external circuit. The value of the current is a function of the anode–cathode potential V_{AK}. The electrons leave the cathode with relatively low velocity, since most of the energy imparted to them thermally is used up in breaking through the potential barrier at the surface of the cathode. They immediately come under the influence of the electric field set up by the positive potential on the anode, and are thereby accelerated towards the anode. On striking the anode and being absorbed by it, their kinetic energy is converted into heat at the anode. In small valves this heat may be directly dissipated by radiation through the glass envelope but in large transmitting valves several kilowatts of power may need to be conducted away; this is achieved either by the use of radiating fins on the anode, often cooled by a fan, or by circulating distilled water around the anode.

6.4.1 Thermionic triode

In order to produce a device which is capable of amplification or control, a third electrode needs to be inserted in the valve in order to modulate the flow of electrons between cathode and anode. This takes the form of a fine grid of wires in cylindrical form placed between the cathode and the anode, termed the control grid (or usually just the grid). A negative potential applied to the grid will set up an electric field at the cathode which is opposite to that caused by the anode; this will result in a reduction in the total current flowing to the anode.

The anode characteristics of a triode are shown in Fig. 6.11 for various values of grid–cathode voltage, v_{GK}, both positive and negative. In high power valves it is essential to obtain the maximum available power from the device and so the grid is usually driven positive to values which may, in cases, approach the anode voltage. It may, therefore be expected that for positive excursions of v_{GK} grid current will flow and this is approximately linearly related to grid voltage.

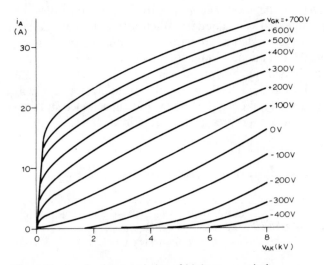

Fig. 6.11 Anode characteristics of high power triode

It is important to remember that the static characteristics of a device give no information whereby the transient behaviour may be predicted. We may obtain a conceptual insight into the dynamic response by appreciating that the anode current is controlled by the charge on the grid; as a result of the capacitance between grid and cathode C_{gk}, the grid charge will take a finite time to build up when the applied voltage is changed. This will thus impose a limit on the maximum frequency at which the valve may be operated. In addition the capacitance between anode and grid may cause a variation of grid voltage directly dependent on anode voltage. This again will impose a limit on the frequency of operation and under some conditions may even give rise to instability in a circuit.

Typical values of these parameters for a tetrode are

$$C_{gk} = 45 \text{ pF}$$

$$C_{ag} = 33 \text{ pF}$$

Since the anode and cathode are effectively screened from each other by the grid, the anode–cathode capacitance C_{ak} is relatively small, about 1·0 pF.

6.4.2 Tetrode and pentode valves

One great shortcoming of the triode is the relatively large capacitance between anode and grid. This capacitance may be considerably reduced in the tetrode by interposing a screen in the form of a cylindrical mesh between anode and grid which reduces the electrostatic coupling between these electrodes. The valve is normally operated with the screen held at a fixed potential; thus the cathode, grid, and screen act rather like a triode with a fixed anode voltage; the electron stream is not collected by the screen but passes through it to the anode. Thus the current collected by the anode remains fairly constant regardless of variations in the anode potential. The anode characteristics of a tetrode are similar to those shown for the triode but are flatter, indicating that the anode current is almost independent of anode voltage for anode voltages greater than screen voltage. The tetrode will not operate in a linear manner if the anode voltage falls too low. A third grid, at earth potential, the suppressor grid, will cure this defect and, incidentally, reduce the anode–grid capacitance still further. A pentode is a valve containing three grids as described above.

References

BECK, A. H. W. and AHMED, H., *An Introduction to Physical Electronics*, Arnold, 1968, Chapters 6 and 8.

ALLISON, J., *Electronic Engineering Materials and Devices*, McGraw-Hill, 1971, Chapter 10.

COMER, D. J., *Introduction to Semi-conductor Circuit Design*, Addison Wesley, 1968, Chapter 10.

ATKINSON, P., *Thyristors and their Application*, Mills and Boon, 1972, Chapter 1.

MAZDA, F. F., *Thyristor Control*, Newnes-Butterworth, 1973, Chapters 1 and 2.

Problems

6.1 Two Zener diodes A and B are available. A has a breakdown voltage of 12·0 V and a reverse saturation current of 10 nA, B has a breakdown voltage of 7·5 V and a reverse saturation current of 20 nA. Sketch the characteristic relating supply voltage to current flowing for the following combination:

(a) A and B in parallel and both in the same direction

(b) A and B in parallel and back to back

(c) A and B in series and back to back.

6.2 A Zener diode has a reverse saturation current of 100 nA, a breakdown voltage of 12·0 V and when conducting the voltage rises by 0·03 V for an increase in current of 1 mA. It is connected in series with a resistor of 100 Ω to a sinusoidal supply of r.m.s. value 10 V. Sketch and dimension the voltage across the Zener diode.

6.3 A Zener diode is used to provide a fixed output voltage to a variable load resistance. The circuit used is shown in Fig. P6.1. The 40 V supply is available from a power pack having an internal resistance of 100 Ω. The Zener diode has a breakdown voltage of 20 V and an incremental resistance when conducting of 10 Ω and a maximum current of 40 mA. The load resistor R_L may vary from 1 kΩ to 10 kΩ. Select a value for R_S and determine the variation of output voltage over the complete range of load resistance. What is the minimum value of R_S which will allow the output voltage to be stabilized?

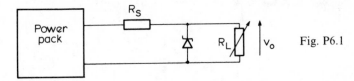

Fig. P6.1

6.4 An SCR in series with a load resistor connected to a 250 V alternating supply is switched on by applying a direct voltage to its gate. When $v_{GK} = 0\cdot1$ V, the device will conduct for anode voltages exceeding 350 V and when the gate voltage exceeds $0\cdot8$ V conduction occurs for all positive anode voltages; between these values assume a linear relationship between gate voltage and anode firing voltage. Sketch the relationship between conduction angle and gate voltage.

6.5 The load resistor in Problem 6.4 has a value of 50 Ω. Sketch the power in the load as a function of gate voltage.

6.6 An SCR is switched ON by a pulse applied to the gate controlled by a variable voltage. The current flowing in the load is required to be linearly related to the control voltage. Plot the relationship between time of occurrence of pulse and control voltage to achieve this.

6.7 The photodiode whose characteristics are shown in Fig. 6.10 is connected across a resistance of value 40 Ω. Determine the power dissipated in the resistance and in the diode when illuminated at (a) 10 mW/cm^2 and (b) 20 mW/cm^2.

6.8 Determine the value of load resistance such that maximum power is delivered at an illumination of 10 mW/cm^2 by the diode of Problem 6.7 to the load. What is the maximum power available?

6.9 What total area of photocell surface at 10 mW/cm^2 illumination would be needed to deliver 2 kW to a load at 250 V? If the surface area were divided into cells of 1 cm^2 area, how would you interconnect these cells?

7
Integrated Circuit Fabrication

7.0 Introduction

The devices which we have considered in the preceding six chapters have been fabricated as discrete units, each one separately packaged and individually provided with external leads for connection into a circuit.

The ability to construct transistors occupying a very small volume has forced component designers to reduce the physical size of other components, notably resistors and capacitors; inductors have been reduced in size but not to the same extent. A logical development of this miniaturization process is to fabricate a complete circuit including resistors, capacitors, transistors, diodes, and interconnecting leads by a single manufacturing process. Such a circuit is known as an integrated circuit.

We shall look at methods whereby the passive components, resistors, capacitors, and inductors, may be produced using semiconductor material and how they may be incorporated in circuits involving transistors and other active devices. Constraints will be imposed on the values of the various components which are obtained using the various types of manufacturing process; we shall then be in a position to see what limits this imposes on the circuit design using an integrated form of fabrication.

7.1 Fabrication techniques

The most usual form of integrated circuit construction, the monolithic form, is to manufacture all components on one single chip of silicon by a diffusion process. The various components on the chip are isolated from each other and this is achieved either by using a diffusion isolating technique or by dielectric isolation.

An alternative method, known as the hybrid process, is to construct the resistors and capacitors in the network by a thin film technique on a ceramic substrate and to bond discrete transistors on the upper surface of this. Although thin film components are relatively simple and cheap to construct, the tendency in integrated fabrication is to use monolithic techniques as these lend themselves more readily to large scale integration, involving perhaps several hundred components in one circuit. We shall therefore mainly study the monolithic form of construction of integrated circuits.

In the monolithic form two methods are available for isolating the various circuit elements. The simplest isolating technique is to fabricate all the elements on a substrate of, usually, p-type silicon within islands of n-type silicon which have been diffused into the substrate. All components will therefore have a junction diode between the n-type island and the p-type substrate. If the substrate is always kept at a more negative potential than any other part of the circuit, these diodes will always be reverse biased and there will be negligible leakage current between any of the components; however it must be remembered that there is a finite capacitance across the reverse biased diode and so additional parasitic elements will appear in the model of the system.

An alternative isolation technique is to construct all the elements on a ceramic substrate, usually of silicon dioxide. This not only reduces the stray capacitance and leakage conductance but also permits freedom of choice of transistor type, allowing the use of complementary pairs of transistors in the output stage. It is, however, more expensive to fabricate circuits using this technique. Although superficially similar to the hybrid process, mentioned above, all circuit elements are fabricated using monolithic techniques, the ceramic substrate serving only to isolate them.

The first step in monolithic fabrication is to decide on the number of isolated parts needed in a circuit. As a first approach we may assume that each component—resistor, capacitor, transistor—may be considered as a separate unit; however, as we shall see on further study of the techniques, certain groups of components having one common terminal may be lumped into one unit manufactured on a single isolation island.

The p-type substrate, perhaps 200 µm thick is first prepared by growing an n-layer of thickness about 10 µm over the whole surface. This n-type layer is now divided up into isolation islands. First a mask is formed on the surface of the n-layer using a photoresistive material which is dissolved away in the regions where the isolation zones are to be formed. p-type impurity is now diffused into the surface through this mask converting the n-type material into the p-type and penetrating down to the underlying substrate. The n-type layer beneath the mask is unaffected. This results in a number of n-type islands separated by p-type isolation. The form of the basic chip will now be similar to the cross-section shown in Fig. 7.1 showing three islands ready for subsequent fabrication processes.

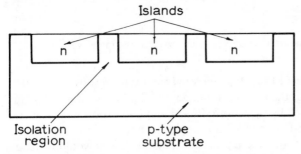

Fig. 7.1 Cross section of monolithic integrated circuit showing isolation islands

Thin film components are constructed on a ceramic substrate by vacuum deposition or electrodeposition of metals on the surface. Since the materials used are not restricted to semiconductors an added degree of flexibility in design is introduced, although the technique is not, at present, suitable for the construction of transistors.

7.2 Resistors

The usual method of construction of resistors in a monolithic circuit is to diffuse a controlled amount of p-type impurity into the selected n-type island, through a mask formed, as discussed previously, from a photoresist; the length and width of the p-diffusion is determined by the mask; the depth of the diffusion is controlled by the length of diffusion process. As we shall see in Section 7.6, transistors require p-diffusion to form the base and, in order to reduce the number of controlled processes in IC fabrication, the same diffusion process is usually used for the formation of resistors; thus the thickness of the layer of resistive p-type material is limited by the length of diffusion needed to construct the base of the transistors on the chip. An alternative to using this p-type diffusion is to use the n-type diffusion which is used subsequently to form the emitters of transistors. Since the impurity density in the emitter is considerably higher than that in the base, resistors having higher values may be produced by this process; this may be verified from equation 1.5. The final process is the oxidization of the surface of the semiconductor chip to provide insulation, and the connection of aluminium leads, usually deposited by a sputtering process in a vacuum. The final form of such a resistor using p-type material is shown in Fig. 7.2.

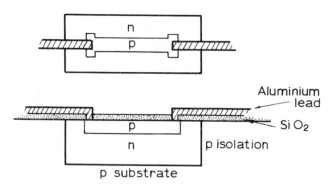

Fig. 7.2 Construction of monolithic resistor (a) plan (b) cross section elevation

The depth of penetration of the p-type material for a typical resistor is of the order of 5 μm, the width of the resistor about 25 μm. The value of the resistor is then primarily determined by the length. It is possible to produce resistances whose values range from 1 Ω to 100 kΩ although the extreme values are more difficult to achieve satisfactorily and a practical range of values would be from about 100 Ω to 30 kΩ.

Since careful control over depth of penetration is necessary in order to obtain accuracy of resistor values, it is generally not possible to maintain absolute values more accurately than $\pm 10\%$. However, when a batch of resistors is made by one diffusion process, it is possible to maintain their ratios to better than 1% since this is controlled solely by the accuracy of construction of the mask. Thus integrated circuits tend to be designed around networks in which the ratio of resistors controls the performance rather than their absolute value.

Resistors may be made using thin film techniques by depositing a thin film of resistive material on the surface of either an insulating substrate or the protective oxide layer on the surface of a monolithic chip. Materials other than p-type silicon, such as nichrome, aluminium, tantalum, etc., may be used to give a wider range of resistivities. The deposited layer is etched to give the desired profile and protected by an oxide layer, ohmic contacts being made to the two ends through this layer. Resistance values from $20\,\Omega$ to $50\,k\Omega$ may be achieved using nichrome as the resistor material. Although thin film resistors may be made with more precision, however, they are more expensive to fabricate.

7.3 Capacitors

In Section 1.9 we noted that a p-n junction diode when reverse biased acts like a capacitor shunted by the resistance of the reverse biased diode. Such diffused junction capacitors may be constructed with values of the order of $3\,pF/mm^2$ with breakdown voltages of the order of 10–20 V. The capacitance will be a nonlinear function of the biasing voltage, the functional form of the nonlinearity depending upon the method of fabrication of the junction; it has also a relatively high temperature coefficient. Thin film capacitors may be constructed by depositing a layer of metal then a layer of oxide dielectric and finally a second layer of conductor on a ceramic substrate. Capacitor values ranging from 0.5 to $10\,pF/mm^2$ may be obtained with typical breakdown voltages of about 10 V using thin film techniques.

A capacitor using a combination of the two techniques is shown in Fig. 7.3.

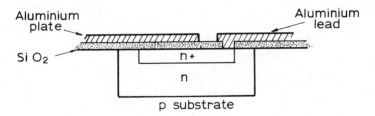

Fig. 7.3 Metal-oxide-semiconductor capacitor

The n-type island is used as one plate of the capacitor, silicon dioxide as the dielectric and an aluminium plate vacuum deposited on the surface as the second plate; usually some highly doped semiconductor is diffused in beneath the oxide layer to provide a good lower conducting plate. Capacitor values of the order of 10 pF/mm^2 with breakdown voltages up to 50 V may thus be obtained.

The diffused junction type of capacitor is the easiest and least expensive to fabricate, but it is associated with a relatively large parasitic capacitance to substrate and also a parasitic resistance in series with the junction capacitance. The oxide layer capacitance is more expensive since additional processing is required; it has the advantage that no polarizing voltage is needed, its value is not dependent on applied voltage, and higher breakdown voltages, up to 200 V, may be obtained.

7.4 Inductors

No satisfactory method of fabricating inductors in either monolithic or thin film form has been discovered to date. Inductors with values of a few nano-henries may be made by constructing a spiral conductor on an insulating substrate, but these have exceedingly limited utility.

It is for this reason that inductors are normally avoided in integrated circuit design. It will be shown in Section 13.9.4 that a simulated inductor may be constructed using high gain amplifiers in conjunction with a capacitor.

7.5 Parasitic elements

It should be noted that since all the components are constructed on a p-type silicon substrate one must always be aware of the possibility of parasitic effects; in particular the reverse biased diode between island and substrate will give rise to a stray capacitance.

In Fig. 7.2, for example, we may note that between resistor, island, and substrate, a potential transistor exists and, depending on the biasing voltages, this may give rise to parasitic currents. An equivalent circuit model of the resistor shown in Fig. 7.2 is given in Fig. 7.4. R represents the desired resistance; the

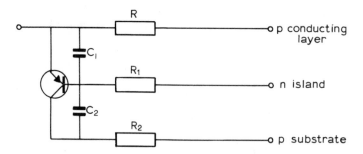

Fig. 7.4 Equivalent circuit model of monolithic resistor showing parasitic elements

transistor and capacitors C_1 and C_2 represent the effects of the junctions between the three types of semiconductor; these effects are distributed along the length of the resistance, but may be approximately modelled by the lumped circuit shown. R_1 and R_2 represent the lateral resistance of the island zone and the p-substrate. In order to maintain this parasitic transistor nonconducting, the substrate is connected to the most negative potential in the circuit; the island is usually left unconnected.

7.6 Bipolar transistors

The most usual form of bipolar transistor used in integrated circuit construction is of the n-p-n variety. This utilizes the n-type island zone as the collector and is fabricated by diffusing two further zones, one p-type for the base and one n-type for the collector into this. The cross-section of such a device is shown in Fig. 7.5.

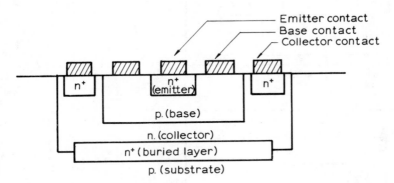

Fig. 7.5 Monolithic n-p-n transistor

It should be noted that the device suffers certain disadvantages compared with a similar discrete transistor. The necessity to make contact with the collector on its upper surface introduces a considerable resistive path between the collector contact and the collector–base junction. This may be reduced by introducing an n^+, highly doped and hence highly conductive, layer along the interface between the substrate and the n-type collector layer. The effect of this collector resistance will be a deterioration in the performance of the transistor at high frequencies. Against this disadvantage may be set the ability to construct many transistors in close proximity and hence to achieve better thermal stability. Beneath the ohmic contacts to the collector, a small n^+ region is usually diffused. This ensures that the contact with the external lead, usually made of aluminium, an acceptor impurity, is a truly ohmic contact and does not form a p-n junction.

The construction of p-n-p transistors may be simply accomplished using the substrate as the collector, but such a transistor is very limited in application. A more useful type is obtained by utilizing the n-type island as a base and diffusing both collector and emitter into the surface giving the lateral p-n-p transistor shown in Fig. 7.6. This transistor suffers the grave disadvantage that, in addition to the desired p-n-p transistor, there is another parasitic transistor having the substrate as its collector. Overall the current gain of such a transistor is limited to about 5, since the current flows laterally between emitter and collector across the base and the inaccuracies in location of the diffusing masks necessitates a wide base.

The substrate used in the construction of integrated circuits is normally p-type silicon. It is possible to use n-type substrate, in which case the most easily fabricated transistor is a p-n-p type. In many circuits it is desirable to have complementary pairs of transistors, one an n-p-n and the other p-n-p. We have seen that p-n-p transistors may be constructed in lateral form using the same number of diffusion processes as conventional n-p-n transistors but having very low values of current gain.

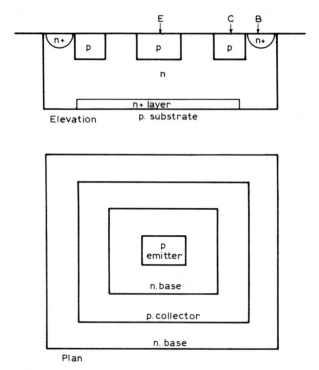

Fig. 7.6 Lateral p-n-p transistor (a) cross section (b) plan

p-n-p transistors with high current gains may be constructed but at the cost of additional diffusion processes. One technique is to use an extra isolation region of n-material within the chosen island and then to construct the p-n-p transistor within this double island. An alternative technique is to use the principal of dielectric isolation in which each component or group of components is isolated from its neighbours by a silicon dioxide layer. Separate transistors may be constructed in these islands of either p-n-p or n-p-n form. Either of these methods entails additional diffusion processes above those needed for a single type of transistor.

7.7 Diodes

The diodes used in integrated circuit technology are frequently transistors with one pair of electrodes shorted together; a common configuration is to connect collector and base as the anode of the diode. Alternative connections are possible or the final p diffusion may be completely omitted. The diode formed by connecting the base and collector of a transistor together is preferred as less charge is stored in the device in this configuration and hence it has a higher switching speed. Diodes having common cathodes may all be constructed in one n-type island whereas those with common anodes each require a separate island region.

7.8 Field-effect transistors

One of the simplest devices to produce by integrated circuit technology is the mosfet. It is constructed as shown in Fig. 7.7 directly on the p-type substrate;

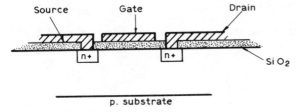

Fig. 7.7 Cross section of monolithic mosfet

this gives an n-channel mosfet identical in operation to the device described in Section 2.8. The fabrication is essentially much simpler than that for a bipolar transistor since fewer operations are needed. In addition, since, under operating conditions, the substrate to source, channel and drain all form reverse biased diodes, there is no need for isolation regions between adjacent mosfets and the n-type islands may be omitted completely for an n-channel mosfet. One result of this is that the packing density of mosfets is much higher than that of bipolar transistors. A bipolar transistor occupies an area of the order of 5×10^{-8} m^2 compared with a mosfet which occupies only about 2×10^{-9} m^2.

Another related advantage is that the area taken up by a mosfet is considerably smaller than that of a monolithic or thin film resistor. Mosfets are, therefore, frequently used at very low drain voltages as discussed in Section 2.3 as a substitute for conventional integrated circuit resistors. A more usual configuration for a mosfet as a resistor is obtained by connecting the gate of an enhancement type mosfet directly to the drain. It may be shown that the small signal incremental resistance of such a device is approximately $1/g_m$. Since the drain current in an integrated circuit may be as low as say 10 μA, the transconductance, g_m, may be only 10 μS corresponding to a resistor of 100 kΩ. The reduction in chip area required for such an element is about 70 over that needed for a similar monolithic resistor. This allows the construction of entire circuits using mosfets alone; this is particularly relevant to certain digital circuits.

References

ALLISON, J., *Electronic Engineering Materials and Devices*, McGraw-Hill, 1971, Chapter 13.
MILLMAN, J., and HALKIAS, C. C., *Integrated Electronics*, McGraw-Hill, 1972, Chapter 7.

WARNER, R., *et al.*, *Integrated Circuits—Design Principles and Fabrication*, McGraw-Hill, 1965, Chapters 5–10.
HIBBERD, R. G., *Integrated Circuits*, McGraw-Hill, 1969, Chapters 2 and 3.
MEYER, C. S., *et al.*, *Analysis and Design of Integrated Circuits*, McGraw-Hill, 1968, Chapters 1–5.

Problems

7.1 Determine the capacitance of a mos integrated capacitor of area 10^{-6} m^2 if the silicon oxide thickness is 80 nm. The relative permittivity of silicon dioxide may be assumed to be 6. What is its value if titanium dioxide is used as dielectric (relative permittivity = 100)?

7.2 Determine the resistance of a diffused resistor 1 mm in length and 2×10^{-3} mm in width. It is constructed by diffusing an acceptor impurity into the surface of the silicon so that the surface impurity density is 5×10^{24}/m^3 falling to zero at a depth of 2 μm. (Make a simple approximation for the variation of impurity density with depth.)

7.3 A resistor is manufactured using the same mask as in Problem 7.2 but donor atoms are diffused such that the surface impurity density is $1{\cdot}2 \times 10^{27}$/m^3 falling to zero at a depth of $1{\cdot}1$ μm. Determine the resistance.

8
Amplifiers—
Operating Point and
Biasing

8.0 Introduction

One of the most important applications of bipolar transistors is in the amplification of small alternating signals; field-effect transistors are also used for this purpose, although to a much smaller extent. We shall look at a very simple form of amplifier using a bipolar transistor and consider the limitations which are imposed on such an amplifier by the various nonlinearities in the characteristics and the power dissipation constraints of the transistor. Since the voltage fluctuations which we wish to amplify will be superimposed on a steady bias voltage, we shall also study the way in which we may fix the operating point so that it does not change too much despite variation of temperature or between transistors of one type.

8.1 Small signal voltage amplifier

We may see from the collector characteristics of a bipolar transistor shown in Fig. 4.9b that, if we keep the collector–emitter voltage constant at, say, 4 V, and change the base current from 20 µA to 60 µA, then the collector current will change from about 3 mA to 9 mA. A transistor used in this manner is therefore a current amplifier having a current gain of 150, since a change of 40 µA of base current results in a change of 6 mA in collector current. Current amplifiers such as this are sometimes used in electronic circuits but voltage amplifiers are much more common; it is relatively straightforward to convert this current amplifier to a simple voltage amplifier. From Fig. 4.10 we note that a change of base–emitter voltage from 0·47 V to 0·53 V will cause the base current to change from 20 µA to 60 µA. As we have seen above this results in a change in the collector current from 3 mA to 9 mA. If we now allow this current to pass through a resistor of value, say, 400 Ω a voltage will be developed across the resistor which will vary from 1·2 V to 3·6 V. (This presupposes that the collector voltage is kept constant.) We have thus obtained a simple voltage amplifier having a voltage gain of 40. If we intend to use this to amplify small alternating signals, we need to choose a bias voltage at the input upon which to superimpose the input voltage variations; the output voltage fluctuations will similarly appear superimposed on a fixed output bias voltage. In the amplifier we have been discussing we could choose an operating point at $V_{EB} = 0.5$ V, $I_B = 40$ µA, $I_C = 6$ mA.

In the preceding discussion we have assumed that the input voltage is applied between base and emitter and the output voltage taken between collector and emitter. The emitter terminal is common between input and output and the circuit is thus known as a common emitter amplifier. Alternatively we may have a common collector amplifier, often called an emitter follower, with the collector terminal common to input and output circuits; another more rarely used form is the common base amplifier. Similarly a field-effect transistor may be connected in common source, drain, or gate.

The simple common emitter amplifier may be closely realized by the circuit of Fig. 8.1. The input voltage changes are applied between base and emitter,

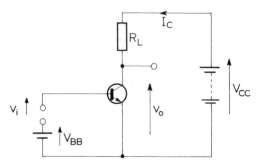

Fig. 8.1 Basic common emitter amplifier

and a resistance R_L is connected in series with the collector and battery. The output voltage fluctuations are observed between the collector and emitter, rather than directly across R_L; although these two total voltages are not identical, the voltage variations are equal but of opposite sign. In order to establish the desired operating point we need the two batteries V_{CC} and V_{BB}. One assumption we made previously was that the collector voltage remained constant at V_{CC}. We see that in this circuit this is not valid but that at all times the collector–emitter voltage is given by

$$v_{CE} = V_{CC} - i_C R_L \tag{8.1a}$$

and the base–emitter voltage by

$$v_{BE} = V_{BB} \tag{8.1b}$$

We now need to solve these equations with the added constraints which are imposed by the transistor on i_C, v_{CE}, v_{BE}; these may be expressed most conveniently graphically by Figs. 4.9b and 4.10, reproduced in Figs. 8.2a and b. Let us assume that $V_{BB} = 0.5$ V and $V_{CC} = 5$ V with $R_L = 400$ Ω. Then from Fig. 8.2b we may read directly that $I_B = 40$ µA. Our problem now reduces to the solution of equation 8.1a with the collector characteristic of Fig. 8.2a corresponding to $I_B = 40\ \mu A$; this is done by plotting equation 8.1a on Fig. 8.2a and noting the point of intersection of the two graphs. This occurs at point Q where $I_C = 6$ mA, $V_{CE} = 2$ V and this is known as the operating point. The graph of equation 8.1a is known as a load line. All values of i_C and v_{CE} for different values of v_{BE} must lie on this line.

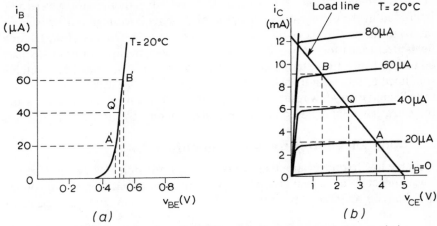

Fig. 8.2 Graphical determination of amplifier gain from transistor characteristics

We may now refine the crude analysis which we performed above to determine the voltage gain of the amplifier. When the input voltage between base and emitter is 0·47 V, the base current is 20 μA and thus the collector current and voltage are given by point A on Fig. 8.2, the intersection between the load line and the 20 μA characteristic; the collector–emitter voltage is thus 3·8 V. When the input voltage rises to 0·53 V, the base current becomes 60 μA and the collector voltage is given by point B, namely 1·4 V. Thus for an increase in base–emitter voltage of 0·06 V the collector voltage falls by 2·3 V corresponding to an incremental voltage gain of −40. The negative sign indicates that an increase of input voltage results in a fall in collector voltage. Although we obtain the same magnitude of incremental voltage gain as earlier, this is only because the collector characteristics of a bipolar transistor are approximately parallel horizontal lines in this region of operation. The latter analysis is more correct since it takes into account the variation of collector voltage with variation of collector current. The magnitude of voltage gain which we have obtained is not a realistic one for a transistor amplifier as voltage gains of several hundred are quite common; the low value is entirely due to the low value of load resistor which was chosen.

The amplifier we have studied is suitable for small increments of voltage and thus is readily applied to small alternating voltages. These may be applied to the base of the transistor through a capacitor whose impedance is low at the frequency of operation, and removed from the collector through a similar capacitor. These isolating capacitors are necessary in order that the steady bias voltages which determine the operating point are not affected by the connection of the signal source at the input or the load resistance at the output. Another point to note about this circuit is the need for two separate batteries, one of them having a voltage of about 0·5 V. The difficulties associated with these batteries will be studied later in this chapter.

We now have the basic concept of a small signal amplifier, in which signal voltages and currents are small in comparison with the corresponding steady voltages and currents in the transistor. If, as is frequently the case, we are only interested in the voltage or current gain of the circuit to small signals we may model the device by one of the small signal models discussed in Section 3.4 for a field-effect transistor, or in Section 5.4 for a bipolar transistor; this model may be inserted into the external circuit of the amplifier which has been suitably simplified to represent its behaviour to small signals. The usual simplifications arise from the very low resistance of all batteries to small voltage variations and these are generally replaced in the small signal model by short circuits.

8.2 Common emitter transistor amplifier

A simple low frequency amplifier using bipolar transistors is shown in Fig. 8.3. It is connected to a signal generator giving an open circuit alternating voltage v_i and having an internal resistance R_1. The bias voltage to establish the correct

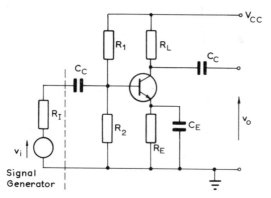

Fig. 8.3 Practical form of common emitter amplifier

operating voltage on the base is determined by the potential divider R_1 and R_2 across the supply V_{CC}. The magnitude of the collector current is controlled by R_E which is shunted at the frequencies of the input signal voltage by a capacitor C_E. The two capacitors C_C prevent the flow of direct current through the signal generator or the flow of collector current through the output voltage measuring instrument; their impedances are small at the frequency of the input signal. As long as we restrict ourselves to frequencies that are low compared with the cut-off frequency of the transistor, we may use the small signal model of the transistor given in Fig. 5.6. The capacitors C_C and C_E may be replaced by short circuits, as they are large in magnitude and therefore have a low impedance. The batteries may be ideally considered as short circuits to alternating voltages. This circuit is shown in Fig. 8.4.

By using Thevenin's theorem on that part of the network to the left of B′E and by combining r_o and R_L in parallel, Fig. 8.4 may be redrawn as in Fig. 8.5, where

$$R'_I = r_x + \frac{R_I R_B}{R_I + R_B} \tag{8.2a}$$

$$v'_i = \frac{R_B}{R_I + R_B} v_i \tag{8.2b}$$

$$R'_L = \frac{r_o R_L}{r_o + R_L} \tag{8.3}$$

where $\quad R_B = \dfrac{R_1 R_2}{R_1 + R_2}$ \hfill (8.4)

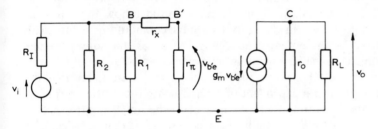

Fig. 8.4 Small signal model of common emitter amplifier

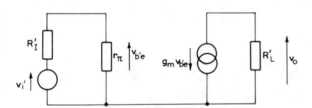

Fig. 8.5 Equivalent model to Fig. 8.4

The voltage gain, k_{v0}, of the amplifier may be directly written down from inspection of Fig. 8.5 as

$$k_{v0} = \frac{g_m R'_L r_\pi}{r_\pi + R'_I} \frac{v'_i}{v_i}$$

$$= \frac{g_m R'_L r_\pi R_B}{(r_\pi + r_x)(R_I + R_B) + R_I R_B} \tag{8.5}$$

$$= g_m R'_L a \tag{8.5a}$$

where $\quad a = \dfrac{r_\pi R_B}{(r_\pi + r_x)(R_I + R_B) + R_I R_B}$

It is of interest to note how the gain may be made as large as possible when the circuit is being fed from a given signal source. The only circuit elements available to the designer, having chosen a transistor, are R_B (namely R_1 and R_2) and R_L. It is obvious, both from equation 8.5 and intuitively, that an increase in R_B will increase the gain since it will shunt less of the signal current outside the transistor. This is obviously a first step in the design and if R_B is made large compared with R_1 (say $R_B = 10\ R_1$) then

$$a \simeq \frac{r_\pi}{r_\pi + r_x + R_1}$$

It may also be thought that an increase in R_L will cause the gain to increase towards the limiting value obtained when $R_L = \infty$ and $R_L' = r_0$. However, equation 5.7a shows that the transconductance g_m is directly proportional to the steady collector current I_C. Thus we must consider how the quiescent currents and voltages vary with change of R_L. In the circuit of Fig. 8.3 the base voltage V_B is fixed by the potential divider R_1, R_2 across the supply voltage V_{CC}. If we assume that the base–emitter voltage is constant at about 0·5 V (it will not vary by more than about 100 mV either way) we observe that the emitter voltage V_E is also independent of R_L and hence the emitter current is given by V_E / R_E and is sensibly fixed. We now make the simple approximation that $r_0 \gg R_L$ so that $R_L' = R_L$. Recalling from equation 5.7a that $g_m = e|I_C|/kT$ we may write equation 8.5a as

$$k_{v0} = e|I_C|R_L a/kT$$
$$= e|V_L|a/kT$$

where V_L is the direct voltage drop across the load resistor. Note that the gain is independent of the transistor type. We must now consider one other aspect of voltage amplifiers, namely the maximum available alternating voltage which may be obtained. This is known as the maximum voltage swing. We have seen in Section 8.1 that the collector voltage varies approximately equally either side of the quiescent voltage. The maximum deviation will be obtained when the quiescent voltage is exactly midway between the supply voltage V_{CC} and the voltage at the knee of the characteristics where the collector current ceases to be approximately constant; this latter value may be taken as zero to a first approximation. Thus a choice of quiescent collector voltage close to $V_{CC}/2$ will give maximum collector swing.

Combining this with the above we shall see that the maximum voltage gain which one can obtain from a given transistor, if we wish to get maximum output voltage excursion, will be $e|V_{CC}|/2kT$ by assuming, rather optimistically, that in equation 8.5a both r_π and R_B are large compared with R_1 and r_x, and hence $a = 1$.

We have therefore shown that for this particular circuit there is a limit on the gain, for maximum output voltage swing, determined solely by the magnitude of the supply voltage. If smaller voltage swings are acceptable then the quiescent point may be at a lower voltage; even in this case, however, the maximum gain is limited to $e|V_{CC}|/kT$.

8.2.1 Common emitter transistor amplifier—input and output resistances

The small signal input resistance of the common emitter amplifier r_{in} may be determined from the equivalent circuit of Fig. 8.4. The input resistance of the circuit to the right of terminals BE is

$$r_x + r_\pi$$

and the input resistance of the amplifier as a whole is thus $r_x + r_\pi$ in parallel with R_B (the parallel combination of R_1 and R_2). Thus,

$$r_{in} = \frac{(r_x + r_\pi) R_B}{r_x + r_\pi + R_B} \tag{8.6}$$

and this obviously has its maximum value when R_B is large. If $R_B > 10(r_x + r_\pi)$, a condition which usually holds in a good design, we may write

$$r_{in} = r_x + r_\pi \tag{8.6a}$$

The small signal output resistance r_{out} may similarly be determined by obtaining the Thevenin equivalent to the network to the left of the terminals CE in Fig. 8.4. This shows that the output resistance of the amplifier without load resistance is

$$r_{out} = r_o \tag{8.7}$$

These expressions indicate that the low frequency input resistance is independent of R_L and the output resistance is independent of R_I.

8.2.2 Common emitter transistor—current gain

The low frequency current gain of the common emitter transistor amplifier, k_{i0}, is defined as the ratio v_o/i_i, where i_i is the input signal current to the base of the transistor. From Fig. 8.4 we see that

$$i_i = v_i/(R_I + r_{in})$$

and using equations 8.6 and 8.5 we obtain

$$k_{i0} = \frac{g_m R_L' r_\pi R_B}{r_x + r_\pi + R_B} \tag{8.8}$$

8.3 Common source fet amplifier

The circuit of a fet amplifier similar to the bipolar transistor amplifier of Fig. 8.3 is shown in Fig. 8.6. An n-channel depletion type mosfet has been selected, although the circuit would be suitable for both junction fets and mosfets of either channel kind and operating in either the enhancement or depletion modes; changes would be required in the component values but not in the essential topography of the circuit.

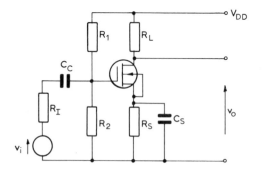

Fig. 8.6 Common source fet amplifier

In a manner similar to that used for the bipolar transistor we may draw an equivalent circuit to this amplifier using the circuit model of the fet given in Fig. 3.9, if the operating frequency is sufficiently low that we may ignore the charge storage effects in the transistor. This circuit model is shown in Fig. 8.7.

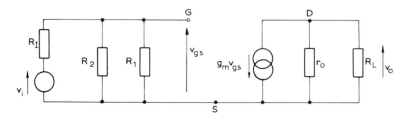

Fig. 8.7 Small signal model of common source amplifier

The voltage gain may be written down by inspection as

$$k_{v0} = \frac{v_o}{v_i} = \frac{g_m R_B R_L'}{R_B + R_I} \tag{8.9}$$

where R_L' and R_B are defined in equations 8.3 and 8.4. It is again immediately seen that the voltage gain is increased by making R_B large; if $R_B > 10R_I$, the voltage gain is given approximately by

$$k_{v0} = g_m R_L' \tag{8.10}$$

It is apparent from equation 8.10 that if g_m remains constant the gain may be increased up to a maximum value of $g_m r_o$ which occurs when $R_L = \infty$; however, since g_m is a function of I_D, the steady current flowing in the drain, this can only be achieved by increasing V_{DD} indefinitely.

We have seen from equation 3.6a that $g_m = (2\beta I_D)^{1/2}$. In order to make any further inferences regarding the variation of gain with load resistance we shall assume that the supply voltage is fixed at V_{DD} and that the transistor is biased so that the drain–source voltage is $V_{DD}/2$; this allows the maximum swing of incremental voltage at the drain, although it is unlikely that this maximum excursion would be utilized in a fet since a large amount of distortion would be introduced. However, under these conditions

$$I_D = V_{DD}/2R_L$$

and the voltage gain becomes

$$k_{v0} = (2\beta I_D)^{1/2} R_L' = (\beta V_{DD} R_L)^{1/2} r_o/(r_o + R_L)$$

It is possible to obtain an expression for the maximum value of voltage gain as we did in the case of the bipolar transistor in Section 8.2. However, the result is more complex and gives less insight into design considerations. To attempt to extract any more from the above expressions would be pointless; the value of r_o is known to be a function of drain current and to pursue the derivation any further analytically would require a knowledge of the functional form of this dependence. It is sufficient to note that here again the voltage gain is dependent upon the supply voltage available, but in this case the parameters of the transistor allow us some freedom in choice of transistor to optimize the gain.

The above analyses have shown that in no case is it permissible, even at low frequencies, to model a transistor by a simple circuit and to expect the parameters of the circuit model to remain invariant when the other components in the amplifier are varied to give optimum performance. This is only permissible if the operating point of the device is not affected by the variation of external components.

8.4 Choice of operating point

We have seen above that the choice of operating point is a critical factor in obtaining optimum performance from the device. The transistors themselves impose certain restrictions on the operating voltage which may be applied and hence restrict the operating region of the device within certain bounds.

Let us consider first the characteristics of bipolar transistors. For minimum distortion of signal we wish to choose a region where the characteristics are most nearly linear.

Referring to Fig. 4.10 we may note that the emitter–base voltage should exceed about 0·4 V in order that base current may flow. However, as it is customary to determine the emitter current and not base–emitter voltage, this constraint does not usually concern the circuit designer.

From Fig. 4.9b we see that the transistor will only operate on the flat part of the collector characteristic if V_{CE} is never allowed to go below about 0·5 V. Again the collector voltage may not exceed a certain prescribed maximum value. When the collector voltage becomes large a high field exists in the collector depletion layer; this results in a large increase in the current for small increments of voltage caused by avalanche multiplication in the collector depletion layer.

A limit is also imposed on the maximum collector current. This is in order to restrict the current to a reasonable value and may not be directly related to any limiting operating conditions.

Finally, a power dissipation constraint is imposed to control the heat developed in the transistor; excessive rise in temperature would be likely to cause failure of the bonded connections to the chip of the transistor, particularly in the vicinity of the collector. Since power is proportional to $v_{CE}i_{C}$ this boundary line is a hyperbola.

The permitted operating region for a bipolar transistor is shown in Fig. 8.8.

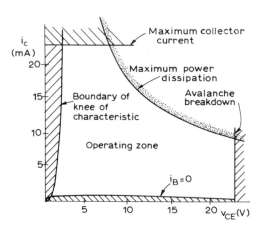

Fig. 8.8 Limitations on linear amplification of bipolar transistor

To obtain maximum performance from the device it is usually operated at about half the maximum permitted voltage for a voltage amplifier or close to the power dissipation limitation for a power amplifier.

A field-effect transistor is also subject to similar constraints. The power limitation curve is similar to the above, as also is the maximum voltage discussed in Section 2.15. The part of the drain characteristics outside the saturation region is also excluded as the drain current is not independent of drain voltage in this region. A further limitation is dictated by the necessity to restrict the gate voltage; in a junction fet the gate–channel junction should not become forward biased, otherwise the gate–source resistance will fall to a low value; in a mosfet the gate voltage is limited so that the oxide layer is not ruptured. These boundaries are shown in Fig. 8.9.

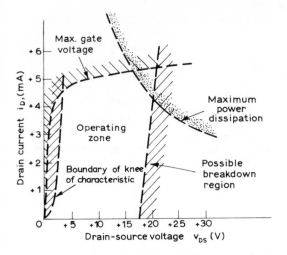

Fig. 8.9 Limitations on linear amplification of field-effect transistor

8.5 Bias network for transistors

Networks for mass production need to be capable of operating with devices randomly selected from a manufacturer's production line. Transistor manufacturer's data indicate that many of the device parameters have a tolerance of $\pm 50\%$ around the mean value.

Again even in custom designed systems the environmental conditions may vary widely. (A network in a satellite may need to operate over a very wide range of temperature.)

It is, therefore, necessary to design biasing networks such that the circuit will still operate even if the transistor is replaced by a similar device with different parameters or if the temperature changes considerably.

8.5.1 Biasing of silicon bipolar transistors

As we have seen in Section 4.5 temperature variation principally affects the base–emitter threshold voltage, V_0, and current gain, β_F, in silicon bipolar transistors. It is just these same parameters which are liable to variation between devices of any given type. We must, therefore, design a biasing network which will ensure that the transistor continues to operate in the active region despite large variations of β_F and V_0 (see Fig. 4.9).

A suitable circuit to achieve this is shown in Fig. 8.10. We may now draw an equivalent circuit to this using the large variable model for the transistor of Fig. 4.6, where the diode has been replaced by a short circuit in series with the battery V_0 since we assume that the emitter–base junction is forward biased. This is shown in Fig. 8.11 where the bias potential divider R_1 and R_2 from the supply voltage V_{CC} has been replaced by the Thevenin equivalent R_B and V_{BB}. Both V_0 and β_F are liable to variation either between transistors of the same manufacturer's type or between different temperatures for a given transistor.

Analysing this circuit and solving for I_C we obtain

$$I_C = \frac{\beta_F(V_{BB} - V_0)}{R_B + (\beta_F + 1)R_E}$$

(8.11)

Manufacturers will normally quote the tolerance on values for β_F and for V_0 or the variation of these parameters with temperature. As an example, a transistor type 2N3114 gives a spread of β_F from 30 to 120 and of V_0 from 0·8 V to 0·9 V at 25°C. For a typical transistor β_F may vary from 40 at −55°C to 100 at 100°C and V_0 from 1·0 V at −55°C to 0·78 V at 100°C.

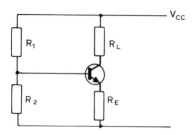

Fig. 8.10 Biasing network of common emitter amplifier

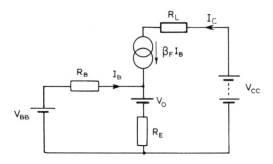

Fig. 8.11 Network model of Fig. 8.10

We may now determine the maximum and minimum values of I_C by inserting maximum and minimum values of β_F and V_0 in equation 8.11. (Note that maximum I_C is associated with maximum β_F and minimum V_0.)
Hence

$$I_{C(max)} = \frac{\beta_{F(max)}(V_{BB} - V_{0(min)})}{R_B + (\beta_{F(max)} + 1)R_E}$$

(8.12a)

and

$$I_{C(min)} = \frac{\beta_{F(min)}(V_{BB} - V_{0(max)})}{R_B + (\beta_{F(min)} + 1)R_E}$$

(8.12b)

If we now assume an upper and lower bound to be placed on the permitted value of I_C, these equations give two relationships between the three unknown quantities V_{BB}, R_B, and R_E. Other constraints may now be invoked to obtain a solution; one important condition is that R_B must be large compared with the small signal input resistance r_{in} so that the input signal is not shunted by the two resistors R_1 and R_2. This then permits an intelligent choice to be made of all the resistors in the design.

8.5.2 Biasing of field-effect transistors

We have noted in Section 2.13 and 2.14 that the transfer characteristics of both junction fets and mosfets are strongly dependent on temperature and also that the spread of such parameters over a batch of transistors is very wide.

For a typical junction fet, quoted values of V_{th} range from -1 to -6 V and I_{DSS} from 5 to 15 mA. Also I_{DSS} varies from 12·5 mA at $-50°C$ to 8 mA at at 100°C. No figures are given for variation of V_{th} although these may be inferred from the quoted variation of g_m assuming a parabolic law for the transfer characteristic; the variation of V_{th} so determined is too small to be significant.

A circuit which may be used to constrain the drain current within given bounds is shown in Fig. 8.12, and it will be seen that this is effectively identical

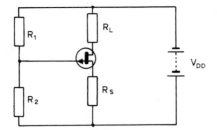

Fig. 8.12 Biasing network of common source amplifier

with the circuit used for the bipolar transistor. This circuit is applicable to both junction fets and mosfets of both channel types. Certain simplifications may be effective in some cases. Since it is not easy to produce a linear or piecewise linear model to a fet we shall use a graphical form of analysis based on the two

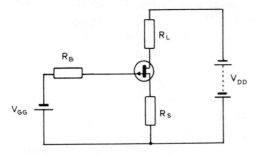

Fig. 8.13 Equivalent network of Fig. 8.12

extreme transfer characteristics. The circuit of Fig. 8.12 may be reduced via Thevenin's theorem to that of Fig. 8.13, where

$$R_B = \frac{R_1 R_2}{R_1 + R_2}$$

$$V_{GG} = \frac{R_2}{R_1 + R_2} V_{DD}$$

We may now plot the two extreme transfer characteristics which we are likely to encounter and mark the upper and lower bounds, $I_{DQ(max)}$ and $I_{DQ(min)}$, of the drain current as in Fig. 8.14. A simple analysis of Fig. 8.13 will give a relationship between gate source voltage and drain current of the form

$$V_{GG} = V_{GS} + R_S I_D \tag{8.13}$$

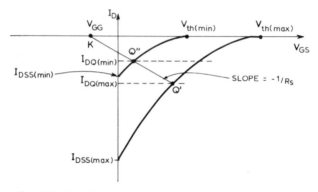

Fig. 8.14 Graphical construction for choosing bias network elements for common source amplifier

It is now necessary to solve this analytic expression with a second functional relationship between I_D and V_{GS} which is imposed by the characteristics of the transistor. We could use equation 2.10 but we would be confronted with a set of nonlinear equations. It is simpler to solve these two equations graphically by plotting them and obtaining the point of intersection. Equation 2.10 is the transfer characteristic of the transistor plotted in Fig. 2.14b. Equation 8.13 is the load line corresponding to the gate voltage. The point of intersection of these two lines will give the operating point Q of the transistor and the drain current may be read directly from it.

Consider first a circuit in which the transistor has maximum and minimum values of I_{DSS} and V_{th} as in Fig. 8.14 and the two relevant transfer characteristics are drawn. Let us assume that we have chosen a bias network giving us a value of equivalent gate supply voltage V_{GG} and equivalent resistance R_B. The load line given by equation 8.13 may be plotted as $KQ''Q'$ intersecting the 'maximum' transfer characteristic in Q' and the 'minimum' in Q''. Thus the drain current if the 'maximum' transistor were used will be $I_{DQ(max)}$ and if the 'minimum' transistor were used will be $I_{DQ(min)}$.

The same procedure may be used in reverse. Draw the two extreme transfer characteristics and mark in the upper and lower limits of drain current permitted by the specification. Identify the two points Q' and Q" and construct the gate load line to pass through them. The intersection on the voltage axis will give the value of V_{GG} and the slope of the load line will be $-1/R_S$. The ratio of R_1 and R_2 is determined from the ratio of V_{GG} to V_{DD} and their absolute magnitudes are chosen so that they do not shunt the applied signal voltage appreciably.

The foregoing analysis has been specifically applied to a depletion mode p-channel junction fet. The same analysis is valid for any n-channel or p-channel depletion fet. It may be appreciated that in cases where the parameter variation is small or where the tolerance on the specification is wide, the required value of V_{GG} may be zero and thus R_1 may be omitted. (Note that the maximum value of V_{GG} is limited to V_{DD} and so a solution may not be possible in some cases.)

The same analysis may be applied to enhancement type mosfets. Here the limitation on V_{GG} may be more serious. It is, however, possible in some of these cases for the bias line KQ"Q' to be vertical and in this case no source resistance R_S is necessary. Such an arrangement permits very simple cascading of stages without the need for blocking capacitors; the gate of the second stage may then be connected directly to the drain of the preceding stage.

8.6 Emitter follower

An alternative method of connection of a transistor between a generator and a load is with the collector common between the input and output circuits. The circuit of Fig. 8.15 shows a common collector amplifier (commonly known as

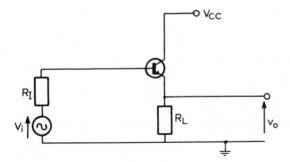

Fig. 8.15 Simplified emitter follower circuit

an emitter follower) in which the input is between base and earth and the output between emitter and earth; the collector is earthed to small variations of voltage directly through the battery, V_{CC}, which may be considered to present negligible impedance to fluctuations of current.

The small signal, low frequency, model of this circuit is shown in Fig. 8.16 and from this we may derive expressions for voltage gain, current gain, input,

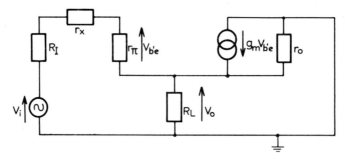

Fig. 8.16 Low frequency model of emitter follower

and output resistances in a manner similar to that adopted in Section 8.2. We obtain for the voltage gain

$$k_{v0} = \frac{(\beta_0+1)R'_L}{R_1+r_x+r_\pi+(\beta_0+1)R'_L} \tag{8.14}$$

where $R'_L = R_L$ in parallel with $r_o = R_L r_o/(R_L+r_o)$. If $\beta_0 \gg 1$ and R_1 and r_x are small compared with r_π

$$k_{v0} \simeq \frac{1}{1+g_m R'_L} \tag{8.14a}$$

If R_1, r_x, and r_π are small compared with $\beta_0 R'_L$, then $k_{v0} \to 1$, but is always less than unity. The current gain is

$$k_{i0} = (1+g_m r_\pi)\frac{r_o}{r_o+R_L} \tag{8.15}$$

and if $r_x \ll r_\pi$ and $r_o \gg R_L$, a fairly common situation, then

$$k_{i0} \simeq \beta_0 \tag{8.15a}$$

If we make the above approximations, the input and output resistances are

$$r_{in} = r_x+r_\pi+(\beta_0+1)R_L \tag{8.16}$$

$$\simeq \beta_0 R_L \tag{8.16a}$$

and $\quad r_{out} = \dfrac{R_1+r_x+r_\pi}{\beta_0+1} \tag{8.17}$

$$\simeq 1/g_m \tag{8.17a}$$

when R_I and r_x are small compared with r_π and $\beta_0 \gg 1$. The value of r_{out} defined above is the output resistance of the transistor circuit measured between emitter and earth. The output resistance measured across R_L will be r_{out} in parallel with R_L.

Using the values for the transistor parameters given in Section 5.5 and assuming a generator resistance $R_I = 1$ kΩ, and a load resistance $R_L = 500$ Ω, we obtain

$$k_{v0} = 0.985$$
$$k_{i0} = 80$$
$$r_{in} = 40.6 \text{ k}\Omega$$
$$r_{out} = 152 \ \Omega$$

The emitter follower is thus a very useful circuit if a high input resistance and a low output resistance are required. The voltage gain is always less than unity but an appreciable current gain is available and thus the circuit can deliver more power to the load than it takes from the source.

It may be thought that the output resistance may be made very small by increasing g_m. However using equation 5.7a we may rewrite equation 8.17a as

$$r_{out} \simeq kT/e|I_C|$$

which shows that for a low output resistance we need a large steady collector current I_C and this, in turn, demands a low value of R_L. As the resistance between emitter and earth consists of r_{out} and R_L in parallel, a limit is placed on the minimum output resistance obtainable for a given supply voltage. It may be noted that the input resistance of an emitter follower is a function of the load resistance varying, for the given transistor, from 225 Ω for small values of R_L to 81 R_L (since $\beta_0 = 80$) for large R_L. Similarly the output resistance ranges from 3 Ω for small values of R_I to $R_I/81$ for large values of R_I. (These figures assume that the transistor parameters do not vary as R_L is changed.)

A similar connection to the emitter follower may be made using a field-effect transistor; in this case it is termed a source follower. It has the same general properties as the emitter follower, namely approximately unity voltage gain, finite current gain, high input resistance and low output resistance. The high input resistance which is obtainable in theory with either of these follower circuits is limited in practice by the necessity to have external resistors connected to the base or gate to provide suitable bias voltages. The circuit arrangements for biasing an emitter follower or a source follower are very similar to those for the common emitter or common source amplifier and will not be discussed in detail.

8.7 Common base or common gate amplifiers

The third possible method of connection of a bipolar transistor is with the base common between input and output giving the common base amplifier. This has a current gain of approximately unity, a finite voltage gain approximately the same as the common emitter amplifier, an input resistance which is relatively low, and a relatively high output resistance. The circuit is not much used except in situations in which the capacitance between output and input needs to be minimized.

A similar circuit, the common gate amplifier, may be constructed using a field-effect transistor. Its properties are generally similar to the common base circuit.

8.8 Summary of amplifier properties

The low frequency properties of the three configurations of transistor amplifiers are summarized in Table 8.1 for a bipolar transistor and in Table 8.2 for a fet. Typical values are given in brackets, although considerable variations from these are likely, depending upon transistor type and values of generator and load resistances.

Table 8.1

	Common emitter	*Common collector*	*Common base*
Voltage gain	high (400)	low (<1)	high (400)
Current gain	high (50)	high (50)	low (<1)
Input resistance	medium (1 kΩ)	high (200 kΩ)	low (50 Ω)
Output resistance	high (50 kΩ)	low (100 Ω)	very high (2 MΩ)

Table 8.2

	Common source	*Common drain*	*Common gate*
Voltage gain	medium (50)	low (<1)	medium (50)
Current gain	very high	very high	low ($\simeq 1$)
Input resistance	very high	very high	low (50 Ω)
Output resistance	high (50 kΩ)	low (200 Ω)	high (100 kΩ)

Note that the high values of current gain and input resistance for the common source and common drain amplifiers will be reduced by the shunting effect of the biasing resistors. The input resistance and output resistance of a common gate amplifier are very dependent on the load and generator resistances.

8.9 Voltage and current gain

The voltage and current gains and input and output impedances may alternatively be determined in terms of the two-port parameters of the transistor. As an example we may assume that the transistor is specified by means of its y

parameters (see Appendix A) measured in common emitter configuration. Since we are only concerned with low frequency operation we consider all the parameters to be real at the frequency of interest. Thus the common emitter low frequency y parameters may be expressed by the four conductances

$$g_{ie}, \quad g_{re}, \quad g_{fe}, \quad g_{oe}$$

The amplifier is now to be operated between a generator of voltage V_i and resistance R_I and a load resistor R_L. For convenience the generator is converted by Norton's theorem to an equivalent current generator of value $I_i = V_i/R_I$ in parallel with a conductance $G_I = 1/R_I$. The load resistor is similarly designated by its conductance $G_L = 1/R_L$. The transistor amplifier may then be represented by the model of Fig. 8.17.

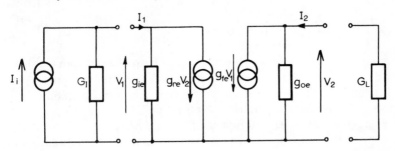

Fig. 8.17 Network model corresponding to low frequency admittance parameters of transistor

We may write the following equations:
For the transistor

$$I_1 = g_{ie}V_1 + g_{re}V_2 \tag{8.18a}$$

$$I_2 = g_{fe}V_1 + g_{oe}V_2 \tag{8.18b}$$

For the source

$$I_i = G_I V_1 + I_1 \tag{8.19}$$

For the load

$$I_2 = G_L V_2 \tag{8.20}$$

These may be solved to give the voltage gain k_{vo} as

$$k_{vo} = \frac{V_2}{V_i} = \frac{V_2 G_I}{I_i}$$

$$= \frac{-g_{fe}G_I}{(g_{ie} + G_I)(g_{oe} + G_L) - g_{re}g_{fe}} \tag{8.21}$$

Similarly the input conductance g_{in} is

$$g_{in} = \frac{1}{r_{in}} = g_{ie} - \frac{g_{fe}g_{re}}{g_{oe} + G_L} \tag{8.22}$$

and the output conductance

$$g_{out} = \frac{1}{r_{out}} = g_{oe} - \frac{g_{fe}g_{re}}{g_{ie} + G_I} \tag{8.23}$$

We may similarly deduce the current gain as the ratio of output current to generator current:

$$k_{io} = \frac{I_2}{I_i} = \frac{g_{fe}G_L}{(g_{ie} + G_I)(g_{oe} + G_L) - g_{fe}g_{re}} \tag{8.24}$$

These expressions are completely general and may be applied to both bipolar or field-effect transistors or any other device specified by its two-port parameters. They may also be used for devices with other terminals common if the appropriate set of conductance parameters are chosen.

Similar expressions may be determined if the transistor or other active element is expressed by any of the other sets of two-port parameters such as the impedance or hybrid set.

It may be seen from equations 8.22 and 8.23 that if g_{re} is very small, the input and output resistances are almost independent of load and generator resistances respectively. This we noted in Section 8.2.1. In the common collector configuration the reverse conductance g_{rc} is not small and hence the input resistance is a function of load resistance as shown previously by equation 8.16a.

8.10 Power gain

We have already computed the voltage and current gains of an amplifier, based either on the device model of the transistor or on its representation as a linear two-port device. In certain circumstances, for example when working from very low power levels, or when driving loads consuming very large power, we shall be concerned with the power gain of an amplifier.

There are three power ratios for an amplifier which are used in different cases.

(a) The *actual power gain* defined as the ratio of output power delivered to the load from the amplifier to input power to the amplifier.
(b) The *transducer gain* defined as the ratio of the output power delivered to the load to the power which is available from the generator.
(c) The *available power gain* defined as the ratio of the power which is available from the amplifier to the power which is available from the generator.

The values of these three power gains may be determined for a common emitter amplifier using the model of Fig. 8.17 where all elements are conductances and the network is represented by the conductance matrix

$$G = \begin{bmatrix} g_{ie} & 0 \\ g_{fe} & g_{oe} \end{bmatrix}$$

g_{re} is assumed zero to simplify the calculations. The network is fed from a current source I_i having a source conductance G_I and feeding a load G_L.

The actual power gain A_P may be directly determined from Fig. 8.17. The input power into conductance g_{ie} is

$$P_i = \frac{I_i^2 g_{ie}}{(G_I + g_{ie})^2} \tag{8.25}$$

The output power from the network is

$$P_o = V_2^2 G_L$$
$$= \left(\frac{I_i}{g_{ie} + G_I}\right)^2 \frac{g_{fe}^2 G_L}{(g_{oe} + G_L)^2} \tag{8.26}$$

Thus the actual power gain is

$$A_P = \frac{g_{fe}^2 G_L}{g_{ie}(g_{oe} + G_L)^2} \tag{8.27}$$

To determine the transducer gain we determine the available power from the generator. This occurs when we maximize equation 8.25 with respect to g_{ie}; namely when $g_{ie} = G_I$. The available power from the generator is thus

$$P_{iA} = \frac{I_i^2}{4G_I} \tag{8.28}$$

The transducer gain, A_T, is thus the ratio of equations 8.26 and 8.28

$$A_T = \frac{4g_{fe}^2 G_I G_L}{(g_{ie} + G_I)^2 (g_{oe} + G_L)^2} \tag{8.29}$$

Finally, to determine the available gain we determine the maximum available power from the amplifier. This available output P_{oA} is obtained from equation 8.26 by setting $G_L = g_{oe}$

$$P_{oA} = \frac{I_i^2}{(g_{ie} + G_I)^2} \frac{g_{fe}^2}{4g_{oe}} \tag{8.30}$$

Hence the available gain of the amplifier A_A is obtained from the ratio of equations 8.30 and 8.28

$$A_A = \frac{g_{fe}^2 G_1}{g_{oe}(g_{ie}+G_1)^2} \tag{8.31}$$

It may be thought that the actual power gain is probably the most useful of these three definitions. It certainly defines the power gain which may be obtained by interposing the given amplifier between a fixed source and a fixed load. However, the amplifier input circuit may be very badly designed so that full use is not being made of the power available from the generator. The transducer gain takes this into account and comparison of the transducer gains of two alternative amplifiers will show that the amplifier with the larger transducer gain will make more effective use of the power available from the generator and deliver a greater power to the load. Finally, the available gain is an optimization of the transducer gain so that the load is matched to the output conductance of the transistor and is thus delivering its maximum power into the load.

8.11 The decibel

In a great many engineering applications the use of a linear scale does not correspond with a subjective assessment of the phenomenon being studied. In many instances it would be preferable to choose a scale in which equal increments were made proportional to the least detectable change in the quantity being measured. The human ear for example can detect changes of sound intensity only when the change is of such a magnitude as to double the energy content of the original sound. This suggests the use of a logarithmic scale in such cases. This idea has been extended to the measurement of power in electrical networks, in particular to the ratio of power entering and leaving a network.

Let the power entering a certain two-port system be P_1 and the power leaving P_2. Then the power gain of the two-port is P_2/P_1 as an absolute dimensionless ratio. Alternatively this quantity may be expressed by its logarithm, in particular as

$$10 \log_{10} (P_2/P_1) \text{ decibels}$$

The decibel is thus seen to be strictly a logarithmic ratio of two powers; for example, 1 dB is defined as the power gain of a network in which the ratio of output to input power is 1·26. If these two powers are being dissipated in identical resistors R, then the power ratio may be written

$$10 \log_{10} (V_2^2/V_1^2) = 20 \log_{10} (V_2/V_1) \text{ decibels.}$$

It should be noted that it has become common practice to quote the ratio of two voltages in decibels even if the two impedances across which these voltages are developed are not identical. There is already in existence a valid unit for the logarithmic ratio of two voltages or currents based on the natural logarithm. Thus a voltage gain may be written

$$\log_e (V_2/V_1) \text{ nepers}$$

The greater convenience of logarithms to the base 10 has caused the decibel to replace the neper in specifying voltage ratios in a great many cases. It would be preferable to define a new unit for the logarithmic ratio of two voltages (or currents) to a base of 10, but custom has sanctioned the use of the decibel in this situation. One should, however, always treat the use of the decibel in specifying voltage (or current) ratios with caution; for example, an attenuator marked in decibels interposed between an arbitrary voltage source and a load will only attenuate the applied voltage by the setting of the attenuator if both terminations of the attenuator are correct. Note that on the assumption of identical terminations 1 neper corresponds to 8·686 dB.

The sign of the gain in decibels is positive when referring to a voltage gain between input and output. The decibel is also used to define the attenuation of a network; in this case the sign convention is that an attenuation of $+20$ dB signifies that $P_o/P_i = 1/100$, and this would correspond to a power gain of -20 dB.

References

THORNTON, R. D., *et al.*, *Characteristics and Limitations of Transistors*, Wiley, 1966, Chapter 2.
COMER, D. J., *Introduction to Semiconductor Circuit Design*, Addison Wesley, 1968, Chapter 3.
SEARLE, C. L., *et al.*, *Elementary Circuit Properties of Transistors*, Wiley, 1964, Chapter 5.
MILLMAN, J., and HALKIAS, C. C., *Integrated Electronics*, McGraw-Hill, 1972, Chapter 9.
GRAY, P. E. and SEARLE, C. L., *Electronic Principles*, Wiley, 1967, Chapter 13.
MALVINO, A. P., *Transistor Circuit Approximations*, McGraw-Hill, 1968, Chapters 8 and 9.

Problems

8.1 A transistor, whose $\beta_F = 150$, is operated in the circuit of Fig. 8.3 with a supply voltage V_{CC} of 12 V. $R_L = 500 \, \Omega$ and $R_E = 200 \, \Omega$. The quiescent base current is 50 μA. Estimate the quiescent collector current and collector–emitter voltage.

8.2 What is the maximum voltage swing which may be obtained in the circuit of Problem 8.1?

8.3 What is the maximum voltage swing in the circuit of Problem 8.1 if R_E is increased to 800 Ω? What is the effect on the voltage swing of a large capacitance shunted across R_E?

8.4 An amplifier whose circuit is shown in Fig. 8.3 uses a silicon n-p-n transistor. The desired operating point is at $V_{CE} \simeq 3$ V, $I_C = 4·5$ mA at room temperature; the collector current must not increase by more than 10% at 100°C. The common emitter current gain, β_F, and the base–emitter threshold voltage, V_0, vary with temperature as below

	25°C	100°C
β_F	80	130
V_0	0·6 V	0·42 V

Estimate values for R_1, R_2, and R_E to achieve this specification when the supply voltage is 10 V.

8.5 What is the maximum available amplitude of output voltage excursion in the circuit of Problem 8.4?

8.6 An amplifier uses an n-channel junction fet in common source configuration to be used at a constant ambient temperature of 25°C. The circuit must be capable of operating at $V_{DS} = 15$ V, $I_D = 4\cdot5 \pm 0\cdot5$ mA from a supply voltage, $V_{DD} = 60$ V when using any transistor from the batch whose parameters lie in the range 15 mA $> I_{DSS} > 5$ mA; $-6\cdot5\ V < V_{th} < -2$ V. Draw a circuit and choose element values.

8.7 A similar amplifier to that of Problem 8.6 is now to be constructed using a p-channel enhancement mosfet. The transistor has been selected and measurements taken on it show that at 125°C, $V_{th} = -3\cdot5$ V, and at $V_{DS} = -10$ V, $V_{GS} = -10$ V, $I_D = -37$ mA; at -55°C, $V_{th} = -3\cdot5$ V and at $V_{DS} = -10$ V, $V_{GS} = -10$ V, $I_D = -65$ mA. The supply voltage is -25 V. Design a circuit so that the quiescent drain current does not vary by more than $\pm20\%$ over this range of temperature.

8.8 The bipolar transistor of Problem 8.4 is used in the circuit of Fig. P8.1. It is to operate at a normal quiescent point where $V_{CE} \simeq 4$ V, $I_C = 4\cdot5$ mA from a supply of 10 V and the load current must not deviate from its nominal value by more than 10% at 100°C. Design the circuit. If this specification cannot be achieved attempt to maintain I_L within 20% of its normal value. Choose $R_L = 500\ \Omega$.

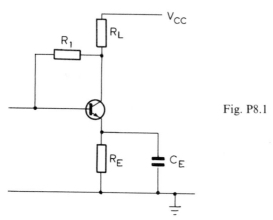

Fig. P8.1

8.9 Design an amplifier having the circuit of Fig. 8.3 and using the transistor of Problem 8.4 to develop a voltage swing of ±3 V across a load resistor of 500 Ω at any temperature from 20 to 100°C from a battery voltage of 10 V. (Note that the operating point may now be selected to satisfy the conditions.) Assume that R_E is shunted by a large capacitor.

8.10 A junction field-effect transistor whose characteristics are given in Fig. 2.14 is connected in the circuit of Fig. 8.12. The supply voltage is -25 V, the load resistor, R_L, is 2 kΩ, and $R_2 = 10$ MΩ, $R_1 = \infty$. The gate reverse saturation current is 3 nA at 25°C. Choose a suitable value for R_S in order that the transistor is operated at $V_{GS} = 0\cdot5$ V.

8.11 What will be the change of reverse saturation current in the transistor of Problem 8.10 if the temperature rises to (a) 45°C, (b) 70°C? How will this change the bias voltage?

8.12 A transistor is specified at low frequencies by the following common emitter parameters

$$y_{ie} = 1\cdot5 \text{ mS}$$
$$y_{re} = 50 \text{ μS}$$
$$y_{fe} = 100 \text{ mS}$$
$$y_{oe} = 100 \text{ μS}$$

Determine the common base and common collector y parameters.

8.13 An amplifier uses the transistor, whose parameters are given in Problem 8.12, fed from a voltage generator of resistance 1·2 kΩ and working into a load of 1·2 kΩ. Determine the voltage gain referred to the open-circuit generator voltage, the current gain, the input and output resistances in common emitter, common base, and common collector configurations. Comment on the fact that some of the values you obtain, for example the common base voltage gain, differ considerably from the magnitude you might expect.

9
Frequency Response of Networks

9.0 Introduction

In Chapter 8 we studied the use of a transistor as an amplifier of small fluctuations of voltage at frequencies where none of the elements in the amplifier had frequency dependent properties. In Chapters 3 and 5 we developed a small signal model for the fet and bipolar transistor which showed that at high frequencies their properties are quite strong functions of frequency. We also noted in Chapter 8 that the input voltage is usually applied to the amplifier through a capacitance and this is likely to place a limit on the amplification available at low frequencies. The main purpose of this chapter is to investigate the frequency response of a transistor amplifier and to correlate the shape of the frequency response with the various reactive elements in the circuit model of the amplifier. As a preliminary we shall review the responses of two simple networks incorporating only one reactive element, to both sinusoidal and step changes of voltage. This will enable us to formulate some simple rules relating the frequency and transient responses of systems.

9.1 Frequency response of simple circuit

The network of Fig. 9.1 is excited by an arbitrary time varying voltage v_i at the input and the response v_o at the output terminals observed. We may analyse

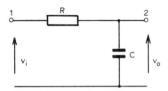

Fig. 9.1 RC low pass network

this circuit by summing currents at node 2 using Kirchhoff's current law to give

$$\frac{v_i - v_o}{R} = C\frac{dv_o}{dt} \tag{9.1}$$

This may be rearranged to give the differential equation relating input and output voltages as

$$\frac{dv_o}{dt} + \frac{1}{CR} v_o = \frac{1}{CR} v_i \qquad (9.2)$$

We may obtain the response of this network to a real sinusoidal signal

$$v_i = V_i \cos \omega t = \text{Re} \left[V_i \exp j\omega t \right] \qquad (9.3)$$

by solving equation 9.2 for this value of input voltage. This gives the frequency response transfer function

$$T(\omega) = \frac{V_o}{V_i} = \frac{1/RC}{j\omega + 1/RC} \qquad (9.4)$$

where V_o is the amplitude of the output voltage given by

$$v_o = \text{Re} \left[V_o \exp j\omega t \right] \qquad (9.5)$$

It is sometimes convenient to substitute $j\omega = s$ and write

$$T(s) = \frac{1/RC}{s + 1/RC} \qquad (9.6)$$

The values of s at which $T(s)$ becomes zero are termed the zeros of the transfer function and the values at which it tends to infinity are the poles. Both poles and zeros may be real or complex.

Equation 9.4 shows how the amplitude and phase of the output voltage is related to that of the input voltage at all frequencies. At any given frequency it gives the relationship between an input phasor and the output phasor at that frequency. It may be plotted in the form of two graphs, one of the magnitude of $T(\omega)$ against frequency and the second of the phase angle of $T(\omega)$ against frequency. These are shown for the network of Fig. 9.1 in Fig. 9.2. For convenience the frequency axis is graduated on a logarithmic scale. The vertical axis on the magnitude graph is also plotted logarithmically and may be graduated in decibels; this has been done in Fig. 9.2.

9.1.1 Response of network to sudden change of voltage

To obtain the response of the network of Fig. 9.1 to a sudden change of applied voltage we need to solve the differential equation 9.2 with $v_i = E$, a suddenly applied steady voltage at time $t = 0$. The solution to this is

$$v_o = E \left[1 - \exp \left(-t/CR \right) \right] \qquad (9.7)$$

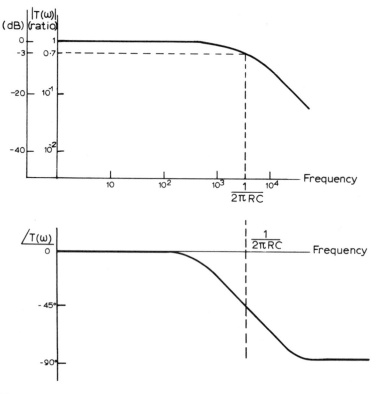

Fig. 9.2 Magnitude and phase response of network of Fig. 9.1

The form of this is sketched in Fig. 9.3. The voltage rises exponentially from zero towards the final voltage E, passing through a value of $0.632\,E$ after a time equal to CR, known as the time constant of the circuit.

It may be noted that networks which have a long time constant tend to respond slowly to a sudden change of input voltage. In addition the response of such a circuit to sinusoidal inputs begins to fall off at relatively low frequencies (its upper cut-off frequency is low). Small time constants correspond to fast responses to step changes of voltage and also to relatively high values of

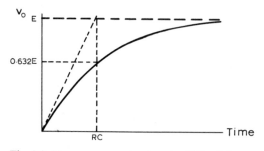

Fig. 9.3 Step response of network of Fig. 9.1

upper cut-off frequency. These conclusions may be extended to much more complex circuits where similar general results are valid.

Thus the time constant of a circuit is an important parameter giving information about both the sinusoidal and step function response of the circuit. It should be noted that for the simple circuit we have been studying the pole of the network function (equation 9.6) occurs at $-1/CR$.

9.1.2 Another network with a single capacitor

A similar analysis of the differential equation relating input to output voltage of the network of Fig. 9.4 leads to a voltage transfer function

$$T(s) = \frac{s}{s+1/CR} \tag{9.8}$$

Such a network has a pole at $-1/CR$ and a zero at the origin.

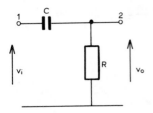

Fig. 9.4 *RC* high pass network

The frequency response transfer function of this network is

$$T(\omega) = \frac{j\omega}{j\omega+1/CR} \tag{9.9}$$

and this is plotted in both amplitude and phase in Fig. 9.5.

The output response of the network to a sudden step change, E, of input voltage is

$$v_o = E \exp\left(-t/CR\right) \tag{9.10}$$

and this is sketched in Fig. 9.6. The time constant RC again determines the form of this response. It may be seen that a network of this form does not affect the speed of response of the output (the output voltage rises instantaneously to the value of the input voltage) but the value so reached slowly decays away to zero.

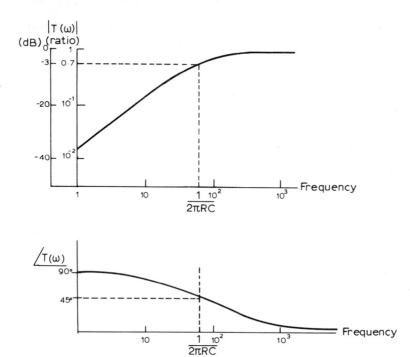

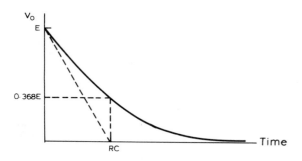

Fig. 9.5 Magnitude and phase response of network of Fig. 9.4

Fig. 9.6 Step response of network of Fig. 9.4

9.2 Frequency and step response

From the above study we may be tempted to make the following generalizations.

(a) Any network having a single capacitor in series between input and output terminals will have a response to sinusoidal voltages which falls off at low frequencies; the frequency at which the response starts to fall is of the order of magnitude of the pole of the network function. The response to a small input step will be instantaneous, but the steady value will droop towards zero after a time of the order of the time constant of the network, namely the reciprocal of the value of the pole.

(b) Any network having a capacitor shunted across the output terminals will have a sinusoidal response which falls off at high frequencies; the frequency at which the fall starts is approximately equal to the value of the pole. The response to a step input voltage will take a finite time to reach its steady value; this rise time will be of the order of the time constant of the circuit.

It should be added that although these generalizations are not true in all networks they can be used to give an insight into the general behaviour of a very large number of fairly simple circuits.

9.3 Rise time

It is often convenient to define the rise time of a network precisely in order to make a comparative estimate of the speed of response of various networks to a sudden change of applied voltage. The rise time is defined as the time taken for the response of the network to a step excitation to rise from 10% to 90% of its final value.

The rise time for the network of Fig. 9.1 is obtained as

$$t_r = t_2 - t_1 = 2 \cdot 2\, CR$$

9.4 Frequency response of amplifier

In Sections 5.1 and 5.5 we have developed the circuit model of a bipolar transistor and in Sections 3.1 to 3.4 the corresponding model for a field-effect transistor. Comparison of these two circuit models shows a great similarity which will enable us to perform an analysis of a circuit involving a bipolar transistor which will be valid for a similar circuit using a field-effect transistor. (In this latter case it may be permissible to simplify the circuit model for the device.)

In Section 8.2 we developed a simple circuit for a single stage transistor amplifier including all those resistors necessary to apply correct bias to the device. A similar circuit could be used for a field-effect transistor. If we wish to use this to amplify small varying voltages, one or two additional components are necessary. A capacitor C_C is needed to couple the signal voltage to the base of the transistor and to block any flow of direct current so that the bias conditions on the transistor are not affected. The capacitor C_E across the emitter biasing resistor R_E is designed to act as a short circuit to the signal currents and thus to prevent any signal voltages being developed across R_E which would be fed back to the input of the circuit; its presence does not affect the bias conditions which have already been established.

The circuit diagram for the complete amplifier is shown in Fig. 9.7 where the signal generator is of magnitude v_i and has a generator resistance R_1. The transistor may be modelled by the circuit of Fig. 5.5, and when this is inserted

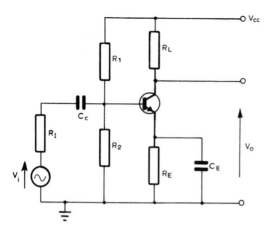

Fig. 9.7 Common emitter amplifier

in the circuit diagram and appropriate modifications made for small signal incremental operation, we obtain the equivalent circuit of the amplifier shown in Fig. 9.8.

The network of Fig. 9.8 now needs to be analysed: this may be done by a variety of methods. One of the most systematic is to obtain the nodal admittance matrix and to reduce this to a two-port matrix relating voltages and currents at input and output ports of the network.

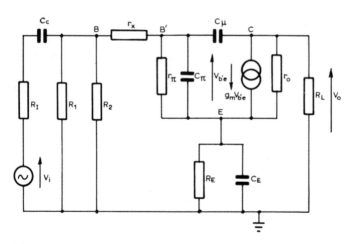

Fig. 9.8 Complete equivalent model of Fig. 9.7

However, although such an analysis is quite straightforward it does not give any insight into the effect that each of the reactive elements in the model has on the overall response of the network.

In order to appreciate the significance of each of the reactive elements, it is better to make the simplifying assumption that the effects of each element may be considered in isolation. This is, of course, only valid if the time constants associated with each reactive element are sufficiently far apart. We shall, therefore, look at the effects on the amplifier response due to (a) the coupling capacitor C_C, (b) the emitter capacitor C_E, (c) the transistor reactive elements C_π and C_μ. First, however, we shall determine the mid-frequency gain where we may assume that all the network is behaving in an idealized manner, namely that C_C and C_E are acting as perfect short circuits to signal currents and that the transistor is not influenced by dynamic effects so that C_π and C_μ are considered as perfect open circuits. This model is valid at mid-frequencies in practice; the term 'mid-frequencies' is left somewhat vague at the moment intentionally as we cannot define the frequency limits of its validity until we investigate the effects of the various capacitances.

9.4.1 Mid-frequency gain

In the mid-frequency range we shall consider that Fig. 9.9 is a satisfactory circuit model of the amplifier. Note that resistor R_B represents the parallel combination of R_1 and R_2. Thus

$$R_B = \frac{R_1 R_2}{R_1 + R_2} \tag{9.11}$$

A further simplification results if we draw the Thevenin equivalent circuit to the left of the points BE. If V_i' and R_i' are the Thevenin equivalent elements then

$$V_i' = V_i \frac{R_B}{R_B + R_I} \tag{9.12}$$

$$R_I' = \frac{R_I R_B}{R_I + R_B} \tag{9.13a}$$

$$= \frac{R_I R_1 R_2}{R_I R_1 + R_I R_2 + R_1 R_2} \tag{9.13b}$$

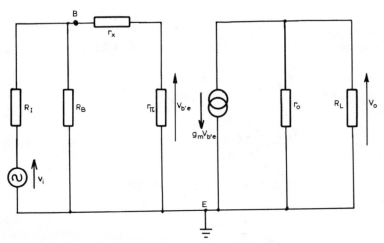

Fig. 9.9 Mid-frequency approximate model of Fig. 9.7

In addition we may combine r_o and R_L in parallel to give R'_L where

$$R'_L = \frac{r_o R_L}{r_o + R_L} \tag{9.14}$$

This modified circuit model is shown in Fig. 9.10.

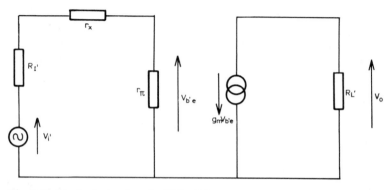

Fig. 9.10 Equivalent network of Fig. 9.9

We may now determine the voltage transfer ratio directly from this circuit. Around the input mesh we obtain

$$V_{b'e} = \frac{r_\pi}{R'_I + r_x + r_\pi} V'_i$$

and in the output mesh

$$V_o = -g_m R'_L V_{b'e}$$

$$= -\frac{g_m R'_L r_\pi}{R'_I + r_x + r_\pi} V'_i$$

The voltage gain (transfer ratio) of the amplifier at mid-frequencies, k_{v0}, is given by

$$k_{v0} = \frac{V_o}{V_i}$$

$$= -\frac{g_m R'_L r_\pi}{R'_I + r_x + r_\pi} \frac{R_B}{R_B + R_I} \tag{9.15}$$

Using the transistor values given in Section 5.5 and taking $R_I = 2\ \text{k}\Omega$, $C_C = 1000\ \mu\text{F}$, $R_1 = 7\cdot5\ \text{k}\Omega$, $R_2 = 2\cdot7\ \text{k}\Omega$, $R_E = 200\ \Omega$, $C_E = 100\ \mu\text{F}$, $R_L = 1\cdot2\ \text{k}\Omega$, we obtain

$$k_{v0} = -38\cdot1\ (\equiv 31\cdot6\ \text{dB})$$

9.4.2 Effect of coupling capacitor

Let us now study the effect on the voltage gain cause by introducing a coupling capacitor C_C. The circuit model which will correspond to this situation is shown in Fig. 9.11. The analysis of this circuit follows the procedure of Section 9.4.1 almost completely.

Thus
$$V_{b'e} = \frac{r_\pi R_B}{(R_B + r_\pi + r_x)(R_I + R_B')} \cdot \frac{1}{1 + 1/j\omega\tau_L} V_i$$

where
$$R_B' = \frac{R_B(r_x + r_\pi)}{R_B + r_x + r_\pi}$$

and
$$\tau_L = \frac{1}{\omega_L} = \frac{1}{2\pi f_L} = C_C(R_I + R_B')$$

Also
$$V_o = -g_m R_L' V_{b'e}$$

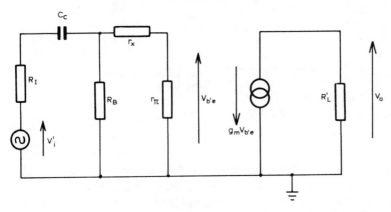

Fig. 9.11 Model of common emitter amplifier including effect of coupling capacitor

Thus the voltage gain is

$$k_v(\omega) = k_{v0} \frac{j\omega}{j\omega + \omega_L} \tag{9.16}$$

If we put $j\omega = s$ we shall obtain the general transfer function for the voltage gain considering this coupling capacitor alone

$$k_v(s) = k_{v0} \frac{s}{s + \omega_L} \tag{9.17}$$

We note that this function has a zero at the origin and a pole at $-\omega_L$.

Referring back to Section 9.1.3 we note that the amplitude-frequency response of this network is shown in Fig. 9.12. Looking at this we may make one simple deduction regarding the meaning of the mid-frequency region which we postulated earlier. At any angular frequencies greater than about $10\,\omega_L$ the value of $|k_v|$ differs negligibly from k_{v0}; below this frequency we must consider the effect of C_C. This will provide us with a design criterion for an amplifier. If the amplifier gain is to remain sensibly constant down to an angular frequency ω_1 we must choose C_C such that $\tau_L > 10/\omega_1$.

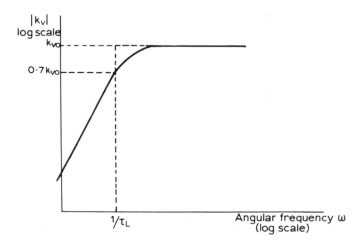

Fig. 9.12 Magnitude response of network of Fig. 9.11

The phase of k_v will be 90° at low frequencies falling through 45° at $\omega = 1/\tau_L$ to zero at $\omega \to \infty$; in a manner similar to that shown in Fig. 9.5.

Again it may be seen that the step response of the amplifier will be an instantaneous rise with amplification k_{v0} falling to zero after a long time, as in Fig. 9.6. For the amplifier whose numerical values have been considered we obtain $f_L = 0.072$ Hz.

9.4.3 Effect of emitter capacitor

If we now study the amplifier considering only the frequency range where the emitter bypass capacitor is significant we shall obtain the circuit model of Fig. 9.13.

A simplifying assumption may be made in this case that the voltage across R_E is small compared with the voltage across R_L. Since r_o is usually large compared with R_L we may approximate the circuit by putting r_o in parallel with R_L to give the simplified circuit of Fig. 9.14.

Analysis of this circuit will lead to an expression for voltage gain of

$$k_v(\omega) = k_{v0}\frac{j\omega + \omega_c}{j\omega + \omega_d} \tag{9.18}$$

or using the s variable

$$k_v(s) = k_{v0} \frac{s + \omega_c}{s + \omega_d} \tag{9.18a}$$

in which

$$\omega_c = 2\pi f_a = \frac{1}{C_E R_E} \tag{9.19}$$

and $\quad \omega_d = 2\pi f_d = \dfrac{1}{C_E R_T} \tag{9.20}$

where $\quad R_T = \dfrac{R_E(r_x + r_\pi + R_1')}{R_E(1 + g_m r_\pi) + r_x + r_\pi + R_1'} \tag{9.21}$

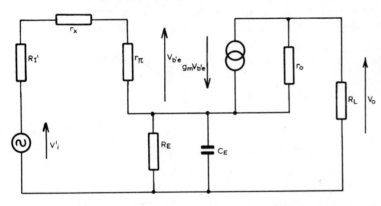

Fig. 9.13 Model of common emitter amplifier including effect of emitter by-pass capacitor

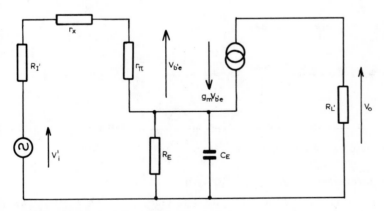

Fig. 9.14 Equivalent network to Fig. 9.13

The effect of this on the frequency response may be seen in Fig. 9.15; the voltage gain falls from k_{vo} at mid-frequencies to a value $k_{vo}R_T/R_E$ at very low frequencies. This might be seen intuitively by appreciating that at high and mid-frequencies, C_E acts as a satisfactory short circuit across R_E and the gain is exactly that given by equation 9.15; whereas at very low frequencies, C_E is

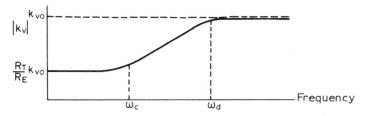

Fig. 9.15 Magnitude response of network of Fig. 9.14

ineffective and feedback is applied across the amplifier reducing the gain appreciably. The phase characteristic of this part of the circuit is shown in Fig. 9.16. The response of the network to a step, E, of input voltage is shown in Fig. 9.17.

For the amplifier considered $f_c = 7.96$ Hz and $f_d = 113.7$ Hz.

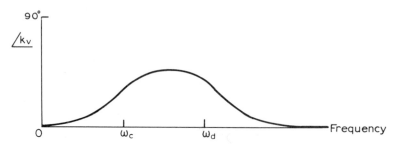

Fig. 9.16 Phase response of network of Fig. 9.14

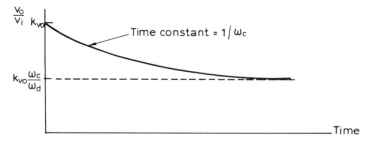

Fig. 9.17 Step response of network of Fig. 9.14

9.4.4 Effect of transistor charging capacitances

We now need to look at the effect on the gain-frequency reponse of the amplifier of the capacitive effects of the transistor, namely C_π and C_μ. The equivalent circuit of the amplifier may now be simplified to that shown in Fig. 9.18. It is convenient at this stage to consider the effects of C_π and C_μ together as they are jointly responsible for a single effect.

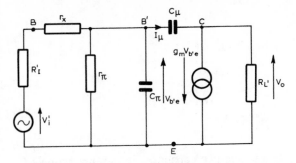

Fig. 9.18 Model of common emitter amplifier at high frequencies

The simplest way both to show this and also to analyse the circuit is to transform the circuit of Fig. 9.18 into that of Fig. 9.19. The equivalence between these two may be seen by the following procedures. The voltage at point B' in Fig. 9.18 is assumed equal to $V_{b'e}$ and hence the voltage at C is approximately equal to $-g_m R_L' V_{b'e}$ where $g_m R_L'$ is the voltage gain between B' and C. This approximation is justified in most cases since the current $g_m V_{be}$ flows almost

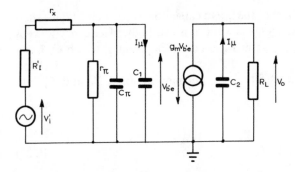

Fig. 9.19 Equivalent network to Fig. 9.18

exclusively through R_L, very little passing into the capacitor C_μ. The current flowing through the capacitor C_μ into node C and out of node B' is $I_\mu = (1+g_m R_L')V_{b'e}j\omega C_\mu$. Now in the equivalent circuit of Fig. 9.19 it is necessary to insert the capacitor C_1 between node B' and earth to give a current I_μ due to a voltage $V_{b'e}$ across it. The required value of capacitance is, therefore, given by

$$j\omega C_1 = \frac{I_\mu}{V_{b'e}}$$

Thus $$C_1 = (1+g_m R_L')C_\mu \qquad (9.22)$$

Similarly we need to insert a capacitor C_2 between node C and earth which will allow a current I_μ to enter node C through C_2 caused by a voltage $-g_m R'_L V_{b'e}$ across it. Thus

$$j\omega C_2 = \frac{I_\mu}{g_m R'_L V_{b'e}}$$

Therefore

$$C_2 = \left(1 + \frac{1}{g_m R'_L}\right) C_\mu \qquad (9.23)$$

We note that on the input side we have two capacitors in parallel, namely C_π and C_1. Consider typical values given in Section 5.5 of $C_\pi = 80$ pF and $C_\mu = 3$ pF. A typical value of internal voltage gain of the transistor would be -500 and hence $C_1 = 1500$ pF which is an order of magnitude greater than C_π. This shows that although the collector–base capacitance, C_μ, may be very small compared with C_π its effects in a high gain amplifier circuit may be more significant than C_π.

If we now look at the output circuit we will note that the equivalent load resistor R'_L is now shunted by a capacitor $C_\mu(1 + 1/g_m R_L)$. Using the above quoted values we obtain $C_2 = 3$ pF.

We can now look at the effect of these two capacitors separately and we shall find that C_2 only becomes of significance at a much higher frequency (except in rather exceptional situations when R_L is small) than that at which C_1 has a significant effect.

Thus considering only the effect of C_1 we may combine C_1 and C_π into C'_1 where

$$C'_1 = C_1 + C_\pi$$
$$= C_\pi + (1 + g_m R'_L) C_\mu \qquad (9.24)$$

Drawing the Thevenin equivalent to the network to the left of C_π in Fig. 9.19 we obtain the circuit of Fig. 9.20 where

$$R''_I = \frac{r_\pi (r_x + R'_I)}{r_\pi + r_x + R'_I} \qquad (9.25)$$

and $$V''_i = V'_i \cdot \frac{r_\pi}{r_\pi + r_x + R'_I} \qquad (9.26)$$

Comparing this with the circuit of Fig. 9.1 we shall expect the response to be similar to that of Fig. 9.2. This is drawn in Fig. 9.21. The low frequency (strictly

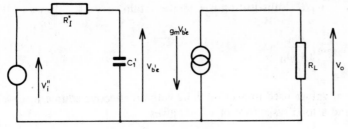

Fig. 9.20 Equivalent network to Fig. 9.19

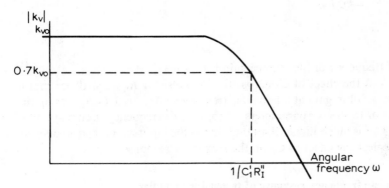

Fig. 9.21 Magnitude response of network of Fig. 9.20

mid-frequency) value of the gain may be determined directly from equation 9.26 since at low frequencies

$$V_{b'e} = V_i''$$

and
$$V_o = -g_m R_L' V_{b'e}$$

$$\frac{V_o}{V_i} = \frac{-g_m R_L' r_\pi}{r_\pi + r_x + R_I'} \cdot \frac{R_B}{R_B + R_I}$$

$$= k_{v0}$$

by comparison with equation 9.15.

At high frequencies the gain becomes

$$k_v(\omega) = k_{v0} \frac{1}{1 + j\omega C_1' R_I''} \tag{9.27}$$

The angular frequency at which the voltage gain has fallen by a factor $1/\sqrt{2}$ is $1/C_1' R_I''$. Rewriting equation 9.27 we obtain

$$k_v(\omega) = k_{v0} \frac{1}{1 + j\omega \tau_H} \tag{9.28}$$

$$= k_{v0} \frac{\omega_H}{j\omega + \omega_H} \tag{9.29}$$

where $\tau_H = 1/\omega_H = C_1' R_1''$. Substituting $s = j\omega$ we obtain

$$k_v(s) = k_{v0} \frac{\omega_H}{s + \omega_H} \tag{9.30}$$

Considering typical values used in Section 9.3.2 with an effective source resistance R_1' of 2 kΩ and a load resistance of 1·2 kΩ gives

$$R_1'' = 557 \ \Omega$$
$$C_1' = 323 \ \text{pF}$$

The cut-off frequency of the voltage gain is thus 0·885 MHz.

If we look at the effect of C_2 across the load resistor R_L' we shall see that it would cause a fall in gain at an angular frequency given by $1/C_2 R_L'$. Using the same values of transistor parameters as above, this frequency occurs at about 44·8 MHz. This is much higher than that due to the input circuit and so we shall normally neglect the effect of C_2 on the frequency response.

9.4.5 Complete frequency response of transistor amplifier

The complete frequency response of the amplifier may be obtained by combining the effects on the voltage gain of the emitter capacitor, the coupling capacitor, and the charge storage phenomena in the transistor. If the frequencies ω_L, ω_c, ω_d, and ω_H are sufficiently remote from each other that their effects do not interact, we may write the overall voltage gain over a very wide frequency range by combining equations 9.17, 9.18a, and 9.30 as

$$k_v(s) = k_{v0} \frac{s}{s + \omega_L} \frac{s + \omega_c}{s + \omega_d} \frac{\omega_H}{s + \omega_H} \tag{9.31}$$

The transfer voltage ratio has zeros at $s = 0$ and $-\omega_c$ and poles at $s = -\omega_L$, $-\omega_d$, $-\omega_H$. Inserting the numerical values we shall obtain for the transfer voltage gain

$$k_v(s) = -212 \times 10^6 \frac{s(s+50)}{(s+0\cdot45)(s+713)(s+5\cdot56 \times 10^6)} \tag{9.32a}$$

Alternatively,

$$k_v(s) = -5\cdot96 \frac{s(1+s/50)}{(1+s/0\cdot45)(1+s/713)(1+s/5\cdot56 \times 10^6)} \tag{9.32b}$$

The above expression ignores the effect of the equivalent capacitor C_2. Including this we obtain

$$k_v(s) =$$
$$-5{\cdot}96\,\frac{s(1+s/50)}{(1+s/0{\cdot}45)(1+s/713)(1+s/5{\cdot}56\times10^6)(1+s/281{\cdot}3\times10^6)}$$
$$(9.32\text{c})$$

It is of interest to note that although equation 9.32c is based on the assumption that the effect of each pole and zero may be considered in isolation, an accurate determination of the transfer voltage ratio by computer leads to an expression which differs from this by a negligible amount.

The voltage transfer ratio for this amplifier is plotted in both magnitude and phase in Figs. 9.22a and b. This response has been plotted using a graphical

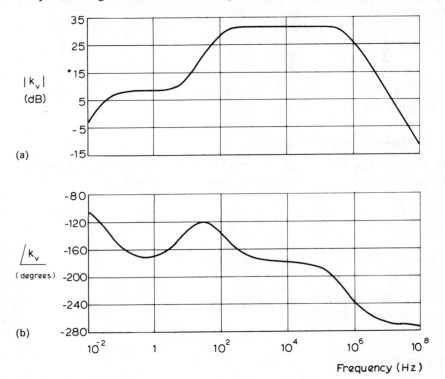

Fig. 9.22 Magnitude and phase response of complete common emitter amplifier of Fig. 9.8 determined by computer

computer program based on a direct analysis of the network. The response curve shows how the magnitude changes its slope at frequencies corresponding to the reciprocals of the time constants of the various parts of the amplifier.

It will be apparent that it would be convenient for the designer to be able to sketch the response of a network in both amplitude and phase quickly from a knowledge of the poles and zeros of the transfer function, since we have seen that these govern both the frequency and transient response. This is readily possible for transfer functions where all the poles and zeros are real, as in this case; it may be extended to the case of complex poles and zeros. We shall now look at methods for the construction of amplitude and phase plots of networks characterized by their poles and zeros. Such graphs are frequently known as Bode plots.

9.5 Network transfer functions

We have seen in Section 9.1 that the ratio of output to input voltage of a simple network is the ratio of two polynomials in the variable s; in both these cases the polynomials are simple. In a more complex circuit, we have seen that the response of a transistor amplifier is a similar rational function in s, as is shown by equation 9.32a. In general the voltage transfer function of a network may be written as

$$k_v(s) = H_0 \frac{s(s-z_1)(s-z_2)(s-z_3)\ldots}{(s-p_1)(s-p_2)(s-p_3)\ldots} \tag{9.33}$$

where the zeros z_1, z_2, z_3, ..., and the poles p_1, p_2, p_3, ..., are either real, imaginary or pairs of complex conjugates.

9.5.1 Magnitude and phase functions

Equation 9.33 may be written in terms of the real angular frequency by making the substitution $s = j\omega$ giving, for a general transfer function

$$T(\omega) = \frac{H_0 j\omega(j\omega-z_1)(j\omega-z_2)\ldots}{(j\omega-p_1)(j\omega-p_2)\ldots} \tag{9.34}$$

The magnitude of the transfer function of frequency is

$$|T(\omega)| = \frac{H\omega |j\omega/z_1-1||j\omega/z_2-1|\ldots}{|j\omega/p_1-1||j\omega/p_2-1|\ldots}$$

where $\qquad H = H_0 z_1 z_2 \ldots / p_1 p_2 \ldots$

Taking the logarithm,

$$20\log_{10}|T(\omega)| = 20\log_{10} H + 20\log_{10}\omega + 20\log_{10}|j\omega/z-1| + \ldots$$
$$- 20\log_{10}|j\omega/p_1-1| + \ldots$$

Now it has been noted in Section 8.11 that $20\log_{10}|T(\omega)|$ is colloquially measured in decibels. Thus, the overall gain in dB is equal to the sum of the gains in dB of each of the constituent factors of the numerator of equation 9.34, reduced by the sum of the gains in dB of the constituent factors of the denominator.

Again, if we consider the phase of the transfer function $T(\omega)$ we obtain

$$\angle\, T(\omega) = \tfrac{1}{2}\pi + \angle\,(j\omega - z_1) + \angle\,(j\omega - z_2) + \cdots$$
$$- \angle\,(j\omega - p_1) - \angle\,(j\omega - p_2) - \cdots$$

and again the phase of the transfer function is the sum (or difference) of the phase angles of each of the component factors.

Since both magnitude or phase of a complex function may be obtained by the addition of the magnitude or phase of the component factors it is therefore desirable to determine the magnitude and phase of a number of these elementary factors. These simple factors include the cases when z (or p) is real or complex (purely imaginary values of z or p rarely occur in practical networks). Thus we must consider the following factors:

(a) A constant term H

(b) $j\omega$ or $1/j\omega$

(c) $j\omega/z - 1$, with z a negative real number

(d) $-\left(\dfrac{\omega}{\omega_0}\right)^2 + 2j\zeta\left(\dfrac{\omega}{\omega_0}\right) + 1$

This does not exhaust all possibilities but we shall not consider repeated factors, nor complex factors given in (d).

When deriving these magnitude and phase plots we shall plot the magnitude in decibels and the phase in degrees and the frequency, f, (or more frequently angular frequency, ω) will be plotted on a logarithmic scale.

9.5.2 Magnitude response

We will now consider the *magnitude* responses due to these elementary factors.

(a) $T(\omega) = H$:

On a dB scale the magnitude becomes $20 \log_{10} |T(\omega)| = 20 \log_{10} H$ and since it is independent of frequency it is a horizontal line at $20 \log_{10} H$ decibels.

(b) $T(\omega) = j\omega$:

Here $20 \log_{10} |T(\omega)| = 20 \log_{10} \omega$.

If we compute this at a frequency ω_1 we obtain a gain in dB of $20 \log_{10} \omega_1$; and at double this frequency the gain is

$$20 \log_{10} 2\omega_1 = 20 \log_{10} \omega_1 + 20 \log_{10} 2$$
$$= 20 \log_{10} \omega_1 + 6$$

Thus for an increase in frequency by a factor of 2 (an octave in musical parlance) the gain increases by 6 dB. Thus a term of this nature represents, when plotted on a logarithmically scaled ω-axis, a straight line whose gradient is unity, corresponding to 6 dB/octave; in addition when $\omega = 1$, the gain is 0 dB. This characteristic is shown in Fig. 9.23. An equivalent expression to 6 dB/octave may be shown to be 20 dB/decade (a decade being an increase of frequency by a factor of 10).

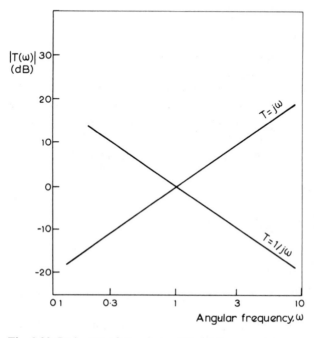

Fig. 9.23 Bode plot of $T = j\omega$ and $T = 1/j\omega$

If $T(\omega) = 1/j\omega$, the plot is a straight line of slope -6 dB/octave passing through the point $\omega = 1$, gain $= 0$ dB. This is shown on Fig. 9.23 also.

(c) *Linear factor:*

$T(\omega) = j\omega/z - 1$ for real negative values of z.

$$|T(\omega)| = \left(\frac{\omega^2}{z^2} + 1\right)^{1/2}$$

and when plotted in decibels on a logarithmic frequency scale has the form shown by the solid line of Fig. 9.24.

It is convenient for quick plotting to consider an approximation to this by using the asymptotes to the graph at low and at high frequencies.

For small values of ω such that $\omega \ll z$,

$$|T(\omega)| \to 1$$

and hence

$$20 \log_{10} |T(\omega)| = 0 \text{ dB.}$$

For large values of ω such that $\omega \gg z$,

$$|T(\omega)| \to \omega/z$$

and

$$20 \log_{10} |T(\omega)| = 20 \log_{10} \omega - 20 \log_{10} z.$$

This is a straight line passing through the point at which $\omega = z$ and the gain is 0 dB and with a slope of 6 dB/octave as in Section (b) above. These two asymptotes are drawn in broken lines on Fig. 9.24 and form a piecewise linear approximation to the exact characteristic.

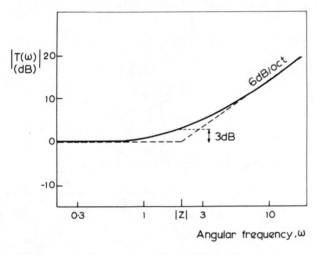

Fig. 9.24 Bode magnitude plot of $T(\omega) = j\omega/z - 1$

It may be noted from the exact expression for $T(\omega)$ given above that the gain at $\omega = z$ is exactly $\sqrt{2}$ which corresponds to a voltage gain 3 dB. It is thus possible to correct the initial two line approximation at this point. Further corrections may be made either side of $\omega = z$ to the asymptotic approximation as given by Table 9.1.

Table 9.1

ω	$z/10$	$z/5$	$z/2$	z	$2z$	$5z$	$10z$
Correction (dB)	$+0{\cdot}05$	$+0{\cdot}17$	$+1{\cdot}0$	$+3{\cdot}0$	$+1{\cdot}0$	$+0{\cdot}17$	$+0{\cdot}05$

It is usually sufficiently accurate to insert the correction of 3 dB at $\omega = z$ and note that the correction is sensibly zero at $\omega = z/5$ and $5z$.

If the linear factor occurs in the denominator of $T(\omega)$ as for example

$$T(\omega) = \frac{1}{j\omega/p - 1}$$

the asymptotes are similar to the above but have negative slopes above the break point at $\omega = p$.

9.5.3 Phase response

The phase response of the elementary factors may be determined in a similar manner.

(a) $T(\omega) = H$ obviously contributes nothing to the phase angle

(b) $T(\omega) = j\omega$

Then $\angle T(\omega) = \tfrac{1}{2}\pi$

For $T(\omega) = 1/j\omega$, the phase angle $\angle T(\omega) = -\tfrac{1}{2}\pi$.

(c) $T(\omega) = j\omega/z - 1$ with z real and negative

$$\angle T(\omega) = \text{arc tan } -\omega/z$$

For $\omega \ll |z|$ $\angle T(\omega) \to 0$
For $\omega \gg |z|$ $\angle T(\omega) \to \tfrac{1}{2}\pi = 90°$ since z is negative
At $\omega = |z|$ $\angle T(\omega) = 45°$

The exact value of $\angle T(\omega)$ is shown by the full line of Fig. 9.25.

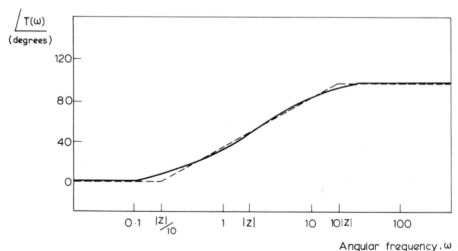

Fig. 9.25 Bode phase plot of $T(\omega) = j\omega/z - 1$

A reasonable approximation to this characteristic is given by the broken line which consists of three segments:

$$\omega < z/10, \qquad \angle T(\omega) = 0$$

$$z/10 < \omega < 10z, \text{ a straight line through } \omega = |z|, \angle T(\omega) = 45°$$

$$\omega > 10z, \qquad \angle T(\omega) = 90°$$

The corrections to be applied to this to obtain the accurate curve are given in Table 9.2.

Table 9.2

ω/z	0·05	0·1	0·2	0·5	1	2	5	10	20
Correction	+1·8°	+5·7°	−2·1°	−4·9°	0	+4·9°	+2·1°	−5·7°	−1·8°

If the factor occurs in the denominator such as $T(\omega) = 1/(j\omega/p - 1)$ the phase curve is similar to Fig. 9.25 but falls from 0° to −90° as ω increases.

9.5.4 Magnitude and phase plots of complex networks

The overall magnitude and phase characteristics of a complex network may be determined by sketching the asymptotic responses of each of the component elements and then summing them over the desired frequency range. Corrections to improve the piecewise linear approximation may be made in the vicinity of each break point.

We shall now use these results to derive the Bode plots of the amplifier whose transfer function is given by equation 9.32c. The break frequencies of the numerator occur at $\omega = 50$ and of the denominator at $\omega = 0.45$, 713, 5.56×10^6, 281×10^6. We may now plot each of the following factors in the transfer function separately

$$T_1 = -5.96 \, (\equiv 15.4 \text{ dB})$$

$$T_2 = s$$

$$T_3 = 1 + s/50$$

$$T_4 = 1/(1 + s/0.45)$$

$$T_5 = 1/(1 + s/713)$$

$$T_6 = 1/(1 + s/5.56 \times 10^6)$$

$$T_7 = 1/(1 + s/281 \times 10^6)$$

These are shown drawn separately by broken lines on Fig. 9.26. The voltage gain is plotted vertically in decibels and angular frequency plotted horizontally on a logarithmic scale.

The characteristics are summed to give the magnitude of the transfer function for the amplifier as a whole and this is shown by the solid line in Fig. 9.26. This may be compared with the computer determined response shown in Fig. 9.22.

The asymptotic phase characteristic may similarly be determined as outlined in Section 9.5.3 and this is sketched in Fig. 9.27 which may be compared with Fig. 9.22. It should be noted that the negative sign of T_1 contributes 180° phase shift; this has been included in Fig. 9.27.

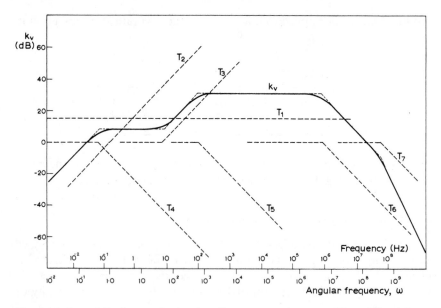

Fig. 9.26 Overall Bode magnitude plot of common emitter amplifier of Fig. 9.8

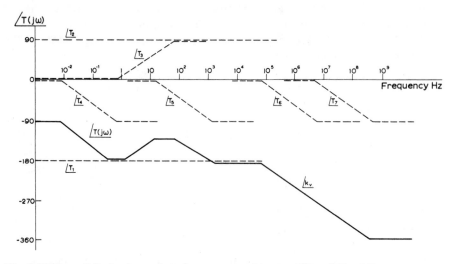

Fig. 9.27 Overall Bode phase plot of common emitter amplifier of Fig. 9.8

References

SEARLE, C. L., *et al.*, *Elementary Circuit Properties of Transistors*, Wiley, 1964, Chapters 6 and 7.
COMER, D. J., *Introduction to Semi-conductor Circuit Design*, Addison Wesley, 1968, Chapters 4, 5, and 7.
LINVILLE, J. G. and GIBBONS, J. F., *Transistors and Active Circuits*, McGraw-Hill, 1961, Chapters 9 and 10.
MALVINO, A. P., *Transistor Circuit Approximations*, McGraw-Hill, 1968, Chapters 10, 13, and 14.
VAN VALKENBURG, M. E., *Introduction to Modern Network Synthesis*, Wiley, 1966, Chapter 9.

Problems

9.1 A field-effect transistor may be modelled at sufficiently low frequencies by a voltage controlled current generator of transconductance $g_m = 6$ mS, shunted by a resistor $r_{ds} = 30$ kΩ. It is used in the circuit of Fig. P9.1. Determine the mid-band gain and the low frequency half power point.

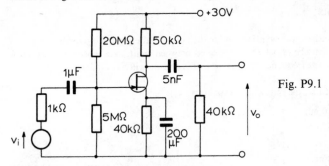

Fig. P9.1

9.2 The circuit of Fig. P9.2 is to be designed as a video amplifier. The lower half power frequency, ω_L, must be at 15 Hz and the upper frequency, ω_H, no lower than 2·5 MHz. It is fed from a source of resistance 1·2 kΩ and drives a load of 400 Ω.

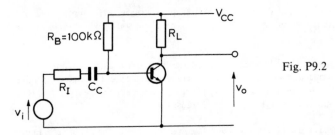

Fig. P9.2

The transistor parameters measured at a collector current of 5 mA are $\beta_0 = 120$, $r_x = 40$ Ω, $r_o = 50$ kΩ, $C_\mu = 3$ pF, and $f_T = 600$ MHz. Determine C_C to meet the specification. What is the mid-frequency gain? Investigate whether the transistor is capable of meeting the design requirements at high frequencies in this circuit. If not, determine the upper cut off frequency. Redesign the circuit using the same transistor to the original specifications. What is the mid-frequency gain of the new circuit?

9.3 The circuit of Fig. P9.2, using the transistor whose data are given in Problem 9.2, is to be designed to amplify rectangular pulses from a generator of internal resistance $R_1 = 50$ Ω. The pulse duration is 10 μs and amplitude 100 mV. Choose values for R_L and C_C such that the output pulse rise time is less than 20 ns and the droop is less than 1%. What is the magnitude of the output pulse?

9.4 A field-effect transistor is used in a source follower having a load resistor of 10 kΩ. The transistor parameters are: $g_m = 7$ mS, $C_{gs} = 5$ pF, $C_{gd} = 0·2$ pF, $r_o = 20$ kΩ. Determine the input capacitance of the amplifier.

9.5 An amplifier has a gain characteristic given by

$$k_v(s) = 4 \times 10^6 \frac{s(s+10)}{(s+30)(s+300)(s+10^5)}$$

Sketch and dimension the asymptotic curves on the Bode plot for $|k_v|$ and $\angle k_v$ against frequency. Use appropriate correction factors to draw in a smooth curve.

9.6 A transistor is represented by the following h-parameters:

$$h_{ie} = 2 \cdot 2 \times 10^3 \ \Omega$$
$$h_{re} = 3 \cdot 6 \times 10^{-4}$$
$$h_{fe} = 55$$
$$h_{oe} = 12 \cdot 5 \ \mu S$$

It is connected to a load resistor of 5 kΩ. Determine the voltage gain and the input resistance. If a signal source having a resistance of 600 Ω is connected between emitter and base via a capacitor of value 0·1 μF, determine the frequency at which the voltage gain will have fallen by 3 dB.

9.7 A transistor is specified by the following parameters:

$$y_{ie} = 5 + j10 \ \text{mS}$$
$$y_{fe} = 50 - j70 \ \text{mS}$$
$$y_{re} = -j4 \ \text{mS}$$
$$y_{oe} = 2 + j4 \ \text{mS}$$

measured at 50 MHz. It is used with a purely inductive load impedance of 0·13 μH to construct a tuned amplifier.

Determine the input admittance of the amplifier at a frequency of 50 MHz. Why would this circuit be unsuitable as a tuned amplifier if it were fed from a current source whose conductance were less than 3·6 mS?

9.8 The transistor of Problem 9.7 is to be used in a tuned amplifier. It is decided to neutralize the feedback susceptance by connecting an inductor between collector and base. Determine the value of this inductor in order that the transistor will be completely unilateral at a frequency of 50 MHz.
Write down the admittance matrix of the neutralized transistor.

9.9 A transistor amplifier has the circuit of Fig. P9.3. The load impedance consists of an inductor of 160 μH tuned with a capacitor to resonance at 150 MHz, and having a Q factor of 30. The transistor may be assumed to have the parameters given below. Determine the input admittance of the amplifier and the voltage gain at 1·5 MHz.

$$r_\pi = 1 \cdot 5 \ \text{k}\Omega$$
$$C_\pi = 60 \ \text{pF}$$
$$C_\mu = 2 \ \text{pF}$$
$$g_m = 12 \ \text{mS}$$
$$r_o = \text{infinite}$$

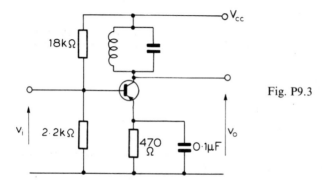

Fig. P9.3

9.10 The emitter follower of Fig. P9.4 using the transistor of Problem 9.2 is required to amplify small rectangular pulses of duration 10 μs from a source of internal resistance 50 Ω. Determine the rise time of the output pulse across R_L and the value of C_C if the droop must not be more than 1%.

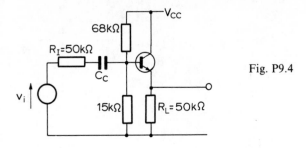

Fig. P9.4

10
Noise

10.0 Introduction

The purpose of most communication and control systems is to process an input signal in order to give an output which is some predetermined function of the input signal. For a simple amplifier this function is simply a constant gain factor over the whole range of frequency. In a radio receiver using amplitude modulation the currents flowing in the loudspeaker are required to be linearly related to the amplitude of the radio frequency waves received. In a digital computer the output should be related to the input by the functional relationship which has been programmed into the computer.

In general there will be some deviation between the idealized output and the output actually obtained from the system. The difference is known by the general term of noise. There are two basic forms of noise, the first is man made noise and may thus, at least in principle, be eliminated; the second is fundamental to the discrete nature of the universe.*

10.1 Man made noise

This may take many forms. One of the most common is mains hum caused by voltages induced in the input circuit of an amplifier by electromagnetic coupling from nearby mains transformers or electrostatic coupling from other parts of the circuit or nearby apparatus at mains potential. Such noise may be eliminated by suitably disposing the circuit elements or by providing electrical screening where necessary.

Other undesired voltages in a circuit may be caused by induction from nearby X-ray apparatus, high voltage power lines, other spark discharges such as motor car ignition, faulty relay contacts, etc.

All these noise voltages may, in principle, be eliminated. Despite this it is often necessary to locate sensitive equipment remote from other electrical apparatus.

The wave shapes of man made noise may vary considerably. Mains hum will

---

* This statement does not preclude an explanation of noise on a wave mechanical basis; but this will not be attempted here.

be restricted to power frequencies (usually 50 Hz). Ignition noise consists of short bursts of voltage usually at very high radio frequencies.

Another form of noise at very low frequencies, often called drift, is caused by the very slow and erratic variations in voltage in a circuit. This may be the result of random fluctuations of supply voltage, random temperature variations and certain very slow electronic conduction variations. The effects of drift are most serious in d.c. amplifiers since no capacitive coupling may be used to filter out these very slow fluctuations.

10.2 Inevitable noise

Since all processes may be considered to be atomic in nature, we should expect all conduction currents to consist of a succession of discrete events. This will result in a very small random fluctuation of the output voltage of an electrical system which was not present at the input. There are a number of causes of electrical noise; the two most important are thermal noise and shot noise and we shall discuss these in some detail. In addition, in semiconductor devices we find that a very low frequency fluctuation of voltage may be detected and that this increases as the frequency of measurement is reduced; this is known as flicker noise, or $1/f$ noise, due to the variation of its magnitude with frequency. It is obvious that this is of significance mainly in d.c. and low frequency amplifiers. With present day transistors it is not important above about 1 kHz.

10.3 Thermal noise

Thermal noise (sometimes known as Johnson noise) is caused by the random movement of electrons as a result of their thermal agitation energies. This phenomenon is common to all conducting media and is a function of the temperature and the resistance of the element only. Since the effect is caused by the entirely random movement of electrons in a crystal lattice, we should expect all frequencies to be equally likely. Thus if we plot the spectrum of thermal noise we will find that it is completely uniform from very low subaudio frequencies up to microwave frequencies and even beyond. It is thus termed white noise because of the similarity of its spectrum to that of white light.

In order to make any meaningful measurements on the noise, we must restrict our measurements of noise power to some limited band of frequencies. If we measure the available power in a resistor, R, at an absolute temperature, T, over a band of frequency, B, with centre frequency, f, we shall find that the power is proportional to BT. It may be shown theoretically that the average power P is given by

$$P = kTB \qquad (10.1)$$

where k = Boltzmann's constant ($= 1.38 \times 10^{-23}$ J/K).

Note that the average available power at a temperature T is independent of the value of the resistor. It is also only proportional to the bandwidth over which measurements are being made and is independent of the centre frequency.

Thus there is exactly the same thermal noise power generated in a band of frequencies from 0 to 10 kHz as between 1·0 and 1·01 MHz.

Since the power is being generated in a resistor of value R, we may model the generation of this power by an equivalent noise voltage in series with the resistor. Since the voltage concerned is of a random nature we must be a little more precise and refer only to average values of power.

Again the power that we have been measuring is the available power from the resistor. This term should now be formally defined. If we consider a sinusoidal signal source of r.m.s. value V_i in series with a resistor R_1, as in Fig. 10.1, the available power from this source is defined as the maximum power which may be drawn from the source by variation of a resistor R_L connected across its terminals.

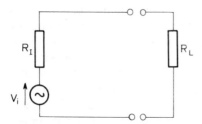

Fig. 10.1 Available power from a generator of resistance R_1

The power dissipated in the resistor R_L is

$$\frac{V_i^2 R_L}{(R_L + R_1)^2}$$

and this is maximum for variation of R_L when $R_L = R_1$. Thus the available power from this source is

$$\frac{V_i^2}{4R_1}$$

Hence if the noisy resistor, R, has an open circuit mean square voltage $\bar{v}_n^2$ the available power $= \bar{v}_n^2/4R$. Equating this to the available power from a noisy

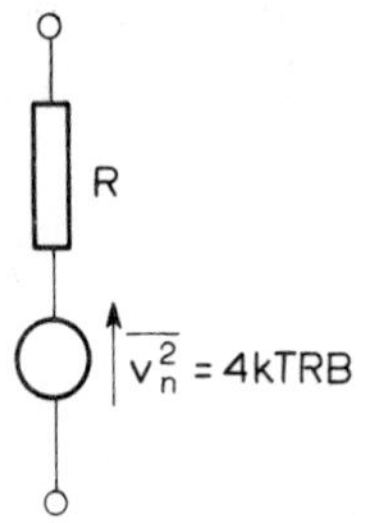

R

$\overline{v_n^2} = 4kTRB$

Fig. 10.2 Noise model of a resistor

resistor given by equation 10.1 we may deduce that a resistor will have a mean square open circuit noise voltage given by

$$\bar{v}_n^2 = 4kTRB \qquad (10.2)$$

An equivalent circuit model of a noisy resistor, R, is thus shown in Fig. 10.2.

An alternative circuit model of the noisy resistor may be obtained by considering the Norton equivalent circuit to that of Fig. 10.2. This is shown in Fig. 10.3 where the current generator has a mean square value

$$i_n^2 = 4kTGB \qquad (10.3)$$

where $G = 1/R$ and is the conductance of the resistor.

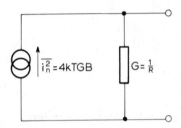

Fig. 10.3 Alternative noise model of a resistor

10.3.1 Noise in inductors and capacitors

We have stated that the cause of thermal noise is the interaction between randomly moving electrons and the crystal lattice through which they move.

An ideal inductor is, by definition, a device in which no collisions occur between the charge carriers and the crystal lattice; if this were not the case, the inductor would have a finite resistance. There can therefore be no thermal noise associated with a pure inductor. Similarly we may see that no noise is generated in an ideal capacitor.

10.3.2 Noise in combinations of resistors

It is easily shown that the noise generated by a combination of two resistors is equal to that which would be generated at the same temperature by a single resistor whose value is equivalent to the given combination of resistors.

Consider the simple case of two resistors in series at temperature T as shown in Fig. 10.4a. The equivalent noise model of this circuit is shown in Fig. 10.4b; the mean square noise voltages of the two resistors are

$$\bar{v}_{n1}^2 = 4kTR_1B$$
$$\bar{v}_{n2}^2 = 4kTR_2B$$

Now we may amalgamate the two noise voltages into a single voltage generator of mean square value $\bar{v}_n^2$ corresponding to a single resistor of value R, where $R = R_1 + R_2$, as shown in Fig. 10.4c.

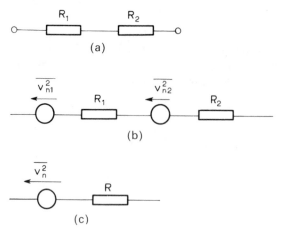

Fig. 10.4 Equivalent noise model of two resistors in series

Since we should hardly expect the noise voltage to be different according to the manner in which the resistor is built up from constituent elements, we should expect

$$\bar{v}_n^2 = 4kTRB$$
$$= 4kT(R_1 + R_2)B$$
$$= \bar{v}_{n1}^2 + \bar{v}_{n2}^2 \tag{10.4}$$

This is a very important result which we have deduced from a rather simple and perhaps naive argument. However, since the events which give rise to noise in resistor R_1 are completely independent of those which give rise to noise in R_2, we could use a statistical argument to show that such voltages must be added quadratically, namely that their mean square values must be added to give the effective total voltage as shown by equation 10.4.*

We may extend this to determine the effective noise voltage due to two resistors R_1 and R_2 in parallel. Considering each resistor in turn, the open

* The summation of thermal noise voltages by adding their mean square values is also applicable to other forms of noise voltage. The only stipulation is that the two sources of noise should have no correlation, namely that the generating mechanisms in the two cases should be statistically independent. When correlation does exist due to the presence of some common factor in the two noise generating processes, a more complicated form of addition of noise voltages must be used incorporating the correlation factor. We shall not consider any such cases in this book.

circuit terminal voltage squared due to generator $\bar{v}_{n1}^2$, representing the noise of resistor R_1, is

$$\bar{v}_{n1}^2 \left(\frac{R_2}{R_1 + R_2} \right)^2 = 4kTB \frac{R_1 R_2^2}{(R_1 + R_2)^2}$$

and due to $\bar{v}_{n2}^2$ it is

$$\bar{v}_{n2}^2 \left(\frac{R_1}{R_1 + R_2} \right)^2 = 4kTB \frac{R_1^2 R_2}{(R_1 + R_2)^2}$$

Superimposing these two voltages by adding their mean squares gives an output voltage

$$\bar{v}_n^2 = 4kTB \frac{R_1 R_2}{R_1 + R_2}$$

and this is exactly the noise which would be developed by a resistor R equal to the parallel combination of R_1 and R_2 at a temperature T.

10.3.3 Frequency spectrum of thermal noise

From equation 10.1 we see that the mean-square thermal noise voltage is proportional to the bandwidth over which it is measured but independent of the frequency at which the measurements are made. This is true up to frequencies in excess of 10 GHz; above this frequency, corrections need to be made to equation 10.1. The spectrum of the amplitude of the mean-square noise voltage is therefore constant from very low frequencies up to microwave frequencies. Such a uniform spectrum is characteristic of white light and so any noise whose spectrum is flat over a very wide band of frequencies is termed white noise.

It is interesting also to consider the waveform of noise. We have seen that it contains all components of frequency and so we should expect the waveform to consist of variations which include fluctuations with extremely long periods up to transients of very short duration. Since all measurements which we can make will be over a limited frequency band we see that some of the very slow fluctuations will be eliminated by the low frequency cut off of the measuring system and the fast transients by the high frequency cut off.

If we now reduce the bandwidth of the measuring instrument, we shall eliminate more of the low and high frequency fluctuations until ultimately we shall find that when the bandwidth is very small the waveform contains only a very narrow band of frequencies and has the appearance of a sinusoidal voltage whose amplitude is fluctuating slightly and having a slight phase 'jitter'. This allows us to consider a noise voltage measured over a narrow band of frequencies as a pseudo sinusoidal voltage and we may therefore analyse the effects of such voltages in a circuit as if they were sinusoidal voltages of corresponding mean square amplitude.

To differentiate between the total voltages over a wide band we shall use capital letters to represent these narrow band noise voltages and currents.

10.3.4 Thermal noise in combinations of resistors, inductors, and capacitors

A one-port network consisting entirely of resistors, capacitors, and inductors will have an impedance, at any arbitrary frequency f, of $Z(f) = R(f) + jX(f)$. The thermal noise voltage which will appear across the external terminals at this frequency in a very narrow band of frequencies δf, due to all the resistors scattered throughout the network, will be equal to the thermal noise voltage generated by the resistive part of the impedance, $R(f)$, in this same bandwidth, namely

$$\overline{V_n^2} = 4kTR(f)\delta f \tag{10.5}$$

The total noise generated by the impedance over the whole frequency band will be obtained by integrating equation 10.5 over the limits of the frequency band.

Consider for example the network of Fig. 10.5a. The noise equivalent circuit is shown in Fig. 10.5b where $\overline{V_n^2} = 4kTR\delta f$. The open circuit voltage is given by

$$\overline{V_n^2} \left| \frac{1/j\omega C}{R + j\omega L + 1/j\omega C} \right|^2$$

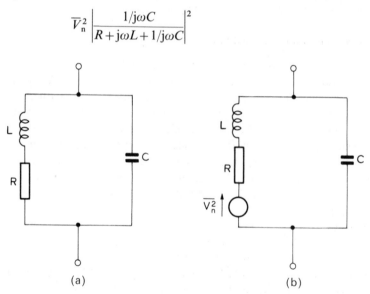

(a)   (b)

Fig. 10.5 (a) Frequency dependent network (b) its noise model

We need to take the modulus of the impedance ratio since we are only concerned with magnitudes of voltages. If we now determine the open circuit noise voltage at the frequency $\omega_0 = (1/LC)^{1/2}$, chosen to simplify the algebra, in a small band of frequencies δf, we obtain

$$\overline{V_n^2} \frac{1}{\omega_0^2 C^2 R^2} = \frac{4kT\delta f}{\omega_0^2 C^2 R} = 4kT\delta f L/CR$$

This is equal to the thermal noise which would be generated in a resistor of magnitude L/CR at a temperature T.

The impedance of the network of Fig. 10.5a is

$$Z(\omega) = \frac{R+j\omega L}{j\omega C(R+j\omega L+1/j\omega C)}$$

At $\omega_0 = (1/LC)^{1/2}$

$$Z(\omega_0) = \frac{L}{CR} - j\omega_0 L$$

The real part of $Z(\omega_0)$ is L/CR and hence we would expect a noise voltage given by

$$\overline{V_n^2} = 4kT\delta f L/CR$$

and this is exactly the expression we obtained previously by a direct analysis from the noise generated in the impedance Z. Thus we have obtained confirmation of the fact that the noise in a complex impedance over a small band of frequencies is equal to the noise which would be developed thermally by the real part of that impedance at the centre frequency of the band.

10.3.3 Magnitude of thermal noise

At room temperature of, say, 300 K, we may compute the available noise power from a resistor in a bandwidth of 1 Hz using equation 10.1 as $4\cdot14 \times 10^{-21}$ W. If we assume that we are using a receiver whose input impedance is 50 Ω and bandwidth is 1 MHz we can compute that this would correspond at the input of the amplifier to an effective noise voltage of $0\cdot91$ μV. Using such a receiver we observe that it would be very difficult, if not impossible, to receive any intelligible signal whose magnitude was less than about 1 μV. There is thus a fundamental limit on the sensitivity of any receiver.

In the preceding discussion we have merely stated that an impedance at a temperature T will give rise to thermal noise given by equation 10.1 or 10.2. It should be stressed that the noise given by these formulae are only valid when the system is in thermal equilibrium with its surroundings, namely that there is no interchange of power between the network and its surroundings. Such a situation exists in any linear resistor; it is not true, however, in nonlinear devices, semiconductor diodes, transistors, etc., where considerably greater noise is observed.

10.4 Shot noise in diodes

The other important noise generating mechanism in modern electronic devices is shot noise. This occurs whenever charge is transferred through a potential gradient by the motion of discrete electrons. Looked at in a very simple manner it may be considered to be due to the random arrival of electrons. It is present in all semiconductor diodes, transistors and all thermionic devices. If we restrict our study to relatively low frequencies (say about 10 kHz), we shall find that the mean square value of the noise current in a semiconductor diode is related to the steady current flowing by

$$i_n^2 = 2eI_DB + 4eI_0B \tag{10.6}$$

where I_D is the steady current through the diode and I_0 the reverse saturation current. This may be written alternatively in terms of the diode conductance g_d. From equation 1.24b we note that the conductance g_d at current I_D is

$$g_d = \frac{e}{kT}(I_D + I_0) \tag{10.7}$$

Substituting equation 10.9 in equation 10.6 gives an alternative expression for the noise in a diode as

$$i_n^2 = 4kTg_dB - 2eI_DB \tag{10.8}$$

10.4.1 Magnitude of shot noise in a diode

A forward biased diode carrying a current of 1 mA connected in a network with a bandwidth of 1 MHz will give rise to a mean square noise current, calculated from equation 10.6, of

$$i_n^2 = 3.2 \times 10^{-16} \ A^2$$

The equivalent current is thus 18 nA. If the diode is connected to a resistor of (or a system whose input resistance is) 50 Ω, the effective noise voltage developed will be 0.9 μV which is of the same order of magnitude as the thermal noise in the resistor, which we calculated in Section 10.3.3.

10.4.2 Noise temperature of a diode

It is sometimes convenient to use the concept of noise temperature of a device. If we assume that the total noise in the device is being produced in the conductance g_d at an artificial 'noise temperature' T_n, then from equation 10.6 we may write

$$2eI_DB + 4eI_0B = 4kT_ng_dB \tag{10.9}$$

and incorporating equation 10.7 we obtain

$$T_n = \frac{I_D + 2I_0}{2(I_D + I_0)} T_0 \tag{10.10}$$

where T_0 is room temperature.

10.5 Noise in transistors

In more complex devices than diodes, noise may be generated as a result of a number of separate processes occurring within or associated with the semiconductor wafer. Some of these physical processes will be completely independent of any other process occurring in the device; others will be dependent either partially or totally on some different mechanism. Many of these processes will give rise to noise although in some cases the noise generated in one part of the device may be correlated with the noise generated in another part of the device, caused by physical mechanisms which are interdependent. Such noise sources are termed correlated and a measure of the degree of interdependence is given by the correlation coefficient. A correlation coefficient of zero indicates that the two noise generating processes are totally independent and a value of unity indicates that they are both the result of a single physical process. We shall consider, in this book, that all noise sources are completely uncorrelated, an assumption which is valid in most devices at relatively low frequencies.

10.5.1 Noise in bipolar transistors

A noise model of a bipolar transistor may be developed by considering the small signal hybrid π model given in Fig. 5.5. If we study the physical mechanisms associated with the circuit elements represented we shall be able to develop an equivalent model for the noise properties of a transistor.

The resistor r_x models the voltage drop caused by the flow of majority carriers from the base terminal to the recombination centres. Since the current is transported by majority carriers, r_x is an ohmic resistor. The noise generated in it is therefore just thermal noise and may be represented by a mean square noise voltage

$$\bar{v}_{nx}^2 = 4kTr_xB \tag{10.11}$$

The resistor r_π is the circuit representation of the flow of minority carriers across the forward biased emitter–base junction. This process is not in thermal equilibrium and so we may not use the thermal noise expression. Since a steady current I_B flows across this forward biased diode, we may model this noise source by a current generator whose mean square value is given by

$$\bar{i}_{n\pi}^2 = 2eI_BB \tag{10.12}$$

In addition a steady current I_C flows from emitter through the base to the collector across the reverse biased collector junction. This current is subject to shot noise and hence we must insert a current generator between collector and emitter given by

$$\bar{i}_{nc}^2 = 2eI_CB \tag{10.13}$$

To complete the model we note that collector saturation current I_{CBO} flows between collector and base and so a current generator should be inserted there of magnitude

$$i_{nco}^2 = 2eI_{CBO}B \tag{10.14}$$

For silicon transistors operating at relatively large values of current gain this may be ignored.

The resistor r_o will not generate any thermal noise since it does not represent an ohmic process but is related to the variation of base width with change of collector voltage and also with the surface leakage of current at the collector.

The complete hybrid π noise model of a transistor is shown in Fig. 10.6. The current generator representing collector saturation current has not been included since we assume it to be small.

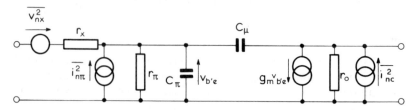

Fig. 10.6 Noise model of common emitter bipolar transistor

10.5.2 Noise in field-effect transistors

The noise in field-effect transistors is considerably lower than that in bipolar transistors. This is principally due to the almost complete absence in fets of semiconductor junctions, and hence the absence of shot noise. Only in junction fets does any shot noise appear, between gate and channel, and this is only dependent, under normal operating conditions, on the reverse current of the gate source diode.

The small signal model for a field-effect transistor is shown in Fig. 10.7, based on the small signal model of Fig. 3.4. The model is applicable to both junction fets and mosfets.

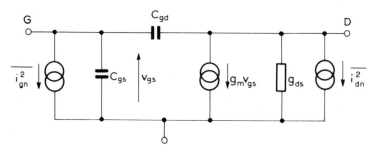

Fig. 10.7 Noise model of common source fet

The current generator i_{dn}^2 is caused principally by the flow of majority carriers along the channel and is therefore almost completely due to thermal noise in the channel resistance. The value of the channel resistance in this situation is the resistance of that portion of the channel up to the point where the inversion layer vanishes (or in the case of a junction fet, where the channel is fully depleted); this has a value which is of the order of $1/g_m$.

The current generator i_{gn}^2 arises from several causes. It contains a component of current which is induced in the gate lead by the random motion of the charge carriers in the channel. In a junction fet a further constituent of gate noise is provided by the shot noise across the reverse biased gate–channel junction; in a mosfet a small amount of noise is generated as a result of charge leakage across the oxide layer; the energy contained in this noise is concentrated at low frequencies and has a spectrum which is approximately inversely proportional to frequency $(1/f)$.

10.5.3 Transistor noise representation as a two-port network

In Appendix A it is shown how a two-port network with no internal independent sources may be represented at any specified frequency by four parameters—the admittance, impedance, or hybrid parameters.

If the network does contain some independent sources, equations A.3 must be modified to read:

$$\left.\begin{aligned} V_1 &= a_{11}V_2 + a_{12}I_2 + V_n' \\ I_1 &= a_{21}V_2 + a_{22}I_2 + I_n' \end{aligned}\right\} \tag{10.15}$$

Although equations 10.15 suggest the use of the a parameters (the cascade two-port parameters), the source-free network may be specified by any convenient set of parameters. In the case of a noisy network, the two independent generators are replaced by equivalent noise generators as shown in Fig. 10.8; these are

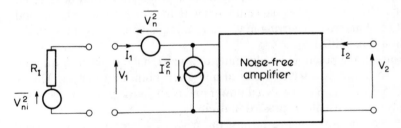

Fig. 10.8 Noisy two-port connected to a noisy source

specified by their mean square magnitude, and, where necessary, by their cross-correlation coefficient in addition.

The advantage of this form of representation is that all the noise generators have been moved outside (in this case, in front of) a noise-free amplifying device. The magnitudes of $\overline{V_n^2}$ and $\overline{I_n^2}$ may be determined from the internal noise generators directly by a straightforward, although in many cases rather tedious, network analysis. The disadvantage of this form of specification lies in the fact that simple expressions like equations 10.11 and 10.14 are no longer applicable and each noise source must be independently specified as a function of frequency, bias currents, etc.

When the input (or output) terminals of the amplifier are connected to an impedance of known value, it is possible to convert the current generator into an equivalent voltage generator and add this to the existing voltage generator. Thus if a noisy source resistor R_I is connected across the input port of the network shown in Fig. 10.8, we shall be able to draw an equivalent network as shown in Fig. 10.9, where

$$\bar{V}_{ne}^2 = R_I^2 \bar{I}_n^2 \tag{10.16}$$

The total input noise to the amplifier may now be determined by summing the mean square noise voltages at the input.

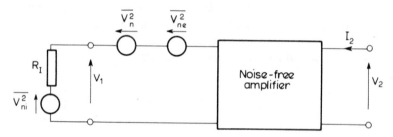

Fig. 10.9 Equivalent noise model to Fig. 10.8

10.6 Signal to noise ratio

We have seen in Section 10.3.3 that because of the thermal noise at the input to an amplifier there will be a certain minimum signal power which may be received intelligibly. The smaller the signal compared with the noise the more likelihood there will be of an error occurring in t.. received signal and hence a loss in the information transmitted.

A measure of the potential intelligibility of a signal may be given by quoting the signal to noise ratio which exists at any given point in a system. This is defined as S/N, where S is the signal power and N the noise power at the same point. This ratio is usually expressed in decibels.

In any system involving transmission networks, amplifiers, frequency changers,* and other linear devices, both signal and noise are equally amplified or attenuated; however, throughout the system the noise power is being progressively increased and thus the signal to noise ratio gets steadily smaller as one progresses through a system.

* A frequency changer is strictly a nonlinear device but if one considers the signal voltage only, it may be regarded as a device which linearly processes a modulated voltage to give a signal voltage.

Thus

$$\frac{S_o}{N_o} < \frac{S_i}{N_i}$$

where the suffixes o and i indicate the output and input respectively.

10.7 Noise figure

We may now define a quantity, the *noise figure*, F, of a system, to indicate this degeneration of the signal purity as it passes through the system:

$$F = \frac{S_i/N_i}{S_o/N_o} \qquad (10.17)$$

We shall be concerned here almost exclusively with the spot noise figure which refers to the noise figure measured over a very narrow band at the selected frequency. The average noise figure is used when considering amplifiers and refers to the integrated spot noise figure over the bandwidth of the amplifier. Although this definition gives a very good physical insight into the significance of noise figure, it needs to be related to the parameters of the system.

Let us define the available gain (see Section 8.10) of the system as A_A over a bandwidth B working from a source resistance R_1 at temperature T_0.
Then

$$S_o = A_A S_i \qquad (10.18)$$

and $\qquad N_i = kT_0 B \qquad (10.19)$

Thus $\qquad F = \dfrac{N_o}{A_A N_i} \qquad (10.20)$

$$= \frac{N_o}{A_A kT_0 B} \qquad (10.20a)$$

Thus a simple method of determining the noise figure of an amplifier is to determine the ratio of the total output noise power to the output noise power if the amplifier contributed no noise. Now N_o is made up of the input noise amplified by the power gain A_A augmented by the noise generated within the system, N_s.
Thus

$$F = 1 + \frac{N_s}{A_A kT_0 B} \qquad (10.21)$$

and we may note that the noise figure of an amplifier must be greater than unity, unless the input resistor is maintained at a temperature less than T_0. Since A_A and N_o are both functions of input termination the noise figure will also be dependent on generator resistance; an optimum value of source resistance may therefore be chosen to give minimum noise figure. Since the noise figure of an amplifier cannot be less than the noise figure of the transistor which is used in it, it is of interest to observe the variation of noise figure with source resistance for both bipolar transistors and field-effect transistors.

Figure 10.10 shows the noise figure of a typical bipolar transistor as a function of source resistance R_I. It may be noted that the noise figure falls to a minimum value of about 1·5 dB at a generator resistance of 5 kΩ.

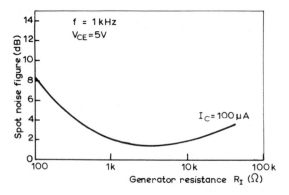

Fig. 10.10 Variation of spot noise figure of bipolar transistor with generator resistance

Figure 10.11 shows the variation of noise figure with frequency for the same transistor at its optimum source resistance. At low frequencies below about 1·0 kHz the noise figure is seen to rise; this is due to the $1/f$ noise mentioned in Section 10.2. The transistor considered is intended mainly for low frequency operation. If the curves for noise figure were continued to higher

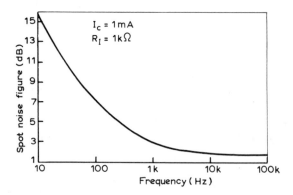

Fig. 10.11 Variation of spot noise figure of bipolar transistor with frequency

frequencies it would be observed that the noise figure would rise again probably in the region of 10 MHz, this rise being caused, in the main, by the loss of gain of the amplifier due to the effects of C_μ and C_π.

It should be added that the noise figure of an amplifier is dependent upon the steady collector current I_C. Lower values of noise figure are generally obtainable at low values of collector current; the curve given in Fig. 10.10 refers to the very low collector current of 100 μA. The optimum value of generator resistance is similarly a function of collector current, being larger for small values of I_C.

A similar dependence of noise figure on generator resistance and frequency occurs with field-effect transistors.

10.7.1 Noise figure from two-port noise parameters

The two-port parameters of Fig. 10.8 are very useful in determining the noise figure of an amplifier. Let us assume that the available gain of the amplifier is A_A. Then in the absence of any noise generated by the amplifier ($\bar{V}_n^2 = \bar{I}_n^2 = 0$) the output noise power $= A_A kTB = A_A \bar{V}_{ni}^2/4R_I$.

If we now consider the effects of the noise in the amplifier, we may convert the circuit to that shown in Fig. 10.9. The total mean square noise at the input is now

$$\bar{V}_{ni}^2 + \bar{V}_n^2 + \bar{V}_{ne}^2$$

and the available noise power at the output of the amplifier is

$$A_A(\bar{V}_{ni}^2 + \bar{V}_n^2 + \bar{V}_{ne}^2)/4R_I$$

The noise figure from equation 10.20 is the ratio of the available output noise power to the available output noise power if the amplifier were noise-free. Thus

$$F = \frac{\bar{V}_{ni}^2 + \bar{V}_n^2 + \bar{V}_{ne}^2}{\bar{V}_{ni}^2}$$

$$= 1 + \frac{\bar{V}_n^2}{4kTBR_I} + \frac{\bar{I}_n^2 R_I}{4kTB} \tag{10.22}$$

which enables the noise figure of an amplifier to be easily determined if $\bar{V}_n^2$ and $\bar{I}_n^2$ are specified over a range of frequency.

From equation 10.22 we may readily determine the value of optimum generator resistance for minimum noise figure.

$$R_{I,opt} = [\bar{V}_n^2/\bar{I}_n^2]^{1/2} \tag{10.23}$$

and thus, given $\bar{V}_n^2$ and $\bar{I}_n^2$, the optimum source resistance is immediately determinable.

10.8 Noise temperature of an amplifier

We have noted that when an amplifier of available power gain A_A is supplied from a fixed source resistance R_I the noise power available from the amplifier

N_o is uniquely defined; this situation is shown in Fig. 10.12a. An alternative circuit model of this situation is shown in Fig. 10.12b in which the output power is referred to the input of the amplifier and is equal to N_o/A_A before passing through a noiseless amplifier of available gain A_A.

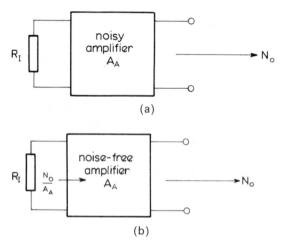

(a)

(b)

Fig. 10.12 Noise properties of an input terminated amplifier transferred to the input resistor

Now again we may consider part of this input noise kT_0B as originating in the source resistor R_1 and the remainder as originating in the amplifier itself. Thus the noise which the amplifier generates referred back to the input terminals is

$$\frac{N_o}{A_A} - kT_0B$$

and, utilizing equation 10.20a, this is equal to

$$(F-1)kT_0B$$

Now since this noise is assumed to exist at the input terminals we may ascribe to the input resistor R_1 a fictitious temperature, the noise temperature, T_n, of the amplifier, to give exactly this noise power.
Hence

$$kT_nB = (F-1)kT_0B$$

i.e. $$T_n = (F-1)T_0 \tag{10.24}$$

It may be seen that the noise temperature is an alternative method of describing the noisiness of an amplifier. It refers exclusively to the additional noise generated within an amplifying system.

References

KING, R. A., *Electrical Noise*, Chapman and Hall, 1967, Chapters 3, 4, 6, 8, and 9.

ALLEY, C. L. and ATTWOOD, K. W., *Electronic Engineering*, Wiley, 1973, Chapter 13.

WALLMARK, J. T. and JOHNSON, H., *Field Effect Transistors*, Prentice-Hall, 1966, Chapter 6.

COBBOLD, R., *Theory and Application of Field Effect Transistors*, Wiley-Interscience, 1970, Chapter 9.

MOTCHENBACHER, C. D. and FITCHEN, F. C., *Low Noise Electronic Design*, Wiley-Interscience.

FREEMAN, J. J., *Principles of Noise*, Wiley, 1958, Chapters 4, 5, and 6.

Problems

10.1 Determine the short circuit mean square noise current developed in a 75 Ω resistor at a temperature of 300 K over a frequency band of 1 MHz.

10.2 Determine the short circuit mean square noise current generated in a bandwidth of 1 kHz by a diode through which a steady current of 10 mA flows. The reverse saturation current of the diode is 150 nA.

10.3 Determine the noise temperature of the diode of Problem 10.2.

10.4 Determine the mean square voltage across the 50 Ω resistor over a bandwidth of 1 kHz in the circuit of Fig. P10.1, when (a) $R = 940\ \Omega$, (b) $R = 9\ \text{k}\Omega$, (c) $R = 90\ \text{k}\Omega$. The capacitor may be assumed to have negligible reactance at the frequencies of interest.

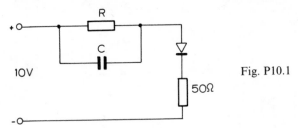

Fig. P10.1

10.5 A resistor of 50 Ω is connected across a capacitor of 220 pF. Determine the mean square noise voltage across the capacitor in a bandwidth of 1 kHz centred at (a) 10 kHz, (b) 15 MHz, (c) 100 MHz.
   Determine the mean square voltage across the capacitor in a very wide frequency band extending from d.c. to over 1 GHz. Show that this is independent of the value of resistor.

10.6 A field-effect transistor has the following low frequency parameters:

$$g_m = 6\ \text{mS}, g_{ds} = 30\ \mu\text{S}$$

It is connected in common source configurations to a generator of resistance 500 Ω. Assuming that gate noise is negligible and the drain noise is given by

$$i_{dn}^2 = 4kTg_mB$$

determine the mean square noise voltage across a 4·7 kΩ load resistor in the drain circuit in a bandwidth of 100 kHz.

10.7 Determine the noise power in the load resistor of the transistor amplifier of Problem 10.6. Hence determine the noise figure of the amplifier and the noise temperature.

10.8 Determine the equivalent two-port noise generators shown in Fig. 10.8 for a bipolar transistor whose low frequency parameters are $r_x = 50\ \Omega$, $r_\pi = 1\cdot2\ \text{k}\Omega$, $r_o = 40\ \text{k}\Omega$, $g_m = 200\ \text{mS}$, $\beta_F = 120$; all other parameters may be neglected. The transistor is operated at a collector current of 10 mA.

10.9 Determine the value of generator resistance to give minimum noise figure at low frequencies and the value of the noise figure for the transistor of Problem 10.8.

10.10 If the transistor of Problem 10.8 is operated at a collector current of 100 μA, how will this affect the minimum noise figure and the value of generator resistance to give this minimum noise figure?

10.11 A resistor at an ambient temperature of 300 K is connected to the input of an amplifier. The noise voltage as measured on a true square law voltmeter at the output is 1·4 V. When the temperature of the source is reduced to 150 K the output voltage falls to 1·15 V. Determine the noise figure of the amplifier.

11
Multistage Amplifiers

11.0 Introduction

The number of circuits involving multiple stages is so great that it is impossible to look at more than a very few of the possible combinations. There are many reasons for resorting to multistage amplifiers; the most obvious is that higher voltage, current, or power gain is required than that which is available using a single stage only. Other reasons for the use of multiple stages are to obtain high (or low) input impedances, low (or high) output impedances, high stability against temperature variation, wide bandwidth, etc. We shall investigate a small selection of circuits involving two or more transistors.

11.1 Cascaded stages

The simplest application of multistage amplifiers is in the cascading of several identical or similar low frequency amplifiers in order to increase the overall gain. We shall look at the situation shown in Fig. 11.1 in which two identical low frequency amplifiers each of gain k_{v0} and upper cut-off frequency f_H, are cascaded. Each amplifier is represented in the diagram as a block having an input resistance r_i, output resistance r_o, and open circuit voltage gain k_v.

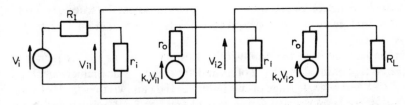

Fig. 11.1 Model of cascade of two transistor amplifiers

For simplicity of analysis we shall assume that R_I the resistance of the generator is identical to the output resistance, r_o, of each stage and that R_L the final load resistor is equal to r_i of each stage.

Initially, we shall neglect the low frequency effects of coupling capacitors and assume that we may write for each amplifier

$$k_v = k_{v0} \frac{1}{1+j\omega/\omega_H}$$

Now the mid-band gain is k_{v0} and hence the overall mid-band gain of the two cascaded stages is k_{v0}^2 and we have thus achieved the desired increase in mid-band voltage gain.

If we now look at the overall voltage gain, k_v', we obtain

$$k_v' = k_v^2 = k_{v0}^2 \frac{1}{(1+j\omega/\omega_H)^2} \tag{11.1}$$

We may draw the asymptotic plot of $|k_v'|$ against frequency and obtain the curve shown by the broken line in Fig. 11.2, having a break frequency at f_H and falling

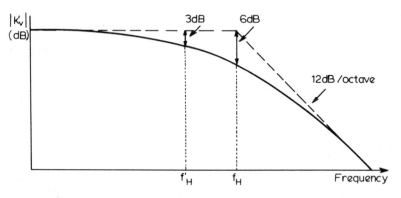

Fig. 11.2 Magnitude response of two amplifier cascade

at high frequencies with a slope of 12 dB/octave. We have noted in Section 9.5.2 that there is an error of 3 dB between the asymptotic amplitude plot and the accurate plot for a transfer function of the form $1/(1+j\omega/\omega_H)$ and hence for the transfer function of equation 11.1 we should expect the true curve to be 6 dB below that given by the asymptotes; this is shown on Fig. 11.2 by the solid line.

Again we have defined the bandwidth of an amplifier as the frequency at which the voltage gain falls to $1/\sqrt{2}$ of its mid-frequency value (namely a voltage drop of 3 dB). For a single stage of this amplifier the bandwidth is f_H. However, for the two-stage amplifier which we are studying, the 3 dB point at frequency f_H' occurs considerably lower than f_H. We may calculate it by determining the value of ω such that

$$\left| \left(1+\frac{j\omega}{\omega_H}\right)^2 \right| = \sqrt{2}$$

This occurs when $\omega = 0\cdot 643\omega_H$; so we observe that for an amplifier constructed of two identical stages the bandwidth is reduced to about 60% of the bandwidth of each stage but that the mid-band gain is squared.

The general expression for bandwidth, B, of a multistage amplifier of n identical stages is

$$B = (2^{1/n} - 1)^{1/2} f_H$$

Thus, in designing a high gain amplifier consisting of a cascade of similar single stages, each stage needs to be designed with a bandwidth considerably greater than that demanded by the specification of the overall amplifier. For example, a five stage amplifier with an overall bandwidth of 1 MHz requires each stage to have a bandwidth of 2·6 MHz.

Although the above example is related to a cascade of identical stages, a similar reduction in bandwidth occurs when the cut-off frequencies of any stages are fairly close. Naturally the overall bandwidth of a cascade will be controlled chiefly by the lowest bandwidth of any stage in the cascade.

It should be noted that a similar bandwidth reduction occurs at the lower end of the spectrum.

11.2 Difference amplifier

Another combination of two transistors in an amplifying circuit is the long tailed pair. This is an extremely versatile circuit and forms the basis of practically all operational amplifiers. It may be used as a single input amplifier in which the first transistor is coupled to the second via a common emitter resistance. It is more customarily used as a difference amplifier in which the output voltage is a magnified version of the difference between the signals applied at the two input terminals.

The basic form of the circuit using bipolar transistors is shown in Fig. 11.3.

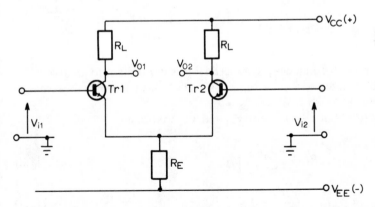

Fig. 11.3 Circuit of differential amplifier

One very great advantage of this circuit is that it tends to be very stable against variations of supply voltage or ambient temperature. If both the input voltages are zero, we note that the circuit is completely symmetrical and thus any variations in temperature which affect the two transistors identically will cause no variation in the output measured between the two collectors. Similarly supply voltage variations will have the same effect on both transistors.

It may be further shown that even if the output is taken between one collector and ground, the variation will be considerably less than that which would occur in a single transistor stage. In a single transistor stage an increase of temperature will cause an immediate increase in collector current which will directly affect the output voltage. In the long tailed pair circuit an increase in temperature causes an increase in both collector currents and this consequently increases the voltage drop across the resistor R_E; thus the base–emitter voltage is reduced tending to minimize the increase in collector current. The same effect occurs in a single stage amplifier with an emitter resistor; however, in the single transistor circuit the gain to a signal is also reduced whereas in the long tailed pair circuit there is no loss of gain to signals applied between the two inputs.

The salient features of the circuit may be shown by means of a small signal analysis based on the transistor model of Fig. 5.7a; we thus obtain the network of Fig. 11.4.

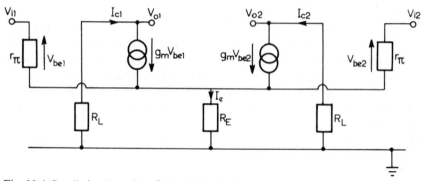

Fig. 11.4 Small signal model of Fig. 11.3

Since the circuit is completely symmetrical it is convenient to analyse it for only one input voltage and then by superposition we may determine the response for both inputs. Thus we shall set $V_{i2} = 0$.

Solving the network of Fig. 11.4 for V_{o1} and V_{o2} we obtain

$$V_{o1} = -g_m R_L \frac{1 + g_m R_E}{1 + 2g_m R_E} V_{i1} \tag{11.2}$$

$$V_{o2} = g_m R_L \frac{g_m R_E}{1 + 2g_m R_E} V_{i1} \tag{11.3}$$

Note that when R_E is very large,

$$V_{o1} = -V_{o2} = -\frac{g_m R_L}{2} V_{i1} \tag{11.4}$$

Thus the amplifier has a gain which is approximately half the gain of a single transistor amplifier; the output is in phase with the input when taken from the collector of the opposite transistor and in antiphase when taken from the same transistor.

We may obtain the output voltages for inputs applied to base 2 by interchanging the suffixes 1 and 2 in equation 11.2 and 11.3. Thus for an input V_{i2} we obtain an output

$$V_{o2} = -g_m R_L \frac{1+g_m R_E}{1+2g_m R_E} V_{i2} \tag{11.5}$$

It is customary to use the amplifier with a single output terminal (say 2) and two input terminals. There is no phase shift between input 1 and output 2 and so input 1 is termed the non-inverting input but there is a phase shift of 180° between input 2 and output 2 which is, therefore, known as the inverting input terminal.

The voltage gain to equal signals at each base is termed the in-phase gain and that to two equal but opposite signals on the bases is termed the antiphase (or differential) gain. The common mode rejection ratio (CMRR) is defined as

$$\text{CMRR} = \frac{\text{differential gain}}{\text{in-phase gain}}$$

This shows the insensitivity of the amplifier to equal signals at both inputs. It is frequently expressed in decibels.

Analytic expressions for these two gains and for the CMRR may be obtained from equations 11.3 and 11.5.

If we set $V_{i1} = V_{i2} = V_i$ and superpose the two output voltages given by equations 11.3 and 11.5, we obtain

$$V_{o2} = -\frac{g_m R_L}{1+2g_m R_E} V_i$$

and thus the in-phase gain, k_i, is

$$k_i = -\frac{g_m R_L}{1+2g_m R_E} \tag{11.6}$$

$$\simeq -\frac{R_L}{2R_E} \text{ for large values of } R_E \tag{11.6a}$$

If we set $V_{i1} = -V_{i2} = V_i$ we obtain the differential gain, k_d:

$$k_d = g_m R_L \tag{11.7}$$

The CMRR thus becomes

$$CMRR = 1 + 2g_m R_E \tag{11.8}$$

It is apparent that a high value of R_E will give a large CMRR. However, it should be noted that an increase in R_E, for a fixed supply voltage, will be accompanied by a reduction in collector current I_C and this will cause a fall in g_m. Thus to achieve a high value of CMRR it is necessary to increase R_E in value and, at the same time, to increase the supply voltage so that the collector current is unchanged. This is usually done by connecting the lower end of R_E to a negative supply voltage. There is, however, a practical limit to the magnitude of negative supply voltage possible.

Consider a circuit with a positive supply of $+15$ V and a negative supply of -15 V, designed for the transistors to operate at a current of 3 mA. The value of g_m is, therefore (from equation 5.7a), 0·12 S. Since we wish to keep the emitters of both transistors at approximately zero volts, the voltage available across R_E is 15 V and the current through R_E is 6 mA. Hence $R_E = 2·5$ kΩ and CMRR $= 600$. Any other choice of I_C will not improve on this value. In fact, we may note that if V_{EE} is the negative supply voltage then

$$R_E = \frac{V_{EE}}{2I_C}$$

Using equation 5.7a we may write equation 11.8 as

$$CMRR = 1 + e|V_{EE}|/kT$$
$$\simeq 40|V_{EE}| \tag{11.8a}$$

showing that the value of the CMRR is entirely governed by the available negative supply voltage.

For this particular design the differential gain is given by $g_m R_L$. Noting that the maximum value of R_L is given by V_{CC}/I_C we see that the maximum possible differential gain is

$$k_d = 40|V_{CC}|$$

and is again limited only by the available supply voltage, in this case the positive supply.

The inference from this is that the choice of transistor is almost immaterial in the design of a long tailed pair if voltage gain and CMRR only are considered; nevertheless, such factors as dynamic range, noise, etc., may influence the choice. Again it should be noted that the collector current of each transistor is defined by the emitter resistor as $|V_{EE}|/2R_E$ and hence the value of the load resistor cannot be greater than $2|V_{CC}|R_E/|V_{EE}|$ if the transistor is not to be bottomed.

The input resistance, r_{in1}, at input 1 may be obtained by solving the network equations for the ratio V_{i1}/I_{i1} where I_{i1} is the current flowing into input terminal 1. We note that

$$V_{be1} = r_\pi I_{i1}$$

and after some algebraic manipulation we obtain

$$r_{in1} = \frac{1 + 2g_m R_E}{1 + g_m R_E} r_\pi$$

$$\simeq 2r_\pi \quad \text{if} \quad g_m R_E \gg 1$$

This is seen to be approximately double the input resistance of a single transistor common emitter amplifier.

11.2.1 Constant emitter current circuit

In order to improve the CMRR without increasing the negative supply voltage we need to simulate a high resistance from a relatively low direct voltage source. Such a circuit is known as a constant current source, or just a current source, in that the current remains constant (or nearly constant) for very wide variations of voltage across it. Such a constant current source may be realized by the circuit of Fig. 11.5, and this may be substituted directly for the resistor R_E in Fig. 11.3.

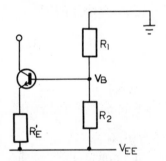

Fig. 11.5 Constant current circuit

The base potential, V_B, of the circuit is fixed at $V_{EE} R_2 / (R_1 + R_2)$ above the negative supply rail and is constant. The emitter potential V_E is $V_B - V_{BE}$ and hence the emitter current is $(V_B - V_{BE})/R'_E$. Now the only fluctuation in current that is possible is that caused by small variations in V_{BE} and if this is kept small compared with V_B we may assume that the emitter, and hence the collector, current is kept constant.

We may determine the incremental resistance of this circuit measured between collector and negative rail by considering the small signal circuit model shown in Fig. 11.6. We may simplify the analysis by assuming that resistors R_1

and R_2 are small compared with r_π for the transistor. This is usually valid since we wish to hold the base potential constant.

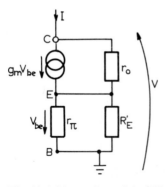

Fig. 11.6 Network model of Fig. 11.5

Assume that an incremental current of magnitude I is flowing into the collector when a voltage V exists across the circuit. We may write the two circuit equations

$$V_{be} = -\frac{R'_E r_\pi}{R'_E + r_\pi} I$$

$$I = g_m V_{be} + \frac{V + V_{be}}{r_o}$$

Whence the incremental resistance of the circuit, R_e, may be found from

$$R_e = \frac{V}{I} = r_o\left(1 + \frac{\beta_0 R'_E}{R'_E + r_\pi}\right) + \frac{R'_E r_\pi}{r_\pi + R'_E} \tag{11.9}$$

If we choose some typical values, say $\beta_0 = 100$, $r_o = 10$ kΩ, $R'_E = 5$ kΩ, and a 'tail' current through the transistor of 6 mA as before we may determine the effective value of R_e as 940 kΩ.

It should be noted that when $R'_E \gg r_\pi$ the value of R_e is closely given by $\beta_0 r_o$ and is independent of the emitter resistor. Thus in a practical design the value of R'_E is chosen to ensure the bias point is such that the maximum variation of voltage across the transistor does not cause it to bottom.

It may be seen from equation 11.8 that the CMRR of the amplifier considered previously has now been increased to over 200 000. The differential mode gain, however, has not been altered. The use of a current source in the tail of a differential amplifier is usually restricted to the first stage in a cascade of amplifiers since the greatest discrimination against noise voltages is needed at the input.

11.3 Large signal characteristics

In Section 11.2 we studied the properties of the long tailed pair amplifier for small signal variations about a mean operating point and saw that it acts as a differential amplifier of small signals. Other applications of the circuit become apparent when we consider large variations of one or both of the input voltages.

Let us consider the circuit of Fig. 11.3 with the amplitude of the small signal input voltage, V_{i1}, etc. replaced by their corresponding total time varying voltages v_{i1}, etc. We may now determine a transfer characteristic relating the output voltage v_{O1} to the input voltage v_{I1} with the second input v_{I2} kept constant at, for convenience, zero volts.

Now we have seen in Section 11.2 that the emitter resistor R_E is usually made so large that the current through it is maintained constant. We may therefore conclude that as v_{I2} varies from negative voltages to positive values, the currents through the two transistors individually will change but their sum will remain constant at I_E, the total 'tail' current through R_E.

When v_{I1} is equal to zero the currents in the two transistors will be identical and equal to $I_E/2$. Consider now v_{I1} taken progressively more negative; the common emitter voltage of the two transistors will be kept at approximately zero volts since Tr2 will be conducting but the current through Tr1 will fall since its base–emitter voltage has been reduced. When v_{I1} has fallen to about $-100\,\mathrm{mV}$, its collector current will have fallen almost to zero. At this state the common emitter voltage will be about $-400\,\mathrm{mV}$; this allows Tr2 to conduct adequately while ensuring that Tr1 is effectively cut off. Any further reduction in v_{I1} will not alter the situation. Thus for $v_{I1} < -100\,\mathrm{mV}$ approximately, $i_{C1} = 0$ and $i_{C2} \simeq I_E$.

Now consider v_{I1} increased slightly above zero volts to say $+100\,\mathrm{mV}$. The common emitter potential will rise slightly to, say $-300\,\mathrm{mV}$, and this makes the base–emitter voltage of Tr1 insufficient to allow appreciable conduction. However, the base–emitter voltage of Tr2 is $400\,\mathrm{mV}$ and thus it will conduct and its collector current i_{C2} will be equal to I_E approximately.

We may thus construct the transfer characteristic of the long tailed pair relating collector current i_{C1} with input voltage v_{I1} when v_{I2} is kept constant. This is shown in Fig. 11.7.

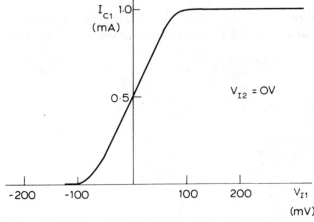

Fig. 11.7 Large signal transfer characteristic of differential amplifier

We may notice from this that a long tailed pair amplifier will act as a linear differential amplifier only over a range of about ± 50 mV differential input voltage. Outside this range the common emitter resistance causes it to act effectively as a limiter. In this connection the circuit may be used as a comparator. In the preceding discussion we considered v_{12} fixed at zero volts; however, if v_{12} is fixed at any reference voltage the form of the transfer characteristic shown in Fig. 11.7 will be retained, the only change being in the magnitude of the input voltage at which transition occurs. Thus if v_{12} is fixed at a given reference voltage V_R, the output voltage taken from the collector of Tr1 will be low if $v_{11} > V_R + 0.1$ and high if $v_{11} < V_R - 0.1$. The output thus switches between two fixed values of voltage for a very small change in input voltage.

11.4 Darlington circuit

A very useful combination of two transistors is the Darlington pair shown in Fig. 11.8. The value of the arrangement may be seen if we consider each transistor to be modelled for small signals by the simple current controlled current generator shown in Fig. 5.7(b). Thus if the input current is I_i, the collector current of Tr1 is $\beta_0 I_i$ and the emitter current is $(\beta_0 + 1)I_i$. This latter is identical with the base current of Tr2 and hence its collector current is $\beta_0(\beta_0 + 1)I_i$.

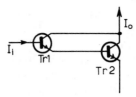

Fig. 11.8 Darlington pair circuit

The total current I_o flowing from the output terminal is $\beta_0(\beta_0 + 1)I_i + \beta_0 I_i \simeq \beta_0^2 I_i$. The whole circuit is thus identical to a single transistor whose current gain is very much greater than that for a single transistor, values of 10^4 or greater being readily achieved.

A more accurate analysis of the performance of the Darlington circuit may be made when it is used as an emitter follower as shown in Fig. 11.9. The small

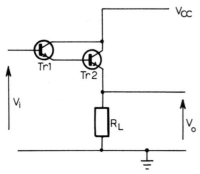

Fig. 11.9 Darlington pair used as an emitter follower

signal model of this may be based on the transistor model of Fig. 5.7(b). This is shown in Fig. 11.10. An analysis of this circuit yields the following expressions for voltage gain, k_{vo}, input resistance r_{in}, and output resistance r_{out}:

$$k_{vo} = \frac{R_L(1+\beta_0)^2}{R_L(1+\beta_0)^2 + r_\pi(2+\beta_0)} \tag{11.10}$$

$$\simeq \frac{1}{1+1/g_m R_L} \quad \text{for} \quad \beta_0 \gg 1 \tag{11.10a}$$

and

$$r_{in} = R_L(1+\beta_0)^2 + r_\pi(2+\beta_0) \tag{11.11}$$

$$\simeq \beta_0^2 R_L \quad \text{for} \quad \beta_0 \gg 1 \tag{11.11a}$$

$$r_{out} = \frac{R_1 + r_\pi(2+\beta_0)}{(1+\beta_0)^2} \tag{11.12}$$

$$\simeq r_\pi/\beta_0 \quad \text{for} \quad R_1 \ll \beta_0 r_\pi \tag{11.12a}$$

$$= 1/g_m \tag{11.12b}$$

where R_1 is the resistance of the generator.

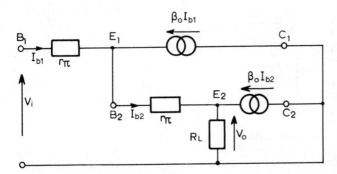

Fig. 11.10 Network model of Fig. 11.9

These expressions may be compared with those of an emitter follower using only one transistor, given by equations 8,14, 8.16a, and 8.17a. Comparison of equations 11.10a and 8.14a shows that the voltage gain of the two circuits is identical, slightly less than unity, and dependent only on the transconductance and emitter resistor. Similarly equations 11.12b and 8.17a are identical for the output resistance. The advantage of the Darlington pair is seen in a comparison of the expressions for input resistance given by equations 11.11a and 8.16a. The input resistance of the Darlington circuit is β_0 times greater than that of the simple emitter follower. It has also been stated above that the current gain is also increased by the same factor.

We may thus consider the Darlington pair as a single three terminal device operating like a single transistor but having a current gain of the order of 10^4. Such devices are now available commercially constructed on a single chip, as single units. It is possible to extend the principle to three or more transistors, but little practical advantage is gained since the circuit performance is now limited by parasitic effects.

References

GIACOLETTO, L. J., *Differential Amplifiers*, Wiley-Interscience, 1970, Chapters 1–6.
MIDDLEBROOK, R. D., *Differential Amplifiers*, Wiley, 1966, Chapters 1–4.
MALVINO, A. P., *Transistor Circuit Approximations*, McGraw-Hill, 1968, Chapter 11.
COMER, D. J., *Introduction to Semiconductor Circuit Design*, Addison Wesley, 1968, Chapter 7.

Problems

11.1 A single stage amplifier has been designed to respond to a small input step with a rise time of 25 ns and to have a voltage gain of 12 dB. Three such amplifiers are cascaded. Determine the overall voltage gain and rise time of the amplifier.

11.2 The circuit of Fig. P11.1 is used as a high impedance voltmeter. The meter has a resistance of 120 Ω and full scale deflection of 10 μA. The transistors may be assumed identical with current gains $\beta_0 = 180$; any other reasonable assumptions may be made. What is the maximum voltage which may be measured?

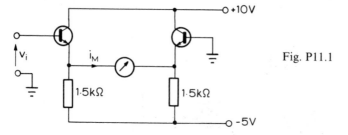

Fig. P11.1

11.3 What is the input resistance of the circuit of Fig. P11.1?

11.4 The Darlington circuit of Fig. 11.8 uses identical silicon transistors having small signal current gain $\beta_0 = 200$ and large signal current gain $\beta_F = 150$. The total quiescent collector current of the circuit is 15 mA. Determine the quiescent collector and base current of the two transistors.

11.5 Obtain the low frequency small signal circuit model for the Darlington pair circuit operating at the conditions given in Problem 11.4.

11.6 The Darlington pair circuit of Problem 11.4 is used as an emitter follower with an emitter resistance of 500 Ω. It is operated at a total quiescent collector current of 15 mA. What is the maximum output voltage swing obtainable? What is the minimum supply voltage needed to ensure that this voltage swing can be achieved in both directions?

11.7 Determine the small signal voltage gain of the Darlington pair emitter follower of Problem 11.6.

12
Feedback

12.0 Introduction

If the output variable of an amplifier or other system is deliberately returned to the input and some fraction is either added to, or subtracted from, the externally applied input, the properties of the system may be radically changed. In many cases a considerable improvement in some aspects can be achieved although this is usually accompanied by a deterioration in some other aspect of performance. This process is termed feedback. We shall study the effects of various types of feedback on amplifier behaviour and how it may be used to introduce an additional element of freedom in design.

12.1 Basic ideas of feedback

Feedback is the term applied to the process whereby a fraction of the output voltage or current of a system is returned and added to the input voltage or current. The definition is usually extended to include electromechanical and other non-electrical systems in which some sample of an output variable is fed back and summed with an input variable.

It is customary to categorize electrical systems according to the output variable sampled and the input variable which is summed. Thus we have

(a) output voltage sampling, input voltage summing;
(b) output voltage sampling, input current summing;
(c) output current sampling, input voltage summing;
(d) output current sampling, input current summing.

In some situations combinations of these simple categories may occur in one amplifier.

Where the feedback variable is added to the corresponding input variable the feedback is termed positive; where the feedback is subtracted it is termed negative. In systems in which alternating variables are fed back, positive feedback refers to situations in which input and feedback variable are in phase and negative when they are in phase opposition.

Many of the properties of feedback systems may be assessed without the need for the above categorization. However, where the properties of impedances are concerned it is necessary to consider the type of feedback configuration.

The four basic feedback configurations are shown in Figs. 12.1a to d. For many purposes these may all be generalized by Fig. 12.2, which shows the connections between the various units. (Note that the interconnecting lines no longer represent wires but indicate the path along which some variable passes from one unit to the next.) A diagram of this form is called a block diagram.

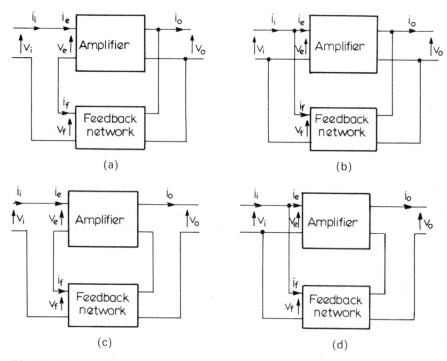

Fig. 12.1 Four methods of connection of a feedback network to an amplifier

The amplifier in Fig. 12.2 may be assumed to have a transfer ratio A; this may represent the ratio of output voltage or current to input voltage or current. The feedback network will be assumed to be passive (containing no amplifying devices) and to have a transfer ratio B; this too may be of any of the four possible ratios of input to output parameter. The feedback network is assumed to have

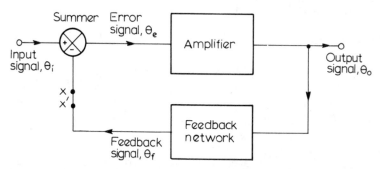

Fig. 12.2 Measurement of loop gain of feedback amplifier

zero input impedance for current sampling and infinite input impedance for voltage sampling, i.e. we assume that it does not load the amplifier. In practice the finite input impedance may be amalgamated with the load impedance of the amplifier when making numerical calculations. Often both A and B will be functions of frequency and where this is essential to the argument they will be expressed as $A(\omega)$ and $B(\omega)$.

The summer in Fig. 12.2 is shown so that feedback signal is subtracted from the input signal; the signs in the summer symbol indicate the operations to be performed. This convention assumes that we are considering negative feedback, in which the feedback variable is subtracted from the input, to be the more usual.

Thus

$$\theta_e = \theta_i - \theta_f \tag{12.1}$$

where θ is a general variable representing current or voltage as appropriate. The subscripts e, i, and f refer to the error, input, and feedback.

12.2 Effect of feedback on gain

The most immediately apparent effect of feedback on an amplifier is in a modification of its gain. This may be seen readily by an analysis of the block diagram of Fig. 12.2.

In addition to equation 12.1 we may also write the transfer relations for amplifier and feedback network as

$$\theta_o = A\theta_e \tag{12.2}$$

$$\theta_f = B\theta_o \tag{12.3}$$

It is implied in each of these equations that each of the blocks is unilateral, that is, it transmits signals in one direction only. The amplifier unit is assumed to have a forward transmission A and zero reverse transmission, an assumption closely realized in practice. The feedback network is assumed to have finite transmission from right to left and zero in the opposite direction; this is in contradiction to our assumption that the feedback network is a simple passive network; nevertheless its transmission from right to left is usually so small compared with that of the amplifier that it may be neglected.

Equations 12.1 to 12.3 may be solved to give the overall gain of the system

$$A' = \frac{\theta_o}{\theta_i} = \frac{A}{1 + AB} \tag{12.4}$$

Equation 12.4 shows that one of the principal effects of negative feedback is to reduce the gain by a factor $1 + AB$. We now consider the four types of feedback shown in Fig. 12.1 and apply the above formula in each case.

In Fig. 12.1a it is obvious that the input variable to the system is a voltage since we are summing input and feedback voltages to give the error voltage applied to the amplifier. Again the output variable is also a voltage as the feedback network is sampling a voltage. Thus the relevant transfer ratio to be considered in this system is the voltage gain, and A becomes the voltage ratio k_v. Similarly the feedback ratio B must also be a voltage ratio. In this situation it is apparent from equation 12.4 that the overall system voltage gain is reduced by the factor $1 + AB$, a dimensionless factor known as the return difference.

In Fig. 12.1b the input variable is a current and the output variable a voltage. Thus A is the ratio of output voltage to input current, namely a transimpedance. Similarly B, a ratio of current to voltage, must be a transadmittance. In this situation the system transimpedance (V_o/I_i) is reduced by the factor $1 + AB$.

In Fig. 12.1c the output is a current and the input a voltage and thus A is a transadmittance and B a transimpedance. In Fig. 12.1d the output and the input are both currents and A and B both represent current ratios.

It is therefore essential that before applying equation 12.4 to any system, we establish the nature of A and B. This is best done by studying the feedback network and identifying the type of sampling and summing which are used in the system. On this basis the input and output variables (current or voltage) of the basic amplifier may be established and hence the nature of A determined, whether voltage or current ratio, transimpedance or transadmittance. Equation 12.4 may now be applied to obtain A', which must be of the same form as A.

Although the majority of amplifying systems are used as voltage amplifiers, the units from which they are constructed may have other basic functional relationships. For example, an operational amplifier with negative feedback (which we shall study in Chapter 13) is essentially a transimpedance amplifier, despite its high input resistance. It is customarily used as a voltage amplifier by supplying the input current from a voltage source through a series resistance. In order to study the effect of feedback it is necessary to convert this voltage source and series resistor into the Norton equivalent of a current source with a shunting resistor. Equation 12.4 may now be applied to obtain the transimpedance of the feedback amplifier and this may then be converted back to a voltage gain.

The return difference $1 + AB$ is seen to occur in all types of amplifier and, as we shall see later, plays a very prominent part in the performance of feedback systems. It is always a dimensionless quantity and is independent of the manner in which it is measured. In particular we may note that if we open the system at some point, say between XX' in Fig. 12.2, and apply a signal θ at X, the output variable of the amplifier for zero input signal θ_i, will be $-A\theta$ and the signal appearing at X' will be $-AB\theta$. There is thus a loop transmission of $-AB$ around the feedback loop. This loop transmission, $-AB$, is termed the loop gain or return ratio and is independent of the point in the loop where the break is made and the signal injected and measured. However it is important that the impedances in the system are not modified when the loop is broken; thus if one wished to make a laboratory measurement of loop gain it would be essential to measure the output voltage (or current) across a load resistance which was identical to that looking forward from the break point in the system.

A single stage amplifier will have a negative loop gain (negative feedback) only over the mid-range of frequencies. At high or low frequencies, as was shown in Chapter 9, the forward gain, $-A$, will be a function of frequency. At some specified frequency, $-A(\omega)$ will be a complex number and thus the loop gain will also be complex. It is therefore no longer possible to categorize an amplifier into positive or negative feedback types on the basis of the sign of the loop gain.

An extended definition may be derived from equation 12.4. We define negative feedback as that situation in which the magnitude of the gain of the feedback system, $|A'|$, is less than the magnitude of the forward gain of the amplifier, $|A|$. This will be seen to agree with the previous definition in the special case in which A is a real ratio.

12.2.1 Sensitivity of amplifier gain with feedback

In most mass produced systems it is desirable to maintain a system performance within a small tolerance of its specification for a wide range of variations in component values; these variations may occur either from the random spread of values within a batch of components or from variation of the parameters of one component with changes in environment, such as temperature.

In an amplifier, variations of gain are very likely to occur as a result of changes of temperature or replacements in transistors. It is of interest to note how the gain sensitivity of a negative feedback amplifier is less than that of the inherent amplifier.

The gain sensitivity of a system may be defined as the fractional (or percentage) change in the gain for a given fractional change in the parameter considered. (Other definitions of sensitivity have been proposed but this is the most convenient for our use.)

The sensitivity of a parameter, say $A(x)$, to a small variation of some element of value x is therefore

$$S_x^A = \frac{1}{A}\frac{\mathrm{d}A}{\mathrm{d}x}$$

Now the sensitivity of the overall transfer ratio, A', of a feedback amplifier to some parameter, x, is

$$S_x^{A'} = \frac{1}{A'}\frac{\mathrm{d}A'}{\mathrm{d}x}$$

If we assume that the parameter x only affects the forward transfer ratio A and that the feedback ratio, B, is independent of x, then

$$S_x^{A'} = \frac{1}{A'}\frac{\mathrm{d}A'}{\mathrm{d}A}\frac{\mathrm{d}A}{\mathrm{d}x}$$

Using equation 12.4 this becomes

$$S_x^{A'} = \frac{1}{1+AB} S_x^A \qquad (12.5)$$

This shows that the variation of transfer ratio A' in a negative feedback system due to change in a parameter x will always be smaller than the corresponding change in the system without feedback. This insensitivity to parameter changes will be more pronounced the greater the value of $1+AB$, namely the greater the magnitude of negative feedback, B, applied across the system.

In the extreme case when the feedback is large

$$A' \simeq 1/B \qquad (12.6)$$

and we see that the transfer ratio of the feedback system is independent of the 'gain', A, in the forward path and thus variation of A has no effect on the system performance.

Equation 12.6 is a valuable expression in the design of a system to a specified overall gain. If the loop gain, $-AB$, has a magnitude greater than 100, equation 12.6 will only be 1% in error. Such values of feedback are quite common in conjunction with high gain amplifiers and so we shall frequently use equation 12.6 to determine the required feedback for a system.

12.3 Effect of feedback on bandwidth

The effect of negative feedback on the bandwidth of an amplifier may be seen by considering a simple amplifier with one pole whose gain may be given by

$$A(\omega) = \frac{A_0}{1+j\omega/\omega_0} \qquad (12.7a)$$

where A_0 is the mid-band gain. This amplifier has a 3 dB bandwidth of $f_0 = \omega_0/2\pi$. If we apply negative feedback of magnitude B we obtain for the transfer function, $A'(\omega)$, of the feedback amplifier

$$A'(\omega) = \frac{A(\omega)}{1+A(\omega)B}$$

$$= \frac{A_0'}{1+j\omega/\omega_0'} \qquad (12.7b)$$

where $\qquad \omega_0' = 2\pi f_0' = \omega_0(1+A_0B) \qquad (12.8)$

and $\qquad A_0' = A_0/(1+A_0B)$

showing that the bandwidth, f_0', is now $1+A_0B$ times greater than the bandwidth without feedback.

The effect of this may be seen by reference to Fig. 12.3 in which the magnitude of the gain of a simple amplifier is shown. The mid-band gain is A and it has one high frequency pole at ω_H and one low frequency pole at ω_L. If feedback of magnitude B is applied across the amplifier the mid-band gain will be reduced by the factor $1+AB$ to A', the high frequency cut-off increased by $1+AB$ to ω_H' and the low frequency cut-off reduced by $1+AB$ to ω_L' as shown. Note that the low and high frequency asymptotes to the amplifier response have not been affected. We shall see later that in amplifiers having more complex transfer functions the effects of feedback are not so simply deduced.

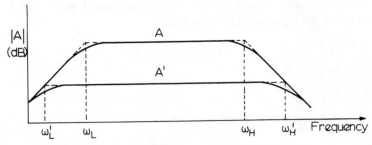

Fig. 12.3 Effect of negative feedback on gain and bandwidth

A concomitant effect is that the range of frequency over which there is negligible phase shift is also extended. This may be appreciated by recalling from Section 9.5 that the phase shift becomes $\pm 45°$ at the corner frequencies of the amplifier gain plot, and with the application of negative feedback these have been extended.

We may now look at the effect of this feedback on the gain–bandwidth product of the amplifier. In most amplifiers the low corner frequency is very small and so we may approximate the amplifier gain by equation 12.7a; the bandwidth is thus f_0. The gain–bandwidth product of the feedback amplifier is

$$A_0'f_0' = \frac{A_0}{1+A_0B}f_0(1+A_0B)$$

$$= A_0f_0$$

Thus the gain–bandwidth product of a feedback amplifier is identical with that of the amplifier without feedback. This reinforces the statement previously made that any attempt to increase the bandwidth of a system using a given amplifying device will inevitably be also associated with a reduction in the gain.

12.3.1 Effect of negative feedback on rise time

We have seen in Sections 9.2 and 9.3 that a network with a transfer function of the form

$$\frac{A_0}{1+j\omega/\omega_0}$$

will give an output, in response to an input step, having a rise time $2{\cdot}2/\omega_0$.

Thus for the feedback amplifier whose transfer function is given by equation 12.7 the rise time will be $2 \cdot 2/\omega_0(1 + A_0 B)$. This shows that negative feedback applied to an amplifier or other system will force it to respond more quickly to a sudden change of input.

12.4 Effect of feedback on impedances

In order to investigate the effect of feedback on input and output impedances it is preferable to look at one specific configuration, say that represented by Fig. 12.1a.

We may consider the system excited by a sinusoidal voltage of magnitude V_i and then write down the equations for this system as

$$V_e = V_i - V_f$$
$$V_o = A V_e$$
$$V_f = B V_o$$

If we define the input impedance of the amplifier as

$$Z_i = V_e / I_e$$

we may obtain the input impedance of the feedback amplifier

$$Z_i' = V_i / I_i$$
$$= Z_i(1 + AB) \tag{12.9a}$$

This shows that a system in which input and negative feedback voltages are summed results in an increase in input resistance. The same analysis applies to the input circuit of Fig. 12.1c.

A similar analysis of the two circuits where input and negative feedback currents are summed as in Figs. 12.1b and d shows that the input impedance of the feedback system, Z_i', is reduced by feedback according to

$$Z_i' = Z_i / (1 + AB) \tag{12.9b}$$

Let us now look at the output terminals of the circuit of Fig. 12.1a. As we are interested in the effect of feedback, we must include the output impedance, Z_o, of the basic amplifier as shown in Fig. 12.4. The output impedance may be determined by applying a voltage V at the output terminals and measuring the current $-I$ which flows into the network with the input terminals shorted. Thus

$$V = A V_e - Z_o I$$

(It is assumed that negligible current flows into the feedback network.)

$$V_e = -V_f$$
$$V_f = B V$$

Solving these gives the output impedance of the feedback amplifier

$$Z'_0 = \frac{V}{-I} = \frac{Z_0}{1+AB} \tag{12.10}$$

Thus when voltage sampling is used at the output the effect of negative feedback is to reduce the output impedance by a factor $1+AB$; this similarly applies to Fig. 12.1b.

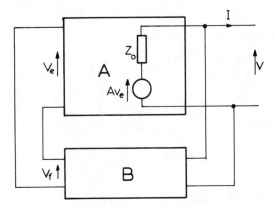

Fig. 12.4 Determination of effect of negative feedback on output impedance

When current sampling is used as in Figs. 12.1c and d, negative feedback increases the output impedance by the same factor.

It may be seen therefore that feedback may be used to control the values of both input and output impedances.

12.5 Effect of feedback on distortion

If the transfer characteristic of an amplifier is nonlinear, the output will contain components having frequencies additional to those present in the input. These additional frequency components will cause the output waveform to be a distorted form of the input waveform. For a single sinusoidal input the distortion components will consist of harmonics of the input signal frequency. For an input consisting of two or more sinusoids of different frequencies, the output will contain, not only the input frequencies and their harmonics, but also cross modulation frequencies at the sum and difference of the input frequencies and their harmonics. For faithful reproduction of a signal it is desirable to minimize the amplifier distortion.

Consider an amplifier with a voltage gain A which introduces a certain distortion. This may be represented by the block diagram of Fig. 12.5a in which

$$v_{o1} = v'_{o1} + v_d$$

$$v'_{o1} = Av_{i1}$$

The fractional distortion in the output is v_d/v_{o1}.

If we now connect feedback B around the amplifier as in Fig. 12.5b we may write

$$v_{o2} = v'_{o2} + v_d$$

$$v_f = Bv_{o2}$$

$$v_e = v_{i2} - v_f$$

$$v'_{o2} = Av_e$$

The above equations lead to the expression for the output voltage

$$v_{o2} = \frac{A}{1+AB} v_{i2} + \frac{1}{1+AB} v_d \qquad (12.11)$$

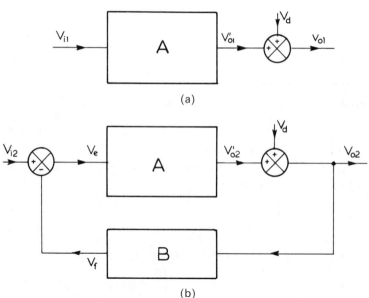

(a)

(b)

Fig. 12.5 Determination of effect of negative feedback on distortion

The first term represents the undistorted output voltage from the amplifier; the second term represents the distortion. The fractional distortion is now

$$\frac{1}{1+AB} \frac{v_d}{v_{o2}}$$

We now set $v_{o1} = v_{o2}$, so that the output voltage swing of the amplifier is the same in each case; this is necessary in order that the conditions causing the distortion are the same. We see that the fractional distortion has been reduced by a factor $1/(1+AB)$ below that of the amplifier without feedback. It should be emphasized that in order to achieve this degree of reduction of distortion it is necessary to preamplify the signal, since the gain to the desired signal has also been reduced. However, at lower signal levels undistorted amplification is much easier to achieve since the voltage does not need to traverse such large ranges of the device characteristics.

The analysis which we have used to derive this result should be considered rather critically and its limitations noted. The amplifier which we are considering here is one which introduces distortion and thus, by definition, is not a linear amplifier; the analysis used is based on the summation of the two components of output voltage and the use thereafter of superposition to obtain the result. The justification for this is that, in a system in which the distortion is small, pseudo-linear conditions may be assumed.

The distortion component will usually be a harmonic of the signal frequency and hence the forward gain to the distortion will be different from that applicable to the signal; we should therefore use a value of A appropriate to the distortion frequency. Again in the above analysis we have ignored the fact that the distortion component of output, on being fed back and once again passed through the amplifier, will be further distorted; for small distortions this may justifiably be neglected.

A more direct approach to the study of the reduction of distortion would be to consider the effect of negative feedback on the transfer characteristic relating output to input. Let us assume that the input–output relationship of the forward amplifier of non-uniform gain A is shown by curve I of Fig. 12.6.

As a result of the negative feedback,

$$v_F = Bv_O$$

where we use total values of voltage, since we may no longer assume that we are operating under small signal linear conditions. Curve II shows this relationship between v_O and v_F. In addition, we have

$$v_I = v_E + v_F$$

and for a given value of v_O we may determine the value of v_I by summing voltages; thus at a value of v_O equal to OP, $v_E = $ PM, $v_F = $ PN, and v_I is given by PQ where PQ = PM + PN. We may therefore construct the transfer characteristic relating v_O to v_I as shown by curve III.

The distortion in the output is dependent upon the deviation of the actual transfer characteristic from the ideal linear characteristic. We may observe that the slope of curve III is less than that of curve I, showing that the gain of a negative feedback amplifier is smaller than one without feedback. However, more significantly, it may be observed that curve III is more nearly linear than I and hence the distortion generated for a given range of v_O is less. If this construction were analysed algebraically we should be able to show that the distortion had been reduced by a factor $1 + AB$.

It should finally be added that distortion which is caused by operating a transistor into cut-off or well into the bottomed region cannot be reduced by any amount of feedback since these are nonlinearities which are imposed by the magnitude of the output voltage swing and cannot be eliminated unless we reduce the magnitude of this voltage swing.

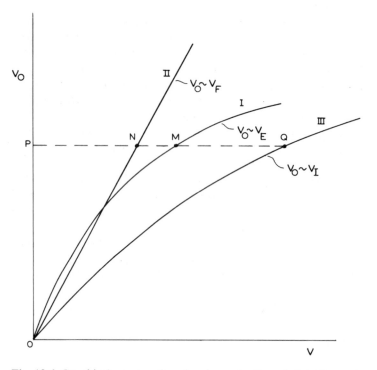

Fig. 12.6 Graphical construction showing reduction of distortion using negative feedback

12.6 Effect of feedback on noise

Since noise consists of an unwanted voltage superimposed on a desired signal and the analysis of the previous section has discussed the effect of distortion in the same terms, it might be thought that feedback would be instrumental in reducing the noise figure of an amplifier. This inference must however be treated with considerable caution.

In Chapter 10 we noted that the noise in the output of an amplifier is due principally to the noise generated in the first stage of the amplifier since this is amplified by the total gain; any noise introduced in subsequent stages will have a much smaller effect on the output since it will not be amplified by the full gain of the amplifier. In addition we note from Section 10.5.1 that the noise generated in a transistor is due principally to the emitter–base diode noise $i_{n\pi}^2$; the effect of collector shot noise and base thermal noise are usually small.

Since the emitter noise generator is at the input terminals of the transistor and the signal is also applied at these same terminals, the amplifier will treat both signal and noise in precisely the same manner, both will be amplified by precisely the same factor, irrespective of whether the amplifier has feedback around it or not. Thus we conclude that there will be no improvement in the signal to noise ratio at the output of a feedback amplifier over that of an amplifier without feedback and hence the noise figure will be unchanged. In fact the noise figure of a feedback amplifier may well be worse since we have noted that the bandwidth of an amplifier is increased by negative feedback and more noise power over a wider frequency band will be included.

The above discussion has centred on the noise induced in the input stage of an amplifier. The other but less common situation is where significant noise is induced in later stages of an amplifier. A classical example of this is the induction of mains hum into an amplifier. Although this is not a random signal we may treat it in the same way as noise. We may draw a block diagram of the system as in Fig. 12.7, where A_1 is the gain of the amplifier before the intro-duction of the noise, v_n, and A_2 is the gain following the point where noise is induced.

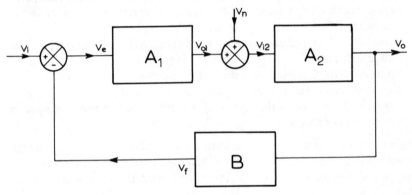

Fig. 12.7 Block model of negative feedback amplifier with internal noise

The equations governing such a system are

$$v_e = v_i - v_f$$

$$v_o = A_1 A_2 v_e + A_2 v_n$$

$$v_f = Bv_o$$

Solving these we obtain

$$v_o = \frac{A_1 A_2}{1 + A_1 A_2 B} v_i + \frac{A_2}{1 + A_1 A_2 B} v_n \qquad (12.12)$$

where the first term gives the signal voltage at the output and the second term gives the noise voltage; the signal to noise ratio at the output is thus

$$\frac{A_1 v_i}{v_n}$$

A valid comparison with this would be to consider the second amplifier of gain A_2 alone; the preamplifier is only needed in the feedback system to replace some of the signal gain lost by the use of negative feedback. The signal to noise ratio of this simple amplifier is v_i/v_n. We see, therefore that the signal to noise ratio of an amplifier in which noise is introduced in later stages may be significantly reduced but that a low noise preamplifier is needed for this to be effective. This is particularly relevant to the high power stages of an amplifier in that it is possible to reduce the hum from poorly filtered power supplies. In the earlier stages it is easier to filter the supply since the drain current is much lower.

12.7 Survey of feedback systems

In the preceding sections of this chapter we have studied some of the properties of feedback amplifiers, principally in connection with negative feedback. We shall now summarize the properties of such simple feedback systems.

(a) Forward transfer voltage (or current) gain reduced by negative feedback.
(b) Gain of feedback system is less dependent on external parameter variations within the forward gain device.
(c) Bandwidth of amplifier increased or, alternatively, rise time to a step change reduced.
(d) Input and output impedances may be varied over a wide range.
(e) Distortion generated within the amplifier is reduced.
(f) Any noise generated in the early stages is unaffected; noise generated in later stages is reduced.

We should note that if B is made negative, the feedback voltage (or current) will be added to the input voltage and we shall have a case of positive feedback. In this case the voltage gain will be increased, and the bandwidth will be reduced.

In general the forward transfer function of the system will not be independent of frequency; the frequency dependence will usually be more complex than that considered in Section 12.3. We need therefore to study the effect which a frequency dependent value of A will have on the performance of a feedback system.

12.8 Frequency dependence of gain on feedback

Section 9.4.5 shows that the voltage gain of a single stage transistor amplifier is dependent on the frequency and, in particular, at high frequencies the phase shift may well approach 360°. At mid-frequencies the gain of the amplifier is $-|k_{vo}|$ or $|k_{vo}| \angle -180°$. If we wish to use this amplifier with negative feedback to improve its properties we require that over the operating range (usually the mid-frequency band) the loop gain $-AB$ is negative. As the frequency increases we observe that an additional phase shift is introduced that may reach a value of 180° in excess of the phase shift at mid-frequencies. (In a single stage amplifier this could only occur at infinite frequency.) At such a frequency the loop gain will become positive and we shall have lost the desirable properties of the negative feedback amplifier.

A more realistic example of an amplifier of this type would be a cascade of three simple common emitter stages. If each amplifier can have a phase shift, relative to mid-frequency value, of up to $-180°$, the overall excess phase shift may approach $-540°$. Certainly at some finite frequency it will have a value of $-180°$ and at this frequency the feedback will become positive.

12.8.1 Significance of positive feedback

Let us consider a simple block diagram of a feedback system, say that of Fig. 12.2, in which the amplifier has a positive gain A and the feedback network has a transfer ratio B which we may vary between $+1$ and -1. The overall gain of the feedback amplifier is given by equation 12.4 as

$$A' = \frac{A}{1+AB} \qquad (12.4)$$

Now so long as $0 < B < 1$, the loop gain, $-AB$, is negative and the feedback is negative and $A' < A$ as discussed in Section 12.1.

Consider now that B is made negative, but is constrained to $-1 < AB < 0$. The loop gain now becomes positive and the feedback is positive. We see that the gain of the feedback amplifier is now greater than the gain without feedback. This application of positive feedback to increase the gain of an amplifier, in particular of a tuned amplifier, is occasionally used with B restricted to the above limits.

As B is made progressively more negative, the denominator in equation 12.4 becomes smaller and the gain, A', steadily increases. When the loop gain, $-AB$, ultimately attains the value of 1, the gain of the feedback amplifier has become infinite. This implies that a finite output is available from the amplifier for zero input voltage. In the absence of any network to limit the bandwidth of the network, any small noise voltage at the input will be greatly amplified, causing the current in the transistor to increase until it is fully bottomed, or alternatively is driven to cut off. The device may stay permanently in this state or possibly oscillate between the extreme values. Nevertheless it will operate well outside its linear region and the circuit is said to be unstable. The instability occurs when the loop gain $-AB$ becomes equal to $1\angle 0°$.

It is pointless to consider what happens when $AB < -1$ since we have seen that the amplifier will now be operating outside its linear region and thus in the context of linear amplifiers the investigation is of no significance.

We saw that in an amplifier which has been designed at mid-band to have negative feedback, at high frequencies an excess phase shift of $180°$ will cause the feedback to become positive. This will not cause serious problems unless at the same value of frequency, $|AB| > 1$.

A simple rule for a stable feedback amplifier may now be formulated. At any frequency at which the phase shift around the feedback loop exceeds $360°$, the magnitude of the loop gain, $|AB|$, must be less than unity for the amplifier to be absolutely stable. In practice we shall restrict ourselves further in order to maintain a margin of safety.

12.8.2 Stability of feedback systems

To illustrate the operation of this phenomenon in another situation let us consider the three-stage common emitter amplifier which is shown in its essentials by Fig. 12.8 (base biasing and emitter resistors have been omitted) with resistor R_F temporarily removed.

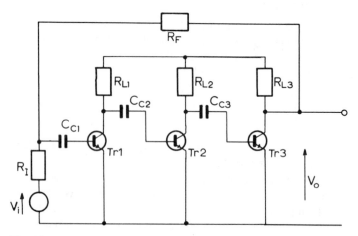

Fig. 12.8 Three-stage transistor amplifier with feedback

At high frequencies this amplifier may be modelled by the network of Fig. 12.9 in which a simplified hybrid π model (neglecting r_x and r_o) is used for each transistor represented by the blocks Tr1, Tr2, and Tr3. If we use the transistor whose parameters are given in Section 5.5 with $R_I = R_{L1} = R_{L2} = R_{L3} = 1 \text{ k}\Omega$ we may use the method of Section 9.5 to obtain graphs of the magnitude and

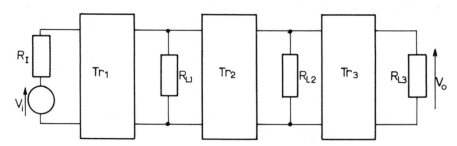

Fig. 12.9 Block diagram of three-stage amplifier

phase of the gain versus frequency shown by the solid line in Figs. 12.10 and 12.11. Now the forward gain of this amplifier, k_v, will have a phase shift of $180°$ at mid-frequencies caused by the three voltage inversions introduced by the three transistors. As the frequency is increased, the phase shift falls until at a frequency of 12·5 MHz the phase shift through the amplifier is zero. At this frequency the forward gain is 70 dB, corresponding to a voltage ratio of 3160.

Let us attempt to apply feedback around this amplifier by connecting a resistor R_F between the output and the input as shown in Fig. 12.8. As the resistor R_F is reduced, thereby increasing the feedback factor B, the magnitude of the mid-band gain is reduced; this agrees with the predictions which could be made from equation 12.4. Also we note that the upper cut-off frequency is increased according to equation 12.8. This is shown by the broken lines on Fig. 12.10 for two values of feedback resistor. For relatively large value of R_F,

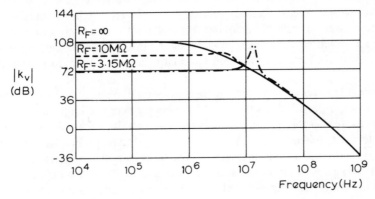

Fig. 12.10 Computed gain magnitude and phase of three-stage amplifier with varying degrees of feedback

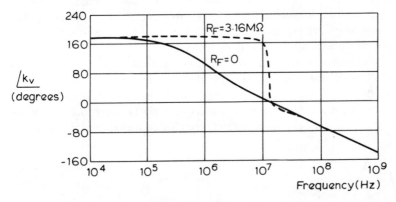

Fig. 12.11 Computed phase of three-stage feedback amplifier

corresponding to small degrees of feedback, the general shape of the gain characteristic is similar to that of the amplifier without feedback, although the gain is reduced and the bandwidth increased. However, as the value of R_F is reduced we notice a peak occurring in the gain characteristic at a frequency of about 12 MHz. Further increase of feedback causes this peak to rise, swiftly becoming infinite when the feedback resistor has a value of about 3·15 MΩ. The feedback ratio B is given by R_I/R_F and thus the amplifier remains stable until $B = 1/3160$. Now the low frequency gain of the feedback amplifier $A' = 1/B$

is then equal to 3160 (= 70 dB) and this agrees with the measured value of low frequency gain obtained above.

A slight further increase in feedback will result in the possibility of an output voltage existing in the absence of any input. Depending on the circuit this may result in a sinusoidal or pseudo-sinusoidal oscillation of the circuit or, alternatively, the circuit may be driven into a nonlinear region by the large currents which flow. In either case the circuit is unstable and of little use as an amplifier. This likelihood of instability at a high frequency thus limits the amount of feedback which may be applied across a multistage amplifier. Since we frequently wish to use operational amplifiers with an overall gain of unity it is apparent that some means of removing this potential oscillation is essential. The amplifier which we have been considering is limited by a maximum feedback factor of 1/3160 and thus the mid-band gain can never be reduced below 70 dB.

In practice it is necessary to keep the feedback down to such a value that the loop gain never approaches unity at any frequency where the loop phase shift is zero. To ensure that the system does not become unstable for small variations of parameters it is customary to introduce a safety factor known as the gain margin. This is defined as the reciprocal of the loop gain of the system at zero phase shift; or alternatively as the negative of this loop gain when expressed in decibels. Thus for an absolutely stable system the gain margin must be greater than unity or positive if measured in dB. A frequently used figure of gain margin is 14 dB, since a simple amplifier with this value of gain margin has a frequency response which is flat over all frequencies up to the cut-off frequency and a transient response to a step which shows negligible overshoot. If we apply this to the feedback amplifier of Fig. 12.8 we shall see that the maximum permitted feedback will allow a closed loop gain of 84 dB.

Study of the transient response of the amplifier will show that with zero feedback the output response to a step input is a monotonically rising voltage as shown in Fig. 12.12 curve A. As the degree of feedback is increased the rise time increases, as shown by curve B, in which a very slight overshoot of voltage may be noticed; such a curve corresponds roughly to a gain margin of about

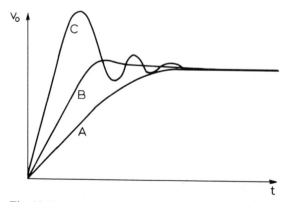

Fig. 12.12 Transient response of three-stage amplifier with varying degrees of feedback

14 dB where the frequency response is approximately flat over the maximum bandwidth. Further increase in feedback will result in the amplifier having a transient response as shown by curve C; the overshoot has increased to such an extent that several oscillations occur before the voltage settles down to its final value. It is obvious that a response such as that of curve C would, in general, be unacceptable and, although the amplifier is not unstable, its performance is unsatisfactory.

The preceding discussion has concentrated on the high frequency response of the amplifier; however, at low frequencies very similar phenomena will occur, caused by the phase shift introduced by the coupling capacitors. In any multistage amplifier incorporating three capacitors which give rise to time constants of the same order of magnitude, it is likely that at that frequency at which the phase of the open loop gain is zero, the magnitude of the gain will be greater than unity. This will result in an unstable amplifier if too much feedback is applied. In a capacitance coupled amplifier this will occur at low frequencies as a result of the phase shift in the coupling capacitors, or at high frequencies as a result of the phase shift in the transistor internal capacitances.

The gain margin has already been defined as the reciprocal of the magnitude of the loop gain at zero phase shift and for a stable amplifier it must be positive when measured in dB. Another measure of the stability of a feedback amplifier is the phase margin; this is defined as the loop phase shift at the frequency at which the magnitude of the loop gain has fallen to zero. The phase margin of the amplifier whose response curves are given in Fig. 12.10, is seen to be $-105°$, occurring at a frequency of 285 MHz.

In order to allow for variations in the gain due to component variability, it is customary to specify a finite phase margin which is always positive for stability; a phase margin of 45° will give an amplifier whose properties are similar to that corresponding to a gain margin of 14 dB. If we consider the amplifier of Fig. 12.8 we shall note from Fig. 12.10 that for a 45° phase margin, which occurs at a frequency of about 3·44 MHz, the magnitude of the open loop gain is 92 dB, signifying that it is possible to apply 92 dB of feedback across the amplifier and it will not only remain absolutely stable, but also will give a good transient response. The figure for permissible feedback based on a 14 dB gain margin is less restrictive than that obtained from using the 45° phase margin criterion; this is not surprising as there is no analytical relationship between these two conditions in the general case. It should be noted that the advantage of phase margin over gain margin occurs in the design of two stage amplifiers which never display more than 180° phase shift; hence the concept of gain margin is inapplicable. Nevertheless excessive feedback, although not causing instability, might result in unacceptable performance; for example the gain characteristic might show a peak towards high frequencies.

It should be stressed that these figures of 14 dB for gain margin and 45° phase margin are in no way sacrosanct and may be adjusted according to the system which is being studied.

We may look at the stability of the feedback amplifier of Fig. 12.8 analytically by making a few simplifying approximations. Let us assume that the amplifier is composed of a cascade of three identical stages each having a voltage gain

$$k_v = \frac{k_{v0}}{1+j\omega/\omega_0}$$

Then the overall open loop gain of the three stage amplifier, assuming that the loading on each stage is the same, is

$$k_{v3} = \frac{k_{v0}^3}{(1+j\omega/\omega_0)^3} \tag{12.13}$$

$$= \frac{k_{v0}^3}{1+3j\omega/\omega_0 - 3(\omega/\omega_0)^2 - j(\omega/\omega_0)^3}$$

Now the phase shift of this amplifier is zero when the imaginary terms are zero, i.e. when

$$3\omega/\omega_0 - (\omega/\omega_0)^3 = 0$$

which is satisfied at $\omega = \sqrt{3}\omega_0$. At this frequency, the magnitude of the gain becomes

$$|k_{v3}| = |k_{v0}|^3/8$$

From this we can see that the maximum feedback before instability occurs is $B = 8/|k_{v0}|^3$, or if we wish to keep a gain margin of 14 dB ($\equiv$ 5 times) $B \not> 8/5|k_{v0}|^3$. A similar analysis is possible if the three stages differ in their high frequency characteristics; this would be a more realistic comparison with the circuit of Fig. 12.11 since the first and second stages are loaded by the input resistance of the following stages.

12.9 Summary of stability consideration

We have developed a stability criterion which states that a system is strictly stable if at the unity gain crossover frequency, ω_0, (that frequency at which the magnitude of the loop gain is 1) the loop phase shift is positive, alternatively, that it is greater than $-360°$. It is assumed that the loop gain falls monotonically from its mid-frequency value thereafter; in some rather rare situations the gain may rise above unity at some higher frequency but we shall not be concerned with such systems. Normally we shall design an amplifier to have a finite phase margin of, say, 45°, or alternatively a gain margin of about 14 dB. The study on an amplifier which we made in Section 12.8.2 used an accurate plot of magnitude and phase of gain against frequency which was obtained graphically from a computer whose input was a suitable model of the network. Such a design aid is not always available and we may normally require to use approximate methods.

We have seen in Section 9.5.1 that it is possible to draw an asymptotic graph of the magnitude and phase of a given network transfer function directly. A cascade of such single stage amplifiers will give rise to a Bode plot of magnitude and phase which will be the sum of those appropriate to each amplifier assuming that one stage does not load a previous stage.

It is also interesting to note a general relationship between the magnitude and phase angle of a network. Consider the network of Fig. 9.1 in which the magnitude falls at very high frequencies at a rate of -6 dB/octave; for the same network we observe that the phase shift tends asymptotically to $-90°$. We may therefore correlate a phase shift of $-90°$ with a gain slope of -6 dB/octave. Similarly a phase shift of $-180°$ may be associated with a gain slope of -12 dB/octave.

From the above we may therefore reformulate the condition for stability of an amplifier with feedback. The amplifier will be unstable if the loop gain is greater than unity (0 dB) at that frequency at which the slope of the magnitude of the loop gain is 12 dB/octave or greater. Such a criterion leaves us with no margin of safety at all. A lower slope than this would allow for the safety factor introduced by having a positive gain or phase margin. Again Bode plots are usually drawn using the straight line asymptotes and the slope of these coincides with the true magnitude plot at only a few points. We may thus make a further refinement to our stability criterion, phrasing it this time not as a condition of instability but as a criterion determining stability of a system. A system is stable if the loop gain is not greater than unity (0 dB) at any frequency at which the slope of the asymptotic magnitude of the loop gain is greater than 6 dB/octave.

Let us consider an amplifier of forward gain A over which we wish to apply feedback of magnitude B in order to design a feedback amplifier whose overall gain is A'.

Now the loop gain of the amplifier is $-AB$ and so the magnitude of the loop gain in decibels is

$$20 \log |AB| = 20 \log |A| - 20 \log (1/|B|)$$
$$= 20 \log |A| - 20 \log |A'| \qquad (12.14)$$

if we assume that $|AB| \gg 1$ and we use equation 12.6. Now A and A' are both functions of frequency and at frequency ω_0, the unity gain cross-over frequency, $|A(\omega_0)||B| = 1$. Thus at ω_0, from equation 12.14,

$$20 \log |A(\omega_0)| = 20 \log |A'(\omega_0)|$$

or $\qquad |A(\omega_0)| = |A'(\omega_0)| \qquad (12.15)$

Thus for stability the magnitude of the forward gain must equal that of the feedback amplifier gain at some frequency where the forward gain is not falling faster than 6 dB/octave. (This is only valid if, as we have assumed, B is independent of frequency.) This is shown graphically in Fig. 12.13. Let us assume that we have a basic amplifier in which

$$A(\omega) = \frac{A_0}{(1+j\omega/\omega_1)(1+j\omega/\omega_2)(1+j\omega/\omega_3)}$$

showing that it has three break frequencies in its asymptotic gain plot at ω_1, ω_2, ω_3 where the slope of the gain changes successively from 0 by 6 dB/octave to 18 dB/octave. This is shown by curve a in Fig. 12.13. Let us assume that we

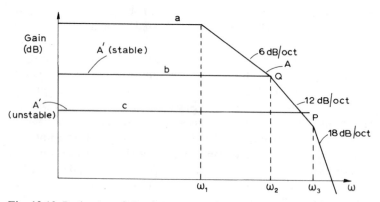

Fig. 12.13 Bode plot of three-stage amplifier showing limit of resistive feedback

wish to obtain an amplifier whose gain at low frequencies is A', given by the line c, over as wide a band of frequencies as possible. We note that the point P where $A = A'$ lies on a part of the A graph which has a slope of 12 dB/octave. Thus an amplifier designed with sufficient feedback to give this value of gain will not be stable. We may see that the lowest gain which we may obtain from this simple amplifier with resistive feedback is given by curve b, where the A and A' curves intersect at Q which is just on the 6 dB/octave asymptote. It is not possible to apply resistive feedback of magnitude greater than this unless we can modify the frequency response of the amplifier in some way.

12.10 Amplifier compensation—phase lag

The simplest way in which we can modify the characteristic of an amplifier so that it will remain stable with greater feedback applied is to reduce the bandwidth of the open loop amplifier of gain A. Suppose that we consider the shunting capacitance which is responsible for the lowest of the high frequency break points, ω_1, and connect a capacitor external to the circuit in parallel with it to reduce the cut-off frequency to ω_1'; the break frequencies ω_2 and ω_3 are unchanged. The open loop gain of this compensated amplifier is given by the solid line a' in Fig. 12.14. We may note that the amount of feedback which may be applied is now greater since it is possible to design a feedback amplifier with gain down to the limit given by curve b' in Fig. 12.14. This is considerably lower than that given by b in Fig. 12.13.

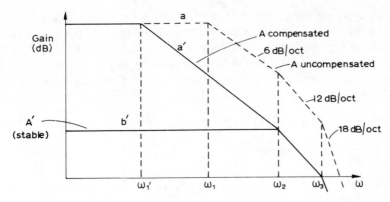

Fig. 12.14 Bode plot of compensated amplifier showing improvement in feedback obtainable

We may apply this to the amplifier of Fig. 12.8 by connecting a capacitor across the load resistor of the second stage. Figure 12.10 shows that the uncompensated amplifier has zero phase shift at a frequency of 12·5 MHz and the magnitude of the gain is then 70 dB. We have seen that we can apply a maximum of 70 dB of feedback before the amplifier becomes unstable. If we connect a capacitor of 8 μF across the second stage the variation of gain with frequency will be as shown in Fig. 12.15. At 12·5 MHz the gain is now about

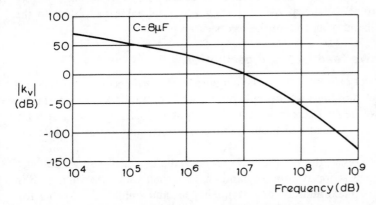

Fig. 12.15 Computed magnitude and phase plot of three-stage amplifier with lag compensation

−8 dB showing that it is now possible to reduce the low frequency gain to unity with a gain margin of 8 dB. If therefore we make $R_F = 1$ kΩ to give unity gain at low frequencies we obtain the closed loop frequency response shown in Fig. 12.16. The low frequency gain is, as desired, 0 dB and the frequency response is flat up to about 3 MHz; since the gain margin is only 8 dB we are not surprised to see the peak in the high-frequency response at about 7 MHz. The overall bandwidth of this compensated feedback amplifier is seen to be about 10 MHz.

Thus in order to achieve stability in a feedback amplifier with large amounts of feedback some form of compensation is necessary and this inevitably reduces the bandwidth of the amplifier on open loop. It also, as may be seen from curve b′ in Fig. 12.14, restricts the bandwidth of the closed loop amplifier to be smaller than that which might be available at the chosen value of A' with an uncompensated amplifier.

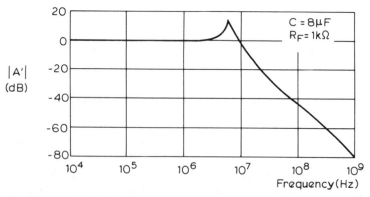

Fig. 12.16 Magnitude of lag compensated amplifier with feedback to give unity gain

Summarizing; in order to design a stable amplifier using resistive negative feedback we must ensure that the magnitude of the gain $|A|$ is not falling more quickly than 6 dB/octave at the frequency where its value has fallen to the specified mid-band gain, $|A'|$, of the feedback amplifier.

12.11 Positive feedback and oscillation

We noted in Section 12.8.1 that if the magnitude of $-AB$ becomes $+1$ the gain of the feedback amplifier will become infinite and an output will be obtainable for no input. If the feedback is resistive the output will usually increase steadily until it reaches a value at which the amplifier saturates and no further increase is possible.

Consider, however, the situation where the feedback consists of a frequency dependent network, for example, a band pass filter network. The value of B in this case will be maximum at some frequency ω_0 and smaller at all other frequencies. Thus, as the amplifier gain is increased the loop gain, $-AB$, will reach the value $+1$ at the frequency ω_0 while its magnitude is less than unity at all other frequencies. Thus we may see that instability can now occur only at the frequency ω_0. We thus obtain an output voltage from the system at frequency ω_0; it has now become an oscillator.

Thus for a sinusoidal oscillator of this type we require a frequency selective network and an amplifier. The significance of these two elements is realized when we consider in detail the conditions for a feedback network to be an oscillator. Firstly the system must be such that at the chosen frequency and, ideally, only at that frequency, the loop gain has zero phase shift (or AB has a phase shift of 180°). Secondly, at that frequency the magnitude of the loop gain,

$|AB|$, must be equal to or greater than unity. These are known as the Barkhausen conditions for oscillation.

Strictly speaking it is only necessary for the loop gain to be equal to unity for oscillation to be maintained. However, in order that oscillation may build up it is essential that the loop gain is slightly greater than unity; when the desired amplitude of oscillation has been reached the loop gain must be reduced to exactly unity to maintain the output at a constant level. We may see therefore that we require a system in which the amplifier gain is variable, being large for small amplitudes of output voltage and falling as the voltage becomes larger. It is therefore an unfortunate requirement of all practical oscillators that they must contain some nonlinear element; this frequently is provided by the non-linearity in the transistor or other device which provides the amplification. One result, of course, of the presence of such nonlinearity is that the output voltage cannot be a perfect sinusoid; good design will keep the nonlinearity to the minimum needed to stabilize the output amplitude.

The two requirements of an oscillator, namely gain and frequency selectivity, may be achieved in several ways. The gain is usually achieved by a transistor or valve amplifier, or an operational amplifier. The frequency selectivity may be obtained by using either networks containing capacitors and inductors, or networks containing resistors and capacitors. We shall consider RC networks first as these will permit the construction of oscillators in integrated circuit form.

12.11.1 Wien bridge oscillator

A circuit of one form of Wien bridge oscillator is shown in Fig. 12.17. The feedback network is shown in the box consisting of R_1, R_2, C_1, C_2. We may now analyse this network to determine whether there is one frequency at which the phase shift between V_b and V_a is either zero or 180°. If the feedback network has zero phase shift at any frequency we shall require a two stage (non-inverting) amplifier to complete the oscillator; if the feedback gives 180° phase shift, a one or three stage (inverting) amplifier will be needed.

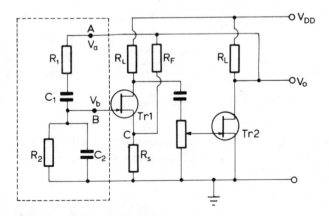

Fig. 12.17 Wien bridge oscillator

If we consider the feedback network as a potentiometer across the output we obtain

$$\frac{V_b}{V_a} = \frac{Z_2}{Z_1 + Z_2}$$

where $Z_1 = (1 + j\omega C_1 R_1)/j\omega C_1$

and $Z_2 = R_2/(1 + j\omega C_2 R_2)$.

For simplicity of analysis we will assume that

$$R_1 = R_2 \quad \text{and} \quad C_1 = C_2 \quad \text{and} \quad \omega_0 = 1/RC$$

Then

$$\frac{V_b(\omega)}{V_a(\omega)} = \frac{j\omega/\omega_0}{1 + 3j\omega/\omega_0 - (\omega/\omega_0)^2} \tag{12.16}$$

The phase response curve of this voltage ratio is sketched in Fig. 12.18. It will be noted that the phase shift becomes zero at a frequency equal to ω_0 and hence we require a two stage amplifier to provide the necessary gain. In the circuit of Fig. 12.17 we have chosen to use field-effect transistors as they have very high input impedances which will not load the feedback network.

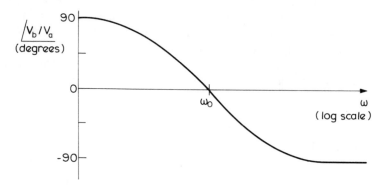

Fig. 12.18 Phase shift through feedback network of Wien bridge oscillator

We now need to determine the attenuation introduced by the feedback network at this frequency and this may be obtained directly by substituting $\omega = \omega_0$ in equation 12.16. This gives

$$\left|\frac{V_b(\omega_0)}{V_a(\omega_0)}\right| = \frac{1}{3}$$

and hence the transistor amplifier must give a gain of at least three to ensure that the magnitude of the loop gain is at least unity at ω_0.

In order to maintain oscillation at a fixed amplitude, a negative feedback gain control is frequently used in addition. This consists of the two resistors R_F and R_S shown in Fig. 12.17. As the amplitude increases the voltage fed back to the source of Tr1 increases and the gate–source voltage at that transistor is thereby less, reducing the amplitude of oscillation. The desired nonlinearity to allow oscillations to build up to a stable value is often introduced at R_S. One common method is to make R_S a temperature-dependent resistor; if the amplitude decreases below the equilibrium value, the average current flowing through R_S will fall, its temperature will drop, and thus its resistance will fall reducing the amount of negative feedback and permitting the amplitude to rise again to its equilibrium value.

An alternative form of Wien Bridge oscillator would incorporate an operational amplifier in place of the two-stage fet amplifier. Other RC networks may be used to give frequency selective properties; some of these only require a single stage amplifier to provide the necessary phase shift.

12.11.2 *LC oscillators*

As noted in Section 12.12, frequency selective networks using inductors and capacitors are frequently used in oscillators. Such circuits may not be constructed in integrated circuit form but are very valuable for use at high frequencies. One example of such an oscillator is shown in Fig. 12.19 where the emitter biasing resistor and capacitor have been omitted.

The ratio of voltages at B and A may be directly written down noting that the transformer is wound so that the secondary voltage is of opposite polarity to the primary

$$\frac{V_b}{V_a} = -\frac{j\omega M}{R+j\omega L}$$

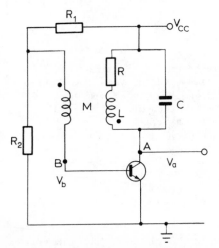

Fig. 12.19 Tuned collector oscillator

If we neglect the value of the small resistance R we note that the feedback will always introduce a phase shift of 180° and hence a single stage amplifier will be sufficient to form an oscillator. With this simplifying assumption we might infer correctly that these phase relationships would hold good at all frequencies and thus we have lost the frequency selectivity of the feedback network. However a more exact analysis incorporating the resistor R will show that the feedback phase shift can never become 180° but that this is corrected for by additional phase shift in the forward gain of the amplifier due to the frequency dependence of the load impedance. Thus the total loop gain is dependent on frequency and the Barkhausen criterion can only be satisfied at one specific frequency. Despite the fact that a tuned circuit is being used in this instance there is no direct connection between the resonant frequency and the frequency of oscillation although they are usually of the same order.

12.11.3 Frequency stability of oscillators

The frequency of an oscillator is, we have seen, solely determined by the frequency at which the loop gain has zero phase shift. Any variations of the components determining this frequency caused perhaps by temperature variation, will cause frequency instability. If we can stabilize the frequency selective network such that its phase shift does not vary from zero by more than say $\pm\theta$ at the oscillation frequency, the most stable oscillator will be one in which the minimum change of frequency is needed to restore the phase shift to zero. Thus the most stable oscillator is one in which $d\theta/d\omega$ at ω_0 is greatest (where θ is the phase angle).

In LC circuits having high Q, i.e. narrow bandwidth, the rate of change of phase shift with frequency is large, increasing as Q is increased. It is, in general, more difficult to construct circuits using RC elements only which have a comparably high rate of change of phase shift and the simple RC oscillators we have discussed have very low values of $d\theta/d\omega$. Thus LC oscillators usually have better frequency stability; this may be further improved by using high Q factor circuits, for example quartz crystals.

References

THORNTON, R. D., *et al.*, *Multistage Transistor Circuits*, Wiley, 1965, Chapter 3.
COMER, D. J., *Introduction to Semi-conductor Circuit Design*, Addison Wesley, 1968, Chapter 8.
ALLEY, C. L. and ATTWOOD, K. W., *Electronic Engineering*, Wiley, 1973, Chapter 14.
MILLMAN, J. and HALKIAS, C. C., *Integrated Electronics*, McGraw-Hill, 1972, Chapter 13.
SMITH, R. J., *Circuits, Devices and Systems*, Wiley, 1966, Chapter 19.
THORNTON, D., *et al.*, *Handbook of Basic Transistor Circuits and Measurements*, Wiley, 1966, Chapter 4.

Problems

12.1 An amplifier has a voltage gain of $-2 \cdot 5 \times 10^4$. What fraction of the output voltage must be fed back to the input to reduce the gain to 100?

12.2 The gain of the amplifier of Problem 12.1 is liable to vary by $\pm 10\%$. What is the variation in gain of the feedback amplifier?

12.3 An amplifier of gain $-2 \cdot 5 \times 10^4$, input resistance of $1 \cdot 2$ MΩ and output resistance of 400 Ω, has negative feedback applied to reduce the output resistance to 25 Ω. How is the feedback connected at the output? The input resistance is to be increased; how would you connect the feedback at the input? What is the magnitude of the input resistance and the overall voltage gain?

12.4 An amplifier of gain -10^4 gives a response to a 1 μs pulse which has a rise time of 40 ns and a droop of 8%. The rise time is to be reduced to 2 ns by application of negative feedback. Determine the gain of the feedback amplifier and the percentage droop. What is the feedback ratio?

12.5 The amplifier A of Fig. P12.1 has voltage gain of -2×10^3 and infinite input resistance. Determine the overall voltage gain of the feedback amplifier shown. If the input resistance is 1 kΩ, what is the voltage gain of the feedback amplifier?

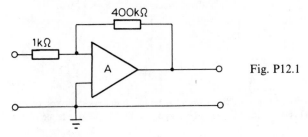

Fig. P12.1

12.6 The amplifier A in Fig. P12.2 has a voltage gain of -2×10^3 and an input resistance of $1 \cdot 5$ kΩ; determine the type of sampling and summation used in this feedback circuit. What amplifying property of the system is modified by this feedback? Identify the magnitude and dimensions of A and B.

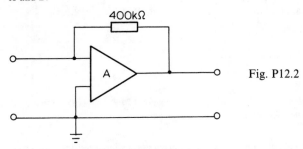

Fig. P12.2

12.7 Determine the overall transresistance (ratio of output voltage to input current) of the amplifier of Fig. P12.2. Find its input resistance.

12.8 The amplifier of Problem 12.7 is modified as shown in Fig. P12.3. Determine the voltage gain and input resistance of this amplifier.

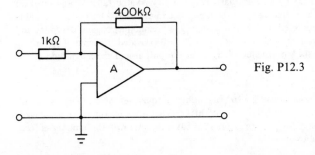

Fig. P12.3

12.9 The amplifier of Fig. P12.4 is operated with a collector current of 5 mA and uses a transistor with $\beta_0 = 200$. Determine the mid-frequency voltage gain and the input resistance.

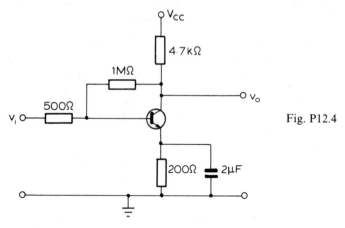

Fig. P12.4

12.10 The amplifier, A, of Fig. P12.5 has a voltage gain of -2×10^3, input resistance of 1·5 kΩ, and output resistance of 2·7 kΩ. Determine the overall voltage gain and input resistance of the feedback amplifier.

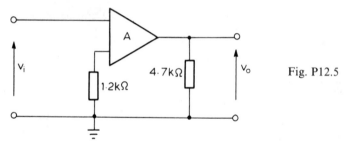

Fig. P12.5

12.11 The amplifier, A, of Problem 12.8 has an output resistance of 2·7 kΩ and works into a load resistor of 10 kΩ. What will be the effect on the overall voltage gain?

12.12 A transistor amplifier, of voltage gain 10^4, when connected to a source resistance of 1 kΩ, gives a noise voltage at the output of 8 mV. Assume that the noise is generated primarily in the base–emitter junction of the first stage. A second amplifier of voltage gain 200 is cascaded after this and feedback provided over both amplifiers to restore the gain to its original value. What is the feedback ratio and what is the output noise voltage?

12.13 A transistor amplifier is fed by a sinusoidal voltage of r.m.s. value 120 mV, and gives an output voltage of 10 V at the same frequency as the input and 15 mV at 100 Hz (twice the mains supply frequency) caused by poor filtering of the power supply. A preamplifier of gain 50 is connected to the main amplifier and overall negative feedback of $-38·5$ dB provided. A signal of 120 mV is applied at the input. What is the r.m.s. value of signal at the output and the magnitude of the 100 Hz component?

12.14 An amplifier has a mid-band gain of 2×10^4, and at high frequencies the gain falls as a result of three poles at $f_1 = 10$ MHz, $f_2 = 15$ MHz, $f_3 = 45$ MHz. What is the maximum resistive feedback which may be applied if stability is to be assured? What is the gain with this amount of feedback and the bandwidth of the amplifier?

12.15 The amplifier of Problem 12.14 is to be lag compensated. Estimate to what frequency the lowest pole must be shifted to ensure stability when feedback is applied to reduce the gain to unity. What is the feedback ratio? What are the closed and open loop bandwidths?

12.16 The amplifier whose gain/frequency characteristics are given in Fig. 9.22 is used in each stage of a three stage amplifier. Assume that the input of one stage does not load the output of the preceding stage. If negative feedback is applied, at what frequencies will the loop phase shift be $0°$? What is the forward gain at these frequencies? What is the minimum gain which can be obtained by resistive feedback if a gain margin of 14 dB is acceptable? What is now the phase margin?

12.17 The transistor used in the circuit of Fig. P12.6 is operated at $I_C = 500\ \mu A$, $V_{CE} = 5\ V$, $V_{CC} = 10\ V$. It has a value of $\beta_0 = 100$, $C_\mu = 2.5\ pF$, $f_T = 120\ MHz$ (r_x may be neglected). Determine the mid-frequency gain, input, and output resistances; also determine the high cut-off frequencies.

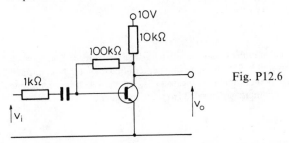

Fig. P12.6

12.18 The circuit of Fig. P12.7 is designed as a voltage amplifier. The transformer is connected to give negative feedback. The transistors have a current gain, β_0, of 100 and a transconductance, g_m, of 100 mS. The turns ratio, n, of the transformer is 10. Determine the overall gain at a frequency of 150 Hz. The reactance of the capacitor C is negligibly small.

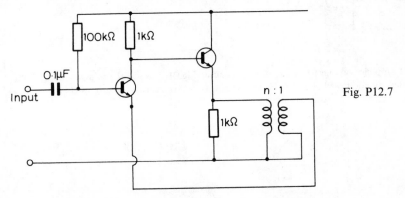

Fig. P12.7

12.19 Determine the voltage transfer ratio of the networks of Fig. P12.8. Sketch these functions in magnitude and phase as a function of frequency when $C = 0.1\ \mu F$ and $R = 4\ k\Omega$.

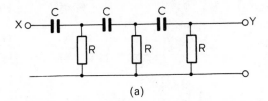

(a)

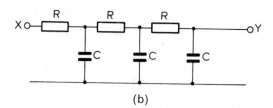

(b)

Fig. P12.8

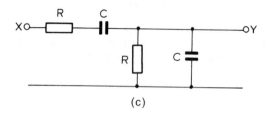

(c)

12.20 The circuit of Fig. P12.9 uses an amplifier of gain A with a feedback network B. The network B is one of the three networks of Problem 12.19. Determine whether terminal Y should be connected to the inverting or non-inverting input of the amplifier to create an oscillator. In each case determine the frequency of oscillation and the minimum gain needed from the amplifier.

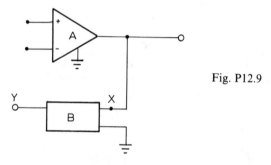

Fig. P12.9

12.21 The amplifier of Fig. P12.9 uses the network of Fig. P12.8(c) in the feedback path. If the amplifier has an additional phase lag of 45°, what gain is now required to give oscillation and at what frequency will the circuit oscillate?

13
Linear Integrated Circuits

13.0 Introduction

The modern techniques for the fabrication of integrated circuits (ICs) permit the construction of complete multistage amplifiers and other linear devices on a single silicon chip. The most common type of IC is the operational amplifier which may be used with different forms of feedback and external circuitry to construct summers, integrators, filters, etc. In addition, audio amplifiers, modulators, frequency dividers may all be constructed in integrated form.

The design techniques used are basically identical to those for the construction of circuits using discrete components. However, some modification in techniques is necessary in some cases because the range of magnitude of the value of some components is limited, and in other cases because new possibilities are presented as a result of the form of construction. We shall look firstly at some of the basic circuits which differ from their counterparts using discrete components, and follow this with a study of the use of these circuits in an operational amplifier.

13.1 Biasing circuits

The problem of biasing a transistor amplifier has been discussed in detail in Section 8.6.1. The essential philosophy outlined there was that the collector current should be maintained constant for all variations of temperature or choice of transistor of a given type. In the case of an IC, we are only concerned with maintaining stability for variations in temperature since no interchange of individual transistors is possible. In the design of discrete circuits this was achieved by the use of negative feedback operative at direct voltages to stabilize the operating point but rendered ineffective at the desired signal frequencies by the use of coupling capacitors in order not to degenerate the a.c. gain inordinately. For a low frequency amplifier this necessitated the use of capacitors of very high value (100 to 1000 μF) which cannot be fabricated by IC techniques.

One alternative method of biasing a transistor amplifier avoiding large capacitors is shown in Fig. 13.1. The amplifying transistor is Tr1 with its load resistor R_1. Transistor Tr2, assumed identical to Tr1 and with similar temperature variations, is used to bias Tr1. If $R_3 = R_4$, then the two transistors have equal

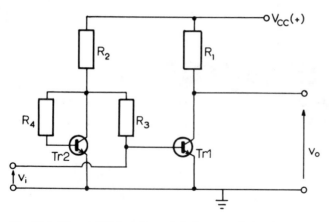

Fig. 13.1 Bias stabilizing network in integrated circuit

base currents and their collector currents will also be identical. The collector currents of Tr2 is given by

$$I_{C2} = \frac{V_{CC} - V_{BE}}{R_2} - (2 + R_4/R_2)I_B$$

Since these are silicon transistors we may assume that I_{CBO} is negligible and hence $I_{C2} = \beta_F I_B$. Thus

$$I_{C2} = \frac{V_{CC} - V_{BE}}{R_2 + (2R_2 + R_4)/\beta_F} \qquad (13.1)$$

For small changes in V_{BE} due to temperature variation we see that

$$\frac{\Delta I_{C2}}{I_{C2}} \simeq -\frac{\Delta V_{BE}}{V_{CC}} \qquad (13.2)$$

For a temperature variation of about 50°C, ΔV_{BE} will not be greater than 0·1 V and hence I_{C2} will be stabilized to about 1% when using a 10 V supply.

The effect of variation of β_F may also be seen from equation 13.1 as

$$\frac{\Delta I_{C2}}{I_{C2}} \simeq \frac{2 + R_4/R_2}{\beta_F} \frac{\Delta \beta_F}{\beta_F} \qquad (13.3)$$

Any change in β_F will therefore affect the collector current by an order of magnitude lower than that of its own variation, assuming reasonable values for R_2, R_4, and β_F. The input may be applied directly to the base of Tr1 and hence there will be no degeneration of signal. However, any generator connected to the input, unless blocked by a capacitor, is likely to upset the balance between the base currents of the two transistors and thus, for a directly coupled signal source, a resistor of value equal to the source resistance should be connected between the base of Tr2 and ground.

13.2 Current source biasing

A similar circuit to that of Fig. 13.1 may be used to maintain the bias on a constant current circuit such as that discussed in Section 11.2.1. It may be seen from Fig. 11.6 that the biasing resistors R_1 and R_2 need to be fairly large in order that the current drain on the supply is not too great. In particular, when applied to the construction of integrated circuits where the collector currents of transistors are of the order of 100 µA, the resistors are usually too large for satisfactory fabrication. One solution to the problem is shown in Fig. 13.2; transistor

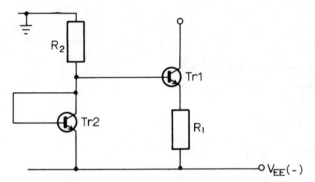

Fig. 13.2 Bias network for constant current source

Tr1 with emitter resistor R_1 forms the current source; transistor Tr2 is connected as a diode and with resistor R_2 acts as a potential divider to provide the base potential. Assuming that the two transistors are satisfactorily matched, variations of temperature will cause little variation in biasing point. However the base–emitter voltage of Tr2 must be equal to that of Tr1 together with the voltage drop across R_1.

13.3 Level shifting networks

Since blocking and coupling capacitors cannot be constructed using integrated circuit techniques, the signal needs to be coupled from one stage to a subsequent stage by resistive networks. In some cases the base of the second stage transistor may be directly connected to the collector of the first stage. This eliminates the need for capacitors but does result in a successive increase in the direct voltage level on which the signal is superimposed. To remove the direct voltage from the signal we might use a potential divider connected between the output and the negative supply rail as shown in Fig. 13.3. Assume that the output of stage 1 consists of a signal v_{o1} superimposed on a direct voltage V_{O1} and resistors R_1 and R_2 are chosen that the output v_{o2} has no superimposed direct voltage. We must choose R_1 and R_2 such that

$$\frac{R_2}{R_1} = -\frac{V_{EE}}{V_{O1}}$$

However the network will also introduce an attenuation of the signal voltage such that

$$\frac{v_{i2}}{v_{o1}} = \frac{R_2}{R_1 + R_2} = \frac{-V_{EE}}{V_{O1} - V_{EE}}$$

and this shows that to reduce the attenuation of the signal $|V_{EE}| \gg V_{O1}$. There is, however, a practical limit to the magnitude of V_{EE} and, if it were used, such a network would introduce a signal attenuation of 25 to 50%.

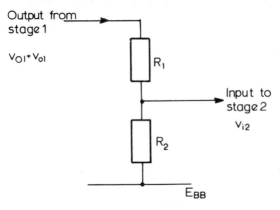

Fig. 13.3 Interstage resistance coupling

An alternative solution when using discrete transistors is to use a complementary pair of transistors, one n-p-n and one p-n-p. Until about 1970 this solution was not available to the designer of integrated circuits since p-n-p lateral transistors could only be constructed having very low values of β_0. Integrated circuits with a complementary pair of transistors in the output stage are now common.

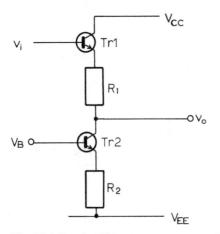

Fig. 13.4 Level shifting interstage network

The solution to this problem using only n-p-n transistors in IC construction is shown in Fig. 13.4. Transistor Tr1 operates as an emitter follower; its equivalent load resistor is provided by R_1 in series with the current source transistor Tr2 which has its base biased at a fixed voltage, V_B, probably by a diode circuit as discussed in Section 13.2. Thus the load resistor of the emitter follower is very high. Resistor R_1 is chosen so that the direct current controlled by Tr1 causes a voltage drop across R_1 to counteract the offset direct voltage which exists at the input. The attenuation to signal voltage is however very small since the signal is only attenuated by the ratio of R_1 to the incremental resistance of the current source. It should be noted that the output resistance of the circuit as it stands is relatively large; this may be substantially reduced by the addition of a further emitter follower to the output.

13.4 Operational amplifier design

A wide variety of operational amplifiers are available, many with specialized functions. In order to illustrate the principles of design of integrated circuits we shall study the 709 operational amplifier. This is a general purpose low frequency amplifier which is in very common use and is probably the simplest to describe and yet which contains a number of techniques peculiar to the design of integrated circuits. There are many more sophisticated circuits such as the 741 which incorporate very complex circuit techniques.

The 709 amplifier consists of a differential input stage followed by a differential to single sided second stage leading to a power output stage. The complete circuit diagram is shown in Fig. 13.5.

The input stage uses the two transistors Tr1 and Tr2 as a long tailed pair. The 'tail' resistor is simulated by the current source Tr11 as discussed in Section 11.2.1; the bias for Tr11 is obtained from the diode connected transistor Tr10 (see Section 13.2). This arrangement provides for a current source of very low value, giving a high input resistance; the collector currents of Tr1 and Tr2 are approximately 20 µA giving a first stage differential gain of about 20.

The second stage is based on the principle considered in Section 13.1. Transistor Tr7 acts as an emitter follower and thus its emitter is at approximately the same potential as the collector of the Darlington pair Tr3, Tr5. The emitter of Tr7 is connected by R_1 and R_2 to the bases of Tr3 and Tr4. Any increase in the current through Tr5 and Tr6 will cause a fall in the potential of the base of Tr7 and hence of its emitter; this will cause a fall in the base currents to both Tr3 and Tr4 and a consequent reduction of the assumed increase of collector currents of Tr5 and Tr6. Thus the bias point of the second stage transistors is effectively stabilized against variation of temperature. Transistor Tr7 is included so that the input stage collector currents do not flow through the load resistor R_5 of the second stage.

A second purpose of the second stage is to enable the full differential output of the first stage to be applied to the base of Tr4. The output of Tr2 is directly connected whereas the output of Tr1 is applied to this base through Tr3 and Tr5 acting as a unity gain inverter.

The advantage of using a Darlington connection in this circuit is that the very high input impedance does not present a serious shunt across the load resistor R_1 and R_2 of the first stage. In addition any variation with temperature of collector saturation current in the first transistor of a conventional Darlington pair circuit will be amplified by the second transistor and cause serious temperature instability. A bleed current is taken from the collectors of Tr3 and Tr4 through R_3 and R_4 and the diode connected to Tr15. This bleed current will increase with temperature as a diode has a positive temperature coefficient and thus any increase in leakage current through Tr3 and Tr4 will be conducted away. This further stabilizes the biasing of the second stage against temperature variations.

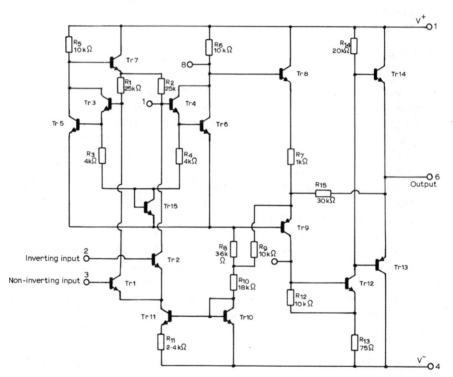

Fig. 13.5 Circuit diagram for 709 operational amplifier

The output stage consists of an emitter follower transistor Tr8 coupled to a common base connected lateral p-n-p transistor Tr9. This latter causes a shift in steady voltage level between the output of the emitter follower and the input to the driver transistor Tr13 and in addition provides a small increase in the voltage gain.

The final power stage consists of a complementary pair of transistors Tr13 and Tr14 which permits a large swing in output voltage and also gives a low output resistance. The p-n-p transistor, Tr13, is fabricated without additional diffusion processes by using the p-type substrate as the collector. Considerable negative feedback is applied through resistor R_{15}, which in conjunction with R_7 determines the gain of the output stage. This negative feedback renders the gain independent of transistor parameters, reduces the output impedance and also minimizes the cross-over distortion introduced in the small range where Tr14 is becoming cut-off and Tr13 starting to conduct and vice-versa.

13.5 Specification of operational amplifier

The typical performance of an operational amplifier is given below; the data are those appropriate to a μA 709 amplifier with $V^+ = 12\,\text{V}$, $V^- = -6\,\text{V}$

Input offset voltage	1 mV
Input offset current	50 nA
Input resistance	400 kΩ
Common mode rejection ratio	90 dB
Large signal voltage gain	45 000
Output resistance	150 Ω

Explanation of some of the parameters mentioned above may be desirable, in particular the offset voltage and current.

13.5.1 Offset voltage and current

Although an operational amplifier is ideally designed to give zero output voltage for zero input signal irrespective of the impedance connected at the input terminals, this idealized situation is not always realized in practice due to differences in the parameters of the two halves of the input stage.

If the input terminals of an operational amplifier are left on open circuit a small voltage will be observable at the output. This voltage could be reduced to zero by connecting an appropriate voltage between the input terminals of the amplifier. The small input voltage necessary to reduce the output voltage to zero is termed the input offset voltage.

Similarly if the two input terminals are shorted together, a small voltage may again be observed at the output. This again may be reduced to zero by connecting a current source in the short circuit across the input terminals. The input current needed to reduce the output voltage to zero is termed the offset current.

These two offset parameters may be represented as shown in Fig. 13.6 as an

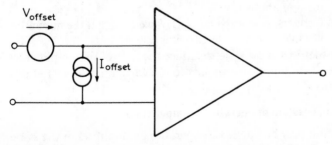

Fig. 13.6 Offset voltage and current in operational amplifiers

equivalent current and voltage generator preceding an ideal operational amplifier. If a source resistance R_1 is connected between the inputs (in practice this will probably comprise two resistors each connected between one input terminal and earth) the offset current will flow through this resistor to produce a voltage which may be added to the offset voltage; the sum of these two voltages will be an equivalent offset voltage appropriate to that particular source resistance. This is occasionally termed the offset voltage, although it must be noted that it is only applicable to a given source resistance.

The input offset voltage may be compensated by applying a small difference in voltage between the two input terminals. This is only of direct relevance in d.c. amplifiers. In a.c. amplifiers, offset voltage may cause such imbalance in later stages that the maximum available voltage swing is not realizable; compensation is thus desirable in this situation also.

13.5.2 Stability

In many applications of operational amplifiers, it is desirable to use amplifiers with very small and stable gains, frequently of the order of unity. This may be achieved by applying feedback between the output and the inverting input terminal. As we have seen in Section 12.8.2 the maximum feedback which may be applied across a multistage amplifier is determined by the possibility of instability occurring, usually at high frequencies. Since most operational amplifiers incorporate at least three stages of amplification, the application of large amounts of feedback will almost inevitably lead to oscillation at high frequencies. This is caused by the summation of the various phase lags which occur throughout the circuit.

Figure 13.7 shows the frequency response in magnitude and phase of a typical operational amplifier. We see from Fig. 13.7b that 180° phase shift occurs at a frequency of 15 MHz for a typical operational amplifier. On the magnitude plot of Fig. 13.7a the open loop gain at this frequency is about 35 dB ($\equiv 56$) and hence the maximum negative feedback which is permissible is 1/56. However it would be preferable to work with some margin of safety. Thus using a 45° phase margin, we note that the phase shift is $(-180° + 45°) = -135°$ at 5 MHz, and the gain at this frequency is 50 dB ($\equiv 316$); hence the maximum permitted feedback is 1/316. Alternatively if we use the criterion of a 14 dB gain margin the maximum feedback ratio permitted is $-35 - 14$ dB equivalent to 1/282. These two alternative criteria give very similar values for maximum permissible feedback ratio of about 1/300.

Since the gain with feedback is approximately $1/B$ we note that the minimum gain which this amplifier may give with feedback is 300.

It has been shown that it is impossible to reduce the gain of this amplifier to unity (a desired gain, A, of unity corresponds to a feedback ratio of 1) unless some form of compensation is used.

13.5.3 Frequency compensation of operational amplifiers

Phase lag compensation may be applied to operational amplifiers in the same manner as that outlined in Section 12.10. A capacitor is connected between the

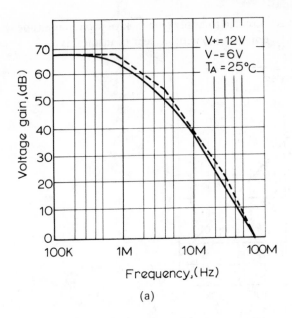

(a)

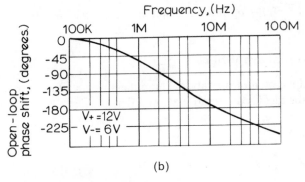

(b)

Fig. 13.7 Frequency response of 709 operational amplifier (a) magnitude (b) phase shift

lag terminal and ground. Reference to Fig. 13.5 will show that transistor Tr6 is operating as an emitter follower having an output impedance of the order of 3 kΩ. Thus a capacitor connected between this point and earth will introduce a new break frequency in the magnitude response curve; this may be placed at such a frequency that the magnitude of the gain will have fallen to zero before the second corner frequency at 0·8 MHz occurs. If we require the gain to fall at 6 dB/octave from a low frequency gain of 70 dB to 0 at 0·8 MHz we need to choose the corner frequency at 70/6 ($\simeq$ 12) octaves below 0·8 MHz namely at $0·8/2^{12}$ MHz, approximately 300 Hz. This response curve is shown in Fig. 13.8 where the new corner frequency is chosen at about 300 Hz to allow sufficient margin of stability. This is implemented using a capacitor of 0·1 μF. If the output resistance of this stage were 3 kΩ we would obtain a break frequency with this capacitor at about 500 Hz which agrees well with the measured value. It may be noted that it is now possible to apply feedback of such magnitude that the gain may be reduced to unity (0 dB) without fear of instability.

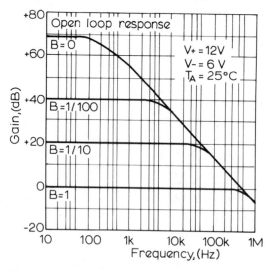

Fig. 13.8 Frequency response of 709 operational amplifier with varying degrees of feedback

13.6 Idealized operational amplifiers

Before considering the application of operational amplifiers we may first summarize the properties of the idealized operational amplifier

(a) input impedance infinite
(b) output impedance zero
(c) voltage gain infinite
(d) infinite bandwidth
(e) zero offset voltage.

The extent to which a practical amplifier approaches this ideal may be seen by reference to the specification for the µA 709 amplifier given in Section 13.5. It will be seen that these requirements are rather poorly approximated. Open loop voltage gains exceeding 100 dB are available in more sophisticated operational amplifiers and values up to 1 MΩ. The output resistance is generally not so important and frequently considerably higher values may be tolerated.

We have seen in Section 13.5 that the bandwidth of an operational amplifier without feedback is constant up to a frequency f_1, the dominant cut-off frequency of the amplifier. The bandwidth of the feedback amplifier is, however, also constrained by the degree of feedback employed.

The zero d.c. offset is not such a serious limitation in performance since it may usually be balanced out by suitable choice of biasing components either at the input or at special offset adjustment points in the amplifier.

Since operational amplifiers work with such high values of gain we may see that the range of input voltage for linear operation is very limited. Considering the μA 709 amplifier specified in Section 13.5 the output voltage swing available is ± 5 V and the gain 4500; hence to ensure that the amplifier does not saturate, the input voltage swing must be restricted to ± 0.1 mV. So long as these conditions are maintained the input terminal of the amplifier deviates negligibly from zero volts and it is considered as a virtual earth point. These conditions of operation are achieved by applying large negative feedback across the amplifier.

13.6.1 Inverting amplifier

An operational amplifier with negative feedback such as was mentioned above may be used to construct a unity gain amplifier whose output is an inverted form of the input; the amplifier has a gain of -1. The circuit is shown in Fig. 13.9. Note that the feedback is connected to the inverting input to give negative

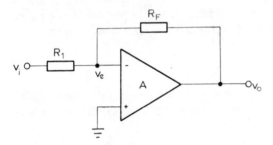

Fig. 13.9 Scaling amplifier

feedback and the noninverting input is not used and is therefore grounded. Assuming that the input resistance and output conductance of the basic amplifier are high, we may assume that no current flows into the amplifier inverting terminal and thus for small signals

$$\frac{v_o}{R_F} = -\frac{v_i}{R_1}$$

giving a voltage gain for the system of

$$\frac{v_o}{v_i} = -\frac{R_F}{R_1} \tag{13.4}$$

If $R_F = R_1$, the voltage gain is -1 as desired.

It is interesting to extend this analysis to see the effects of finite voltage gain,

A, and input resistance, r_{in}. We may now write Kirchhoff's current law at the input node of the amplifier

$$\frac{v_o}{R_F} = -\frac{v_i}{R_1} + \frac{v_e}{r_{in}}$$

also $v_o = Av_e$, thus

$$\frac{v_o}{v_i} = -\frac{1}{R_1(1/R_F - 1/Ar_{in})} \qquad (13.5)$$

Now equation 13.5 may be approximated by 13.4 if $Ar_{in} \gg R_F$. Considering the μA 709 operational amplifier we see that $Ar_{in} = 2 \times 10^4$ MΩ and so for 0·01 accuracy the value of R_F should not be greater than about 1 MΩ.

We may determine the input resistance of the system comprising amplifier and the two external resistors, by noting that the amplifier input terminal is a virtual earth, and thus the input resistance is equal to R_1. A lower bound is therefore placed on the value of R_1 (and hence on R_F) to prevent the amplifier loading the source unduly. A value of 10 kΩ is probably a minimum value for the μA 709 amplifier.

13.6.2 Scaling amplifier

The inverter considered in the preceding section may be generalized by making R_F and R_1 unequal. We now obtain a scaling amplifier in which the output is an inverted form of the input but multiplied by a factor R_F/R_1.

13.6.3 Summing amplifier

A very useful application of an operational amplifier is to add together a number of signal voltages including, if desired, a scaling factor to each input. Expressed algebraically we wish to construct a circuit whose output voltage is

$$v_o = k_1v_1 + k_2v_2 + k_3v_3 + \cdots \qquad (13.6)$$

For three input voltages, the circuit of Fig. 13.10 may be used.

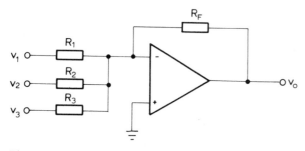

Fig. 13.10 Summing amplifier

Assuming the operational amplifier is ideal we may sum currents at the input node to give

$$\frac{v_1}{R_1} + \frac{v_2}{R_2} + \frac{v_3}{R_3} + \frac{v_o}{R_F} = 0$$

Whence

$$v_o = -\frac{R_F}{R_1} v_1 - \frac{R_F}{R_2} v_2 - \frac{R_F}{R_3} v_3 \qquad (13.7)$$

which realizes equation 13.6 with the constraint that all the multiplying factors are negative. This sign reversal may be rectified by cascading the summing amplifier by a unity gain inverter.

13.6.4 Analogue integrator

If the feedback resistor R_F is replaced by a capacitor C_F as shown in Fig. 13.11 we shall find that the output voltage is the time integral of the input voltage.

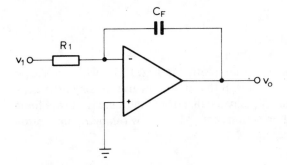

Fig. 13.11 Integrator using operational amplifiers

Assuming an ideal operational amplifier and summing currents at the inverting input we obtain

$$\frac{v_1}{R_1} + C_F \frac{dv_o}{dt} = 0$$

Thus $\qquad v_o = -\frac{1}{C_F R_1} \int v_1 \, dt \qquad (13.8)$

We may consider the effect of finite amplifier gain by assuming that the input voltage is not at virtual earth but has a finite voltage equal to $-v_o/A$. Then summing currents at the input node we obtain

$$\frac{v_1 + v_o/A}{R_1} + C_F \frac{d(v_o + v_o/A)}{dt} = 0 \qquad (13.9)$$

Thus

$$v_o = -\frac{1}{C_F R_1 (1 + 1/A)} \left[\int v_1 \, dt + \frac{1}{A} \int v_o \, dt \right] \tag{13.10}$$

The second term will thus introduce an error. This may be seen by considering the response of the idealized integrator represented by equation 13.8 and the practical integrator of equation 13.9 to a sinusoidal input, V_1, of angular frequency ω.

The ideal integrator will give an output

$$V_o = -\frac{1}{j\omega C_F R_1} V_1$$

We may obtain the response of the practical integrator by solving equation 13.9 for a sinusoidal input phasor $V_1 \exp j\omega t$. We obtain the output phasor V_o as

$$V_o = \frac{-A}{1 + j\omega C_F R_1 (1 + A)} V_1 \tag{13.11}$$

These two frequency response curves are plotted in Fig. 13.12, where A shows the response of an ideal integrator and B that of an integrator using an operational amplifier with finite gain. We may see therefore that there is a lower limit of frequency at which integration action ceases. For a low frequency integrator a very high gain amplifier is needed.

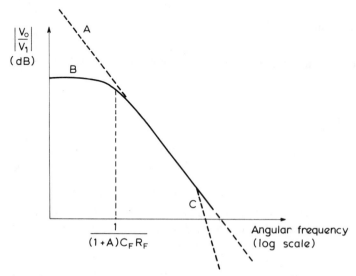

Fig. 13.12 Frequency response of A ideal integrator, B integrator using ideal operational amplifier, C integrator using practical operational amplifier

A further factor needing consideration is the finite bandwidth of the operational amplifier. This will cause the response curve of the practical circuit to have a second break point at a high frequency as shown by the curve C on Fig. 13.12. This frequency occurs at $A\omega_1$ where ω_1 is the lowest break frequency of the operational amplifier and so it is only of significance when working with voltages which change relatively quickly.

13.6.5 Analogue differentiator

An analogue differentiator could be similarly constructed but as its bandwidth is theoretically infinite it is not normally used in analogue computing circuits.

13.7 Analogue computation

The various circuits using operational amplifiers are valuable in building analogue computers. In their simplest form they can give a solution to a linear differential equation for some specified input as a function of time. Thus they may be used to simulate the performance of a mechanical, chemical, or thermal system in the design stage and the effect of modifications to the various parameters may be readily observed. More sophisticated analogue computers incorporate circuits which allow the inclusion of nonlinear functions in the differential equations.

We may consider as an example a system represented by the second order differential equation

$$\frac{d^2y}{dt^2} + a\frac{dy}{dt} + by = x \tag{13.12}$$

in which x is an input function of time and y is the response of the system.

The network blocks which are available are inverters, scalers, summers, integrators, and differentiators. We may see that equation 13.12 may be integrated twice to give an equation involving only integrals of the dependent variable y. We thus have a choice of using either differentiators or integrators as our basic building blocks. All analogue computers use integrators; the reason for this may be seen from the frequency response curves. If we consider the idealized graphs A in each case we shall see that the noise voltage integrated over the complete frequency spectrum will be very much greater for the differentiator than for the integrator, since the noise voltage at any frequency is multiplied by the gain at that frequency and the bandwidth of the differentiator is much greater.

We thus rewrite equation 13.12 in the form

$$\ddot{y} = -a\dot{y} - by + x \tag{13.13}$$

The analogue computer circuit for simulating this differential equation is shown in Fig. 13.13. Starting with $\ddot{y}$ we construct the two other functions of y, namely $-\dot{y}$ and y using two integrators ① and ②, each having unit time constant RC. The right hand side of equation 13.13 is now built up by scaling amplifiers as shown in Fig. 13.13. The output of summer ③ is

$$-\frac{R_1}{R}(-\dot{y})+x = a\dot{y}-x$$

if R_1 is chosen equal to aR.

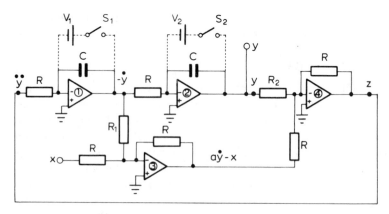

Fig. 13.13 Analogue computer arrangement for solution of a specific differential equation

Similarly the output z from summer ④ is

$$z = -\frac{R_2}{R}y+a\dot{y}-x$$

$$= -a\dot{y}-by+x$$

if $R_2 = bR$.

We have thus constructed the right hand side of equation 13.13 using an assumed input $\ddot{y}$. To satisfy this equation we must now constrain $z = \ddot{y}$ which we do by making a direct connection between the output of amplifier ④ and the input to integrator ① as shown in Fig. 13.13. Thus for any input x, the output y will be the solution of the differential equation 13.12.

Certain practical points should now be considered. In many systems the dependent variable y and its differentials have finite initial conditions. It is therefore essential that the outputs of the two integrators ① and ② are set to a steady value appropriate to the given initial conditions. This may be done by charging each capacitor up to an appropriate voltage before commencing computation, as shown by the batteries V_1 and V_2 in Fig. 13.13; switches S_1 and S_2 are opened immediately prior to integration.

Again, in the majority of commercial analogue computers, the summing amplifiers are restricted to certain fixed multiplying factors, frequently 1/10, 1, and 10. Where multiplying factors other than these are required the amplifier is preceded by a precision potentiometer to give the exact prescribed constant.

In applying the analogue computer to the solution of practical problems it is frequently necessary to scale the time axis. The analogue computer is usually designed to work relatively slowly over a few seconds of time so that output voltages may be plotted graphically, whereas the system being studied may operate at an entirely different speed. Thus, in translating the system into a form suitable for analogue computation, the time variable is usually scaled by an appropriate factor. In addition, the magnitudes of voltages at any of the points in the system must be limited so that the operational amplifiers do not saturate. Since the variables in the real system may be non-electrical quantities, scaling conversion factors may be introduced for all the variables and their differentials.

13.8 Voltage follower

A very useful application of an operational amplifier is as a circuit having a gain of $+1$. This is closely analogous to the emitter follower circuit but the gain can be held even more close to unity. The operational amplifier is provided with unity negative feedback and the input applied to the noninverting terminal as shown in Fig. 13.14. The application of unity negative feedback ensures that

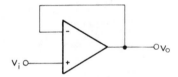

Fig. 13.14 Operational amplifier used as a voltage follower

the voltage between input terminals is sensibly zero (the term 'virtual earth' is no longer applicable as voltages are being applied to the non-inverting input terminal). Hence

$$v_o = v_i$$

and the voltage following action is sensibly perfect.

To obtain the full benefits of a voltage follower special operational amplifiers have been designed having a very high input resistance of the order of 1 GΩ and low output resistances of less than 1 Ω.

13.9 Operational amplifiers in network design

In a great many applications in communication it is necessary to design networks having a specified frequency response. Many desirable frequency response characteristics may be obtained by the use of discrete components involving resistors, capacitors, inductors, and possibly transformers. With the modern trend towards miniaturization, inductors are undesirable components in that they are bulky and cannot be fabricated in integrated circuit form. It is possible using a transistor, or nowadays more usually an operational amplifier, to construct circuits with certain specified transfer functions without the need for inductors, thus allowing complete fabrication in integrated circuit form. In addition, it is possible to construct network elements which are not available using classical passive elements only. The topic of active network design is extremely vast and we shall only study some very simple applications of operational amplifiers in this field.

13.9.1 Negative resistance

A very useful circuit element is one in which the current is proportional to the applied voltage but flows in the opposite direction to that which would flow through a conventional passive resistor. If the reference directions are as shown in Fig. 13.15a, a negative resistor is one in which

$$v = -iR$$

where R is a positive number.

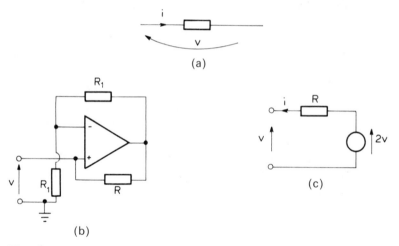

(a)

(b)

(c)

Fig. 13.15 Operational amplifier used to produce a negative resistance

This network element may be realized using an operational amplifier as in Fig. 13.15b. Since we may assume a perfect operational amplifier, the differential voltage between the two inputs is zero and hence the voltage at the inverting input is v. As no current flows into this input we may determine the output voltage of the amplifier directly as $2v$. We may now draw an equivalent model of this circuit as in Fig. 13.15c. The current, i, flowing through resistor R is in the direction shown on Fig. 13.15c and of magnitude v/R. Thus the resistance of the circuit looking in to the input terminals is $-R$.

Such a network is very valuable in that it can be used to cancel out some of the resistive losses in a network and, for example, increase the sharpness of tuning of a resonant circuit.

13.9.2 Voltage controlled current source

An operational amplifier is essentially a device with a high input resistance and a low output resistance. It thus amplifies any voltage applied to the input to give an output voltage which is almost independent of load current flowing. Such devices are termed voltage controlled voltage sources (VCVS). It is sometimes convenient in network design to have available an amplifier with a high input resistance and low output conductance, thus forming a voltage controlled current source (VCCS). A circuit which is capable of doing this using an operational amplifier is shown in Fig. 13.16a and the circuit symbol in Fig. 13.16b.

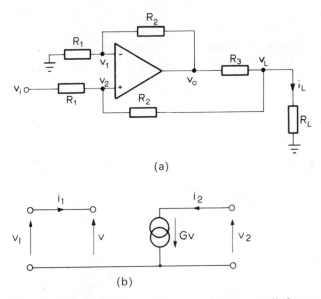

(a)

(b)

Fig. 13.16 Operational amplifier as a voltage controlled current source

Assume that the operational amplifier is ideal and that the feedback resistors R_2 are large compared with the load resistor R_L. Then if v_o is the output voltage of the amplifier,

$$v_1 = \frac{R_1}{R_1 + R_2} v_o$$

Since the differential input voltage must be approximately zero, $v_2 = v_1$, and summing currents at the noninverting input gives

$$\frac{v_2 - v_i}{R_1} = \frac{v_L - v_2}{R_2}$$

At the output

$$v_L = \frac{R_L v_o}{R_L + R_3} = i_L R_L$$

From these equations we obtain

$$i_L = \frac{R_2 v_i}{R_1 R_3} \tag{13.14}$$

In the practical design of such a circuit the input voltage must be limited in order not to exceed the range of linearity of the amplifier.

13.9.3 The gyrator

A useful circuit element in network design is the gyrator. This is usually represented by the conventional symbol of Fig. 13.17a. Its behaviour is described by the pair of equations

$$\left.\begin{array}{l} i_1 = Gv_2 \\ i_2 = -Gv_1 \end{array}\right\} \tag{13.15}$$

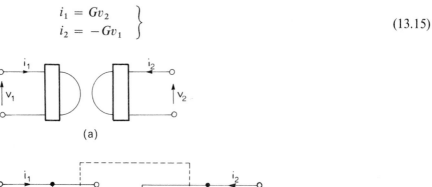

(a)

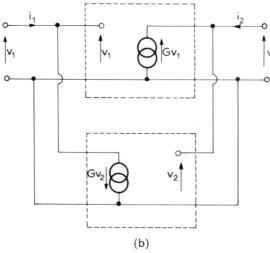

(b)

Fig. 13.17 Construction of a gyrator using VCCS

where G is the gyration conductance. It may be constructed using operational amplifiers in a variety of ways. One very simple form of construction is shown in Fig. 13.17b using two voltage controlled current sources which we discussed in Section 13.9.3. One VCCS is connected directly between input and output and the second with the polarity of the current generator reversed is connected in the opposite direction between output and input. Together they will cause the terminal voltages and currents to satisfy equations 13.15.

One very useful application is a gyrator with a capacitor across its output terminals. We may treat this as a one-port network in which a current i_1 flows as a result of the application of a voltage v_1. The current/voltage relations for the capacitor may be written, for sinusoidally varying quantities,

$$I_2 = -j\omega C V_2$$

Combining this with equations 13.18, we obtain

$$I_1 = j\omega \frac{C}{G^2} V_1$$

Now a one-port network in which the ratio of voltage to current is proportional to $j\omega$ has the properties of an inductor. Thus, a gyrator terminated by a capacitor will simulate an inductor of magnitude C/G^2. Thus it is, at least in principle, possible to dispense with inductors in circuit design and replace them by additional capacitors in conjunction with operational amplifiers.

13.9.4 Network synthesis

The uses of operational amplifiers which we have been studying in the last three sections show that it is possible to design any network which normally requires inductors and capacitors by one which uses capacitors and operational amplifiers only. Furthermore, we have seen that any resistive losses in the network may be cancelled by the addition of an operational amplifier connected to form a negative resistance. Thus a large number of circuits having a very wide range of properties may now be constructed in integrated circuit form.

It is not usually the practice to design circuits using directly simulated inductors and negative resistors as we have discussed. Normally the network is directly designed around the operational amplifier with input, output, and feedback networks composed of resistors and capacitors. In this manner circuits having low-pass, high-pass, band-pass filtering properties may be directly formed. The topic is too vast to be discussed further here.

13.10 Operational amplifier as a comparator

We have restricted our study of the operational amplifier to those applications in which the signals were sufficiently small for it to operate as a linear device. In Section 11.3 we considered the complete transfer characteristic of a long tailed pair amplifier and showed that linear amplification only held over a very small range of differential input voltages of range approximately ± 50 mV. Outside this range the amplifier clamped (or limited) the output voltage either to a low value, approximately zero, or rising almost to V_{CC}, depending on the polarity of the differential input voltage. Since an operational amplifier contains a long tailed pair input stage we should expect the large signal behaviour to be very similar. We should, therefore, be able to use an operational amplifier as a comparator between two input voltages, switching between a high value and a low value as the magnitude of one input exceeds or falls below the other.

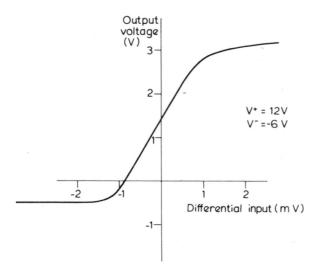

Fig. 13.18 Large input signal transfer characteristic of comparator

Some operational amplifiers are specifically constructed to achieve this property. As an example is the type 710 operational amplifier. The output voltage is normally either at -0.5 V or $+3.0$ V, and only over a very small range of input voltages of the order of ± 1 mV does the output lie between these two values. This small range is achieved by using two long tailed pair amplifiers in cascade as the input stage. The transfer characteristic of the 710 operational amplifier is shown in Fig. 13.18. The output stages of comparator amplifiers are similar to those of linear operational amplifiers, designed to have a low output impedance and thus to provide a large current to an external load.

References

ALLEY, C. L. and ATTWOOD, K. W., *Electronic Engineering*, Wiley, 1973, Chapter 15.
MILLMAN, J. and HALKIAS, C. C., *Integrated Electronics*, McGraw-Hill, 1972, Chapters 15 and 16.
MEYER, C. S., *et al.*, *Analysis and Design of Integrated Circuits*, McGraw-Hill, 1968, Chapters 13 and 15.
GRAEME, J. G. and TOBEY, G. E., *Operational Amplifiers*, Chapters 1–8.
EIMBINDER, J., *Linear Integrated Circuits*, Wiley, 1968, Chapters 3–7.
SMITH, R. J., *Circuits, Devices and Systems*, Wiley, 1966, Chapter 20.

Problems

13.1 The circuit of Fig. 13.1 is used as a small signal amplifier as shown. R_1 is chosen as 820 Ω and the supply voltage is 15 V. The circuit is to be designed so that the maximum voltage swing is available at the output. What is the quiescent collector current of Tr1? Assume that $\beta_F = 120$; design the circuit so that the collector currents of the transistors are equal and that they do not change by more than 0·2% for a 5% change in β_F.

13.2 By how much does the collector current of Tr1 in the circuit of Problem 13.1 change if the base–emitter voltage of both transistors is increased by 0·05 V?

13.3 Estimate the small signal voltage gain of the amplifier of Problem 13.1.

13.4 The circuit of Fig. 13.4 is used to shift the d.c. level of the signal input, v_i. Transistor Tr2 has been biased with a collector current of 260 µA. The supply voltages are $V_{CC} = +10$ V. The input signal v_i is at a mean voltage of 4·5 V. Determine the value of R_1 such that the mean output voltage is zero.

13.5 Determine the quiescent currents and voltages in the circuit of Fig. P13.1. It is not possible to determine the input bias voltage by deduction. Choose this intelligently in order that maximum voltage swing is available across the 10 kΩ resistors.

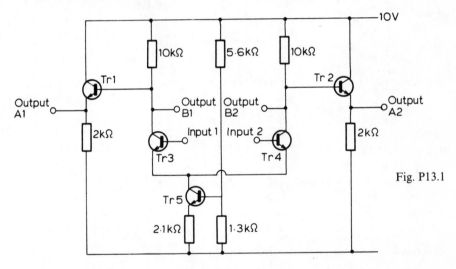

Fig. P13.1

13.6 What is the maximum voltage swing available at outputs A and B of the circuit of Problem 13.5.

13.7 Estimate the order of magnitude of the output impedance at outputs A and B in the circuit of Fig. P13.1. Assume all the transistors have $\beta_0 = 150$.

13.8 Estimate the small signal voltage gain of the circuit of Fig. P13.1 at the output A, if input 2 is earthed and the signal applied to input 1.

13.9 The amplifier of Fig. P13.1 is biased at its input terminals at 3·0 V. Typical values of input offset voltage and current for this amplifier are 1 mV and 150 nA. The amplifier is fed from a source of resistance 200 kΩ. What is the equivalent offset voltage at the input?

13.10 If the offset voltage and current had their maximum values given in Problem 13.9, what would be the quiescent voltages at the four output terminals when the input terminals were connected to a 100 kΩ source? What will be the maximum output voltage swing at B?

13.11 An operational amplifier is used in the circuit of Fig. 13.9. The amplifier has a voltage gain $-2{\cdot}5 \times 10^4$ and an input resistance of $1{\cdot}2$ kΩ. $R_1 = R_F = 1$ MΩ. What is the voltage gain of the circuit?

13.12 An operational amplifier has a voltage gain of 10^4 and is used as an integrator in the circuit of Fig. 13.11 with $C_F = 10$ nF, $R_F = 100$ kΩ. At what frequency will magnitude of the output differ by 1% from the magnitude of a true integral of the input?

13.13 The circuit of Fig. P13.2 is designed as a differentiator. Show that if the amplifier gain is large the circuit performs its required function.

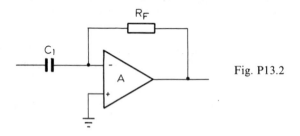

Fig. P13.2

13.14 Show that the transfer function of the differentiator of Fig. P13.2 when the gain A is finite is

$$-\frac{jA\omega/\omega_1}{(A+1)+j\omega/\omega_1}$$

Where $\omega_1 = 1/C_1 R_F$.

13.15 The gain, A, of the amplifier of Fig. P13.2 falls with frequency as

$$A(f) = \frac{A_0}{1+jf/f_0}$$

Show that the overall transfer function of the networks is

$$\frac{A_0 j\omega/\omega_1}{(A_0+1)+j\omega(1/\omega_0+1/\omega_1)-\omega^2/\omega_0\omega_1}$$

Show that at low frequencies the circuit acts as a differentiator but at high frequencies it acts as an integrator.

13.16 Draw a block diagram showing the interconnection of operational amplifiers to form an analogue computer to solve the differential equation

$$\frac{d^2y}{dt^2}+6\frac{dy}{dt}+27y = 3z$$

where z is a given function of time. Use only scalers, summers, and integrators.

13.17 Draw a block diagram of an analogue system which relates two dependent variables x and y with an independent variable of time z, given by

$$3\frac{dy}{dt} + 4y + 6\frac{dx}{dt} = 5z$$

Draw a similar block diagram for solving

$$4y + 3\frac{dx}{dt} = 5z$$

Interconnect these two block diagrams so as to obtain a solution of the above pair of simultaneous differential equations.

13.18 Sketch the transfer function of the network of Problem 13.15 and show that this has a maximum magnitude when

$$\omega = [(A_0 + 1)\omega_0\omega_1]^{1/2}$$

and that at this frequency the phase shift is zero. Show that the maximum magnitude is

$$A_0\omega_0/(\omega_0 + \omega_1)$$

14
Control Systems

14.0 Introduction

So far we have concentrated our attention on systems in which all the parameters studied have been electrical, mainly voltage and current. However such systems can only form a part of any useful engineering system. Essentially the system needs to serve mankind and, with a number of exceptions, man is not designed as a sensor of electrical impulses. It is therefore necessary to convert the electrical signals into those which may be appreciated by man. Thus microphones and loudspeakers act as transducers for sound signals, transforming the sound waves to and from electrical signals. Similar functions are performed by television cameras and receivers for transducing visual signals. A computer requires a line printer to convert the signals into a form intelligible to the average engineer and a card or tape punch and reader at the input of the computer.

In other systems an appreciable amount of work is required from the system in response to an electrical input, the response being a predetermined function of the input. An example of such a system would be the positioning in bearing and azimuth of a transmitting antenna such as that used to relay signals to and from a satellite. Numerous other examples of control systems may be discussed in many other fields; the control of the power to a furnace in order to maintain the temperature at a constant value; the control of water inlet to a tank to keep the head of water constant.

All of these systems involve one common feature, namely the control of some variable of the system in order that some other variable is either kept constant or follows a predetermined path.

Let us focus our attention on a specific problem, a rather simple one, the control of the central heating system in a house in order to maintain the temperature at a satisfactory value, say 24°C. An apparently direct solution to this problem would be to compute the heat losses from the house (an almost impossible task in practice) for an average spring day. From this we could calculate the average energy needed during the day and arrange for the fuel supply valve to be set to deliver this quantity. This is an incredibly crude system and could be considerably improved if we were to include information about outside air temperature, prevailing winds, etc.; a small computer would be needed to process these data and control the fuel supply. Even so it would not take account of such factors as the frequency at which people open external doors and windows and other random events. Such a system is an open loop control system and is shown in block form in Fig. 14.1. It is a perfectly adequate system for a crude control and is still used in simple processes; however even in quite simple situations it is now being replaced by closed loop controllers.

Fig. 14.1 Open loop control system

A closed loop control system is, in principle, considerably simpler. No information is necessary about the various external parameters which influence the performance of the system. To revert to our central heating example, the temperature of the house is measured continuously and this is compared with the desired temperature. The difference between these two values is used to control the flow of fuel to the boiler. This is shown in block form in Fig. 14.2. It may be seen from this that the system is identical with a feedback system in which the feedback function B is unity.

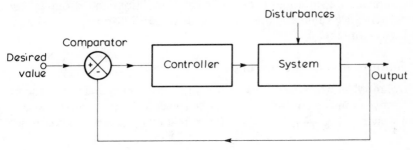

Fig. 14.2 Closed loop control system

Since the output of such a system is frequently a non-electrical quantity and the controller is usually electrically operated, a transducer of some sort is needed between output and input. However this does not affect the nature of the problem. Frequently too the desired input is set by rotation of a shaft and this angular value is conveniently converted to an electrical voltage by connecting a potentiometer to the shaft. It is thus convenient to consider the system as having an electrical input and a non-electrical output. We may therefore define an open loop transfer function for this system as the ratio of the controlled output quantity to the value of the input, and this will no longer necessarily have the dimensions of a voltage ratio or impedance as in the case of a feedback amplifier. In our central heating example the transfer function will be the ratio of house temperature to thermostat setting (the latter will probably take the form of the angular position of a control dial).

One of the significant aspects of a control system is that the output of the system usually requires considerable power for its control. This power may vary from many hundreds of kilowatts as for example in the control of the winding gear on a hoist in a coal mine, to milliwatts for controlling small instruments.

The basic function required of a control system is that it will ensure that the output is equal to the desired reference value for all variations of external conditions. We may thus consider a system as being subject to two inputs, one the desired reference input and the second the variations of external conditions on the system; this latter may take the form of variations of load, environmental temperature, electrical noise, etc. This is shown in Fig. 14.2 as an additional input to the controlled system.

We may observe that all closed loop systems are fundamentally identical to feedback systems. The maintenance of desired performance specifications in the presence of variations in the properties of the controller is analogous to the stabilization of the gain of a feedback amplifier when the internal amplifier is subject to parameter variation caused by external environmental changes. We may therefore apply the majority of feedback theory directly to a study of closed control systems. In particular, in any but the most simple systems, we shall need to study the stability of the closed loop, and this may be investigated using methods identical to those used for feedback amplifiers.

It is thus necessary to use a control system in order to maintain satisfactory performance despite variations in environmental conditions and also to follow any input reference value as closely as desired. We shall take a specific system in order to clarify this.

14.1 Remote position controller

A control problem which frequently arises is the positioning of a shaft at some place remote from the control. This may be achieved by the remote position control system of Fig. 14.3. The armature current of a d.c. motor is controlled

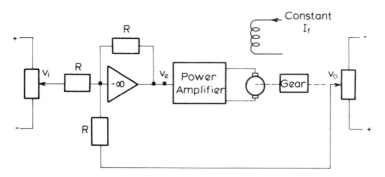

Fig. 14.3 Remote position control system

by a voltage which is the difference between the input position and the output angular position of the shaft; these angular positions may be detected by means of precision potentiometers mounted on the shafts.

We may now write down the equations for this system. The input and output angular positions θ_i and θ_o are directly related to the voltages on the potentiometer sliders by

$$v_i = k_p\theta_i \tag{14.1a}$$

and
$$v_o = k_p\theta_o \tag{14.1b}$$

The error voltage v_e at the output of the unity gain operational amplifier summer is given by

$$v_e = -(v_i - v_o) \tag{14.2}$$

The output voltage of the power amplifier is

$$v_a = -k_a v_e \tag{14.3}$$

where $-k_a$ is the voltage gain of the power amplifier. The torque τ_m of a d.c. motor with constant field excitation is given by

$$\tau_m = k_m i_a \tag{14.4}$$

where k_m is the motor constant, and if R_a is the armature resistance, the induced e.m.f., e, in the motor is

$$e = v_a - i_a R_a \tag{14.5}$$

The speed of the motor ω_m (radians per second) is related to the induced e.m.f. by the same motor constant k_m; thus

$$e = k_m \omega_m \tag{14.6}$$

The gear causes a reduction in speed by a factor n, the gear ratio, and an increase in available torque at the output shaft by the same factor. Thus the output speed, ω_o, and torque, τ_o, are given by

$$\omega_o = \omega_m/n \tag{14.7a}$$

and
$$\tau_o = n\tau_m \tag{14.7b}$$

Finally it should be noted that

$$\omega_o = \frac{d\theta_o}{dt} \tag{14.8}$$

If we now assume that the output load torque is small we may let $\tau_m \simeq 0$ and then using equations 14.1 to 14.8 we may write the differential equation relating output position to input position as

$$\frac{d\theta_o}{dt} + k\theta_o = k\theta_i \tag{14.9}$$

where $k = k_a k_p / k_m n$. Under steady conditions $d\theta_o/dt = 0$ and hence $\theta_o = \theta_i$ and the output shaft takes up a position identical with that of the input shaft.

If the input is given a displacement Θ_I, the output will follow this change in an exponential manner with a time constant $1/k$. Thus

$$\theta_o = \Theta_I[1 - \exp(-kt)] \tag{14.10}$$

However, we have made a number of simplifying approximations in the above analysis. Let us no longer assume that the torque remains constant during a change of angular position. Consider that the output driven shaft has a moment of inertia, J_L, and is subjected to viscous damping, D_L, such that the damping torque is proportional to speed. Finally, let us assume that the static load torque is negligible compared with either of these dynamic torques. Thus in addition to the equations already developed we shall have

$$\tau_o = J_L \frac{d\omega_0}{dt} + D_L \omega_0 \tag{14.11}$$

Solving equations 14.1 to 14.8 with 14.11 we obtain

$$\frac{d^2\theta_o}{dt^2} + 2a \frac{d\theta_o}{dt} + b^2\theta_o = b^2\theta_i \tag{14.12}$$

where

$$a = \frac{D_L + k_m^2 n^2/R_a}{2J_L} \tag{14.12a}$$

$$b^2 = \frac{k_m k_a k_p n}{R_a J_L} \tag{14.12b}$$

As before under steady state conditions the output position θ_o is identical to the input. However the transient response of the system to a sudden change of Θ_I now takes one of the forms shown in Fig. 14.4 obtained by solving equation 13.12 for $\theta_i = \Theta_I$ with zero initial conditions. This may be expressed analytically by

$$\theta_o = \left[\frac{\alpha_2}{\alpha_1 - \alpha_2} \exp(\alpha_1 t) - \frac{\alpha_1}{\alpha_1 - \alpha_2} \exp(\alpha_2 t) + 1 \right] \Theta_I \tag{14.13}$$

where

$$\alpha_1, \alpha_2 = -a \pm (a^2 - b^2)^{1/2}$$

Curve A shows an overdamped response in which the output shaft takes a long time to reach its steady state condition. Curve B shows an underdamped response in which the initial rise is fast but a large initial overshoot occurs followed by several oscillations before settling to its steady value. In any system a compromise in design will be necessary. An overshoot of 20 to 40%

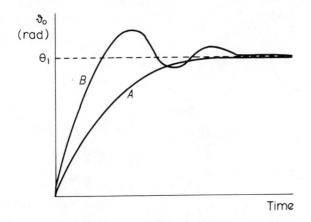

Fig. 14.4 Response of remote position control system to input step

is acceptable in small instrument servos corresponding to a damping ratio a/b between 0·3 and 0·5, but in a large position control system, for example, the aerial positioning system at a satellite tracking station, a much higher damping would be necessary.

14.1.1 Position controller with variable input

The situation we have analysed in Section 14.1 is relatively simple in that we have set the input to a fixed position and investigated the manner in which the output changes and finally coincides with the input. However if we used such a system for controlling an aerial to track a satellite, the input signal, i.e. the predicted position of the satellite, would be continuously variable. As a simple example of a variable input let us consider the situation in which the input angular displacement is a linear function of time (a ramp function)

$$\theta_i = At$$

Solving equation 14.12 for this input, and assuming zero initial conditions gives

$$\theta_o = \left[-\frac{\alpha_2}{\alpha_1(\alpha_1 - \alpha_2)} \exp(\alpha_1 t) + \frac{\alpha_1}{\alpha_2(\alpha_1 - \alpha_2)} \exp(\alpha_2 t) - \frac{\alpha_1 + \alpha_2}{\alpha_1 \alpha_2} + t \right] A$$

$$(14.14)$$

which takes the form shown in Fig. 14.5. We see from equation 14.14 that after a sufficiently long period of time when the initial oscillations have died away

$$\theta_o \simeq A\left[-\frac{\alpha_1+\alpha_2}{\alpha_1\alpha_2}+t\right]$$

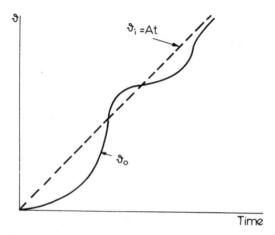

Fig. 14.5 Response of remote position control system to input ramp

Thus, after a considerable time, an error always remains between the output and the input, of magnitude

$$-A\frac{\alpha_1+\alpha_2}{\alpha_1\alpha_2}=\frac{2a}{b}A$$

This steady state error is an inevitable consequence of this type of system and shows that when the input is changing the output cannot follow it accurately but will always lag behind. It may be noted that the error is directly proportional to the damping ratio. Thus a system with a low damping ratio will have a small error to a ramp input but will be subject to a very large transient overshoot. This problem can only be overcome by resort to higher order systems and we shall not consider them here.

14.1.2 Transfer function of system

Although control systems are invariably subjected to time varying inputs, it is often convenient to study them in the frequency domain and obtain the response of the system in magnitude and phase to sinusoidal signals varying over a wide frequency range. The advantage is that frequency response methods have been well studied in relation to feedback amplifiers and the same techniques may be applied here. We should note that in control systems the frequency range extends from d.c. up to perhaps a few hundred hertz since most of the systems involve mechanical, hydraulic, or thermal elements in which the time constants are large compared with those in electrical systems.

It is therefore convenient to solve equation 14.12 for an input

$$\theta_i = \Theta_i \cos \omega t$$

$$= \text{Re} \left[\Theta_i \exp j\omega t \right]$$

We obtain an output

$$\theta_o = \text{Re} \left[\Theta_o \exp j\omega t \right]$$

where
$$\frac{\Theta_o}{\Theta_i} = \frac{b^2}{-\omega^2 + 2ja\omega + b^2}$$

Now the ratio of the output phasor Θ_o to the input phasor Θ_i may be defined as the transfer function of the system $G(\omega)$

$$G(\omega) = \frac{b^2}{-\omega^2 + 2ja\omega + b^2} \tag{14.15}$$

Note that in this example the transfer function is the ratio of two angular positions. The magnitude characteristic of this transfer function is shown in Fig. 14.6. It will be seen that for a small damping ratio a/b there is an appreciable

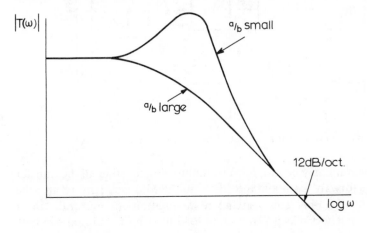

Fig. 14.6 Frequency response of remote position controller

peak in the frequency response curve whereas for large damping ratio (greater than approximately 0·6) the peak vanishes. We may therefore rather loosely correlate a peak in the frequency response curve of the amplifier with an overshoot in the transient response.

Since the control systems we are studying are essentially similar to feedback amplifiers it would be convenient to study them as feedback systems to see whether any insight may be obtained from a knowledge of the open loop system characteristics. To do this it is convenient to draw the system in the form of a block diagram similar to Fig. 14.2.

14.1.3 Block diagram of position controller

This may be obtained by referring to equations 14.1 to 14.8 and 14.11. We notice that the equations for the motor and load are interrelated; let us start by solving equation 14.11 to obtain the transfer function relating the two phasors of output speed Ω_o and torque T_o which represent sinusoidally varying quantities at an angular frequency ω. This gives

$$\Omega_o = \frac{T_o}{D_L + j\omega J_L} \tag{14.16}$$

We may now build up a block diagram for the motor and gearbox as shown in Fig. 14.7. Block (a) may be related with equation 14.5, block (c) with 14.4, (d) with 14.7b, (e) with 14.16, (f) with 14.6, and (g) with 14.7a. Note that since we are dealing with sinusoidally varying quantities, the instantaneous variables indicated by lower case letters have been replaced by capital letters representing phasors.

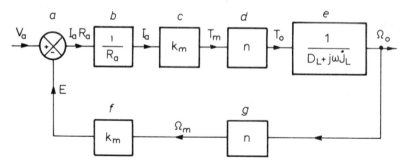

Fig. 14.7 Block diagram of motor and load within remote position controller

This block diagram may be simplified by multiplying together all the transfer functions in the forward path and also those in the feedback path to give the block diagram of Fig. 14.8. This feedback representation of a motor driving a load highlights the effect which variation of load has on the induced e.m.f. as a result of the variation of speed.

Figure 14.8 may be still further condensed by using equation 12.4 identifying A with the forward transfer function and B with the feedback function. Thus the overall transfer function of the motor and load relating speed to applied armature voltage is

$$G_m(\omega) = \frac{k_m n}{R_a(D_L + j\omega J_L) + k_m^2 n^2}$$

and using equation 14.12a this becomes

$$G_m(\omega) = \frac{k_m n}{J_L R_a(2a + j\omega)} \qquad (14.17)$$

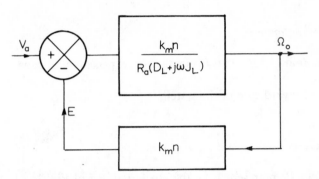

Fig. 14.8 Equivalent block diagram to Fig. 14.7

This is shown in Fig. 14.9. We may now incorporate this in the complete system to give the block diagram of Fig. 14.10. In this diagram we identify block (a) with equation 14.1a, (b) with 14.2, (c) with 14.3. (d) with 14.17, (e) with 14.8, and (f) with 14.1b.

The derivation of block (e) from equation 14.8 is worth elaborating. Consider a variable $x = X \exp j\omega t$. Then if $y = dx/dt$, we have $y = j\omega X \exp j\omega t = Y \exp j\omega t$ and hence the transfer function of a differentiating network is the

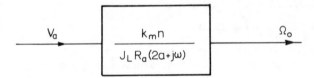

Fig. 14.9 Equivalent block diagram to Fig. 14.8

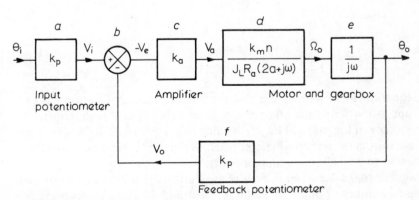

Fig. 14.10 Block diagram of complete remote position controller

ratio of the phasors Y and X namely $G(\omega) = j\omega$. Similarly the transfer function of an integrating network is $G(\omega) = 1/j\omega$.

Condensing the block diagram of Fig. 14.10 as before and using expression 12.4 gives the transfer function for the complete system.

$$G'(\omega) = \frac{b}{-\omega^2 + 2aj\omega + b^2}$$

where b is given by equation 14.12b. This is identical, as expected, with equation 14.15.

We may note that the forward gain of the system is

$$G(\omega) = \frac{b^2}{2aj\omega - \omega^2} \tag{14.18}$$

and the entire system may be represented by the block diagram of Fig. 14.11

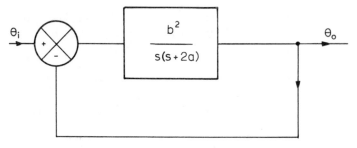

Fig. 14.11 Equivalent block diagram to Fig. 14.10

with unity negative feedback; the transfer function, expressed as a function of the variable $s = j\omega$, is

$$G(s) = \frac{b^2}{s^2 + 2as}$$

$$= \frac{b^2}{s(s + 2a)} \tag{14.19}$$

From equation 14.19 we may see that the forward transfer function has two poles, one at $s = 0$ and the other at $s = 2a$ and therefore at high frequencies the magnitude of the gain will fall off at a maximum rate of 12 dB/octave. Thus the phase shift in the system will never reach 180° at finite frequencies and there will thus be no problems regarding stability.

In higher order systems additional phase shifts may occur and we can then use the techniques outlined in Sections 12.10 and 12.11 to compensate the system and eliminate the possibility of instability.

14.2 Speed control system

As a second example of a control system we shall study a method by which the speed of a motor may be maintained at some preset value despite variations of load. The system consists of a d.c. motor whose armature is supplied with a constant current. The field windings are centre tapped and fed from the output of a power amplifier. The speed of the motor is monitored by a tachogenerator, whose output is compared with a voltage representing the desired speed and the difference is used as input to the amplifier. A circuit diagram is shown in Fig. 14.12.

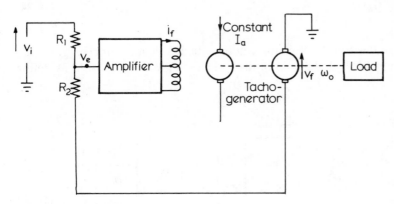

Fig. 14.12 Circuit system of velocity system

Let us draw the block diagram and analyse the system by using the transfer function. First it is necessary to write the equations appropriate to a motor whose field current is variable and whose armature current is held constant. The load will be assumed to consist of a load torque τ_L, which may be variable, inertia torque $J_L \, d\omega_0/dt$ and viscous friction $D_L\omega_0$.

Equating instantaneous developed motor torque to the total load torque we obtain

$$\tau_m = \tau_L + D_L\omega_o + J_L \frac{d\omega_o}{dt} \tag{14.20}$$

Also since the armature current is constant

$$\tau_m = k_f i_f \tag{14.21}$$

where k_f is the torque constant of the motor at the given armature current.

From equations 14.20 and 14.21 we may write the differential equation for the motor as

$$J_L \frac{d\omega_o}{dt} + D_L\omega_o + \tau_L = k_f i_f \tag{14.22}$$

Now, for the moment, consider the load torque to be constant and so we may set the small variations, τ_L, to zero in equation 14.22. We now solve for sinusoidally varying speed and current phasors, Ω_o and I_f, to give the transfer function of the motor as

$$G_m(\omega) = \frac{\Omega_o}{I_f} = \frac{k_f}{j\omega J_L + D_L} \tag{14.23}$$

or $$G_m(s) = \frac{k_f}{J_L s + D_L} \tag{14.24}$$

The units of this transfer function will be radians second^{-1} ampere^{-1}.

The complete block diagram of the system is shown in Fig. 14.13 assuming

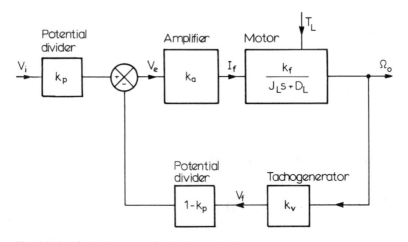

Fig. 14.13 Block diagram of velocity controller

that the load torque is constant. The effect of the potential divider R_1, R_2 is represented by the two blocks involving k_p where $k_p = R_2/(R_1 + R_2)$. Analysis of this system gives a forward transfer function of

$$G(s) = \frac{k_a k_f}{J_L s + D_L} = \frac{k}{s + D_L/J_L} \tag{14.25}$$

where $$k = k_a k_f / J_L$$

Notice that since there is only one pole at $s = D_L/J_L$ the gain fall off will be less than 6 dB/octave and no stability problems will occur.

The closed loop transfer function is easily computed as

$$G'(s) = \frac{k}{s + D_L/J_L + k_v(1-k_p)k} \tag{14.26}$$

$$= \frac{k}{s+a}$$

where $\qquad a = D_L/J_L + k_v k(1-k_p)$

We note by comparison with the network studied in Section 9.1 that the response to a step of input voltage will be an exponential rise with a time constant of $1/a$. Comparison of equation 14.26 with 14.25 shows that the speed of response of the network has been increased by the connection of feedback. Usually in such a system the viscous damping coefficient D_L is small compared to the inertial torque and the load torque.

14.2.1 Speed controlled system with load variation

It is of interest now to see how a variation of load torque will affect the speed of the motor. We shall assume throughout this study that the input voltage V_I is kept constant.

We may now derive from equation 14.22 a relationship between a phasor Ω_o and two variable phasors T_1 and I_f.

$$\Omega_o = \frac{k_f I_f + T_1}{J_L s + D_L}$$

The block diagram of Fig. 14.13 is now redrawn in Fig. 14.14 with phasor variation of input voltage, $V_i = 0$, and the motor block redrawn to include the

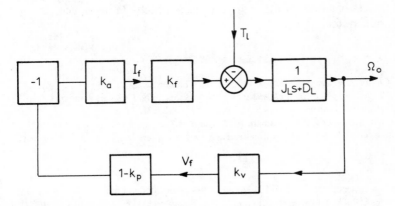

Fig. 14.14 Velocity controller with torque variation

effects of load variation. Analysing this, considering T_1 as input and Ω_o as output, we obtain for the forward transfer function

$$G(s) = \frac{\Omega_o}{T_1} = \frac{1}{J_L s + D_L} \qquad (14.27)$$

and for the feedback function

$$B = k_v k_a k_f (1 - k_p) \qquad (14.28)$$

Thus the closed loop transfer function is

$$G'(s) = \frac{1/J_L}{s + D_L/J_L + k_v k_a k_f (1 - k_p)/J_L} = \frac{1/J_L}{s + a} \qquad (14.29)$$

where a is given by equation 14.26a.

We may observe the improvement on the system response to variation of load torque by comparing equations 14.27 and 14.29.

Under steady state conditions, when all transients have died away, we may obtain the relationship between speed and torque variation by putting $s = 0$.

The open loop system, from equation 14.27, shows that the change in steady output speed $\Omega_o = T_1/D_L$. However, from equation 14.29 we note that under closed loop conditions

$$\Omega_o = T_1/J_L a$$

$$= \frac{T_1}{D_L + k_v k_a k_f (1 - k_p)}$$

This shows that the load variation has a much smaller effect on the speed in a closed loop system than in an open loop system. This is a confirmation of the statement we made in Section 12.6 where we noted that negative feedback can reduce the effects of noise, or other disturbances, which are introduced in the later stages of a system.

References

SMITH, R. J., *Circuits, Devices and Systems*, Wiley, 1966, Chapter 22.
ATKINSON, P., *Thyristors and their Applications*, Mills and Boon, 1972, Chapters 4 and 5.
MAZDA, F. F., *Thyristor Control*, Newnes Butterworth, 1973, Chapter 11.
STOCKDALE, L. A., *Servomechanisms*, Pitman, 1962, Chapters 3 and 5.
KUO, B. C., *Automatic Control Systems*, Prentice Hall, 1962, Chapters 1–6.

Problems

14.1 A direct current generator (Fig. P14.1) has an open circuit rated voltage of 100 V and a full load rated current of 10 A. The open circuit voltage, E, is related to field current, I_f by $E = 100 I_f$; the armature resistance is 1·2 Ω, the field resistance and inductance are 85 Ω and 18 H. Determine the voltage to be applied to the field circuit to obtain rated open circuit voltage at the output. What input voltage must be applied to the amplifier (voltage gain = 125)?

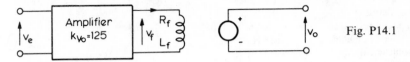

Fig. P14.1

14.2 The generator in Problem 14.1 may be considered as a voltage amplifier between output voltage and field voltage. Write the transfer function of this amplifier on open circuit.

14.3 Determine the time constant of response of the system, generator and amplifier, to a sudden change of input voltage of 10 mV.

14.4 The input voltage is kept fixed to give an output voltage of 100 V. A load of 8 A is now suddenly connected across the output. Determine and sketch the output voltage as a function of time after application of the load.

14.5 The generator in conjunction with the voltage amplifier is used in a feedback loop as shown in Fig. P14.2. Determine the value of reference voltage V_R to give rated output voltage on no load.

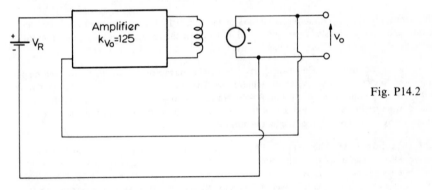

Fig. P14.2

14.6 Determine the variation of output voltage with time of the feedback system of Fig. P14.2 if the reference voltage V_R changes suddenly by 10 mV. Find the time constant of response.

14.7 The reference voltage is kept fixed to give open circuit voltage of 100 V. A load is now suddenly connected across the output. Determine the variation of output voltage with time after application of the load. Compare this with that obtained in Problem 14.3.

14.8 Draw the feedback system of Fig. P14.2 on open circuit in a block diagram and identify the forward transfer network and the feedback network. What are the low-frequency values of A and B?

14.9 Using the expression for reduction of rise time by feedback, confirm the relationship obtained between the two time constants in Problem 14.3 and 14.6. Also confirm the relationship obtained between input voltage on open loop (Problem 14.1) and reference voltage on closed loop (Problem 14.5) required to keep the output voltage at its rated value of 100 V.

14.10 The speed regulator shown in P14.3 has the following parameters

Motor constant	$= 0.12$ N m/A
Motor and load inertia	$= 0.04$ kg m^2
Motor friction	$= 1.6$ N m s/rad
Field circuit resistance	$= 300$ Ω
Field circuit inductance	$= 2.5$ H
Tachometer constant	$= 0.8$ V/rad
Amplifier voltage gain	$= 1000$

Determine the transfer function of the system relating angular speed of shaft to reference voltage V_R.

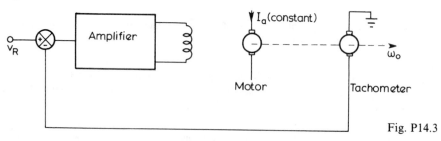

Fig. P14.3

14.11 The feedback link from the tachometer in the system of Fig. P14.3 is broken. Determine the transfer function relating input to output. Sketch the response to a 1 V step.

14.12 Draw a block diagram to represent the system of Fig. P14.3 and determine the low frequency forward and reverse transfer functions A and B. Confirm the reduction in d.c. gain obtained in Problems 14.10 and 14.11 when feedback is used.

14.13 The angular position of a flywheel driven by an electric motor is controlled by the setting of a handwheel. Sketch a suitable system to achieve this. The moment of inertia of the moving parts is 1000 kg m^2; the motor torque constant is 4500 N m/A; viscous friction torque is 1800 N m s/rad. Write down the differential equation of the system. Determine the response of the system to a sudden change in the position of the handwheel by 0.1 rad.

14.14 The control system shown in Fig. P14.4 is designed to keep the speed of a motor constant for load variations. The amplifier will deliver -3.1 mA to the motor field for 1 V at the input; it develops a torque of 0.5 N m/A of field current; the tachometer constant is 62×10^{-6} N m/rad.

The reference voltage V_R is 10 V. Select values for R_1 and R_2 such that the motor runs at a speed of 1000 rev/min when unloaded and that the speed does not fall by more than 5% when the load torque is increased to 0.6×10^{-2} N m.

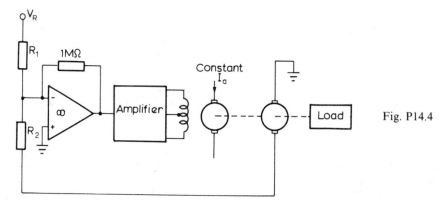

Fig. P14.4

15
Combinational Logic

15.0 Introduction

The circuits which we have considered so far have been, almost exclusively, restricted to linear systems designed to amplify or process signals of small amplitude compared with the supply voltages. Another very large class of circuits are those dealing with large signals in which the relationship between input and output may no longer be assumed linear. A much used group of circuits in this category is switching circuits. Such circuits may be considered as having only two states—when the device is fully conducting with a very low voltage drop across it, and when the device is non-conducting; between these two states the device is in an active condition. Thus in either of its static conditions very little power is dissipated in the device, either the current is small or the voltage drop is low.

15.1 Logic circuits—definitions

The most widely used application of switching circuits is in logic networks. In these only two voltage states are of interest, HIGH and LOW. It is customary to associate the two numbers **1** and **0** with these two logic states. Association of **1** with the high voltage level and **0** with the low voltage level is the convention adopted in positive logic; negative logic is the converse.

The two logic levels are normally defined by stating that logic **1** corresponds to any voltage between V_H' and V_H'' and logic **0** to any voltage between V_L' and V_L''. The upper and lower supply voltages to a circuit usually prevent voltages lower than V_L' and above V_H'' existing. A voltage between V_L'' and V_H' is likely to give an ambiguous output and is therefore excluded. Any circuits which will give steady voltages in this range are considered as unreliable.

15.2 Diode switch

One of the simplest circuits whose behaviour approximates to that of the ideal switch is a diode connected as in the circuit of Fig. 15.1. When the applied voltage is in the forward direction current will flow through the diode and its magnitude will be almost completely determined by the series resistor R, assuming this to be large compared with the resistance of the diode when conducting. The voltage across the diode will be approximately 0·6 V and hence the output voltage v_O will be approximately equal to the input voltage v_I. If the applied voltage is in the reverse direction, the diode will not conduct, apart from the small reverse saturation current of the diode, and the output voltage will be approximately zero. The diode thus acts as a simple switch to control the flow of current through a load resistor. Considered as a device for controlling the power in a load this circuit is unsatisfactory since the switching action is controlled by the supply voltage.

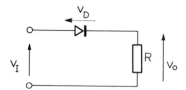

Fig. 15.1 Diode switch circuit

An alternative application of switching is in logic circuits. In order that a circuit shall act as a satisfactory logic element it is essential that the output level is precisely determined as either **1** or **0** according to the input and that the transitional region between these two levels can exist for only a very small range of input voltages. This may best be seen by drawing the transfer characteristic relating v_O to v_I. For the circuit of Fig. 15.1 this is shown in Fig. 15.2.

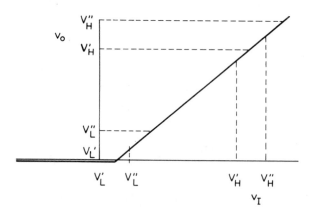

Fig. 15.2 Transfer characteristic of diode switch

We see that so long as the input is within the bounds we have defined by logic **0**, the output will also be within the bounds defining logic **0**. However, if the input is at the lower edge of the band defining logic **1**, the output will fall outside this band and into the indeterminate region. Thus to use diodes satisfactorily as logic elements we shall need to restore the voltage to its correct value at the output; we shall see later how this may be done. With this modification in mind we shall study some of the other properties of diode circuits to investigate their potentialities and limitations.

15.2.2 Dynamic behaviour of diode switch

One very important property of logic circuits is the speed at which they can change state from logic **1** to **0** and vice versa. This switching time is directly related to the dynamic properties of the device, in the present case, a diode.

Consider a diode in the reverse biased condition in which the charge distribution of holes in the n-region is as shown in Fig. 1.7b. Assume now that the diode is forward biased by a steady voltage supply, V_F, through a current limiting resistor R, as shown in Fig. 15.1. If we neglect the small voltage drop across the diode this is equivalent to injecting a forward current $I_F = V_F/R$ into the diode. Now we may see from the model of Fig. 1.18 that the junction voltage cannot immediately take up its steady value given by equation 1.22 since this would require an instantaneous change in the stored charge. We conclude therefore that the junction voltage must rise slowly from zero to its equilibrium value so that the current is at all times given by the rate of change of stored charge.

Since the injected current I_F is approximately constant in the forward direction the gradient of the charge distribution in the n-region of the diode at the edge of the junction is constant (see equation 1.20a). Thus the charge in the diode must rise slowly to its equilibrium value. Since the charge density, $p_n(0)$, at the junction is an exponential function of the instantaneous diode voltage, $v_D(t)$, we shall see that the latter will also rise slowly, as shown in Fig. 15.3b, to its equilibrium value V_D given by

$$V_D = \frac{kT}{e} \ln (I_F/I_0 + 1)$$

The rise time of the diode voltage is thus limited by the equivalent charge storage capacitance of the diode; the changes in charge in the depletion layer have not been considered but they will add to the delay noted here.

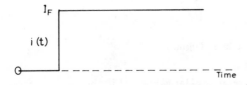

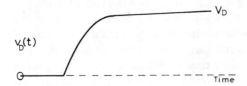

Fig. 15.3 Voltage response of diode switch to a step current change in the forward direction

Let us now reduce the applied voltage to zero thereby reducing the current flow across the junction to zero. From equation 1.20a we see that this requires that the gradient of the charge distribution at the junction falls to zero. Again the charge density at the junction cannot change instantaneously and thus the charge distribution falls slowly towards zero by recombination. Only when all the charge has been removed does the diode voltage fall to zero. As may be seen in Fig. 15.4 this is a relatively slow process and there is a considerable delay time before the diode switches off. This is a serious restriction on the operating speed of diode switches.

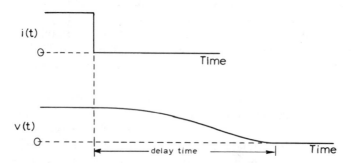

Fig. 15.4 Voltage response for a step current change to the off state

This delay time can be appreciably reduced if the applied voltage is made negative rather than being reduced just to zero. Consider the supply voltage to be switched from V_F in the forward direction to V_R in the reverse direction as in Fig. 15.5. Initially, since the diode is effectively a short circuit, the current

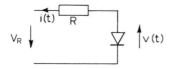

Fig. 15.5 Diode switched off into reverse biased condition

flowing is $i(t) = V_R/R$ and this is a relatively large current in the reverse direction. The charge stored in the diode is now removed partly by recombination but mainly by the flow of this current across the junction. The variation of the diode voltage with time is shown in Fig. 15.6. Only when the charge density at the junction has fallen to approximately zero will the diode become reverse biased; at this point the resistance of the diode will rise and the injected current will fall quickly to the value of the reverse saturation current of the diode. The changes are shown in Fig. 15.6. The delay time in this case is much reduced in comparison with Fig. 15.4b since the charge is removed directly across the junction rather than by relying on recombination only for its reduction. The time of switching from ON to OFF is thus greatly reduced.

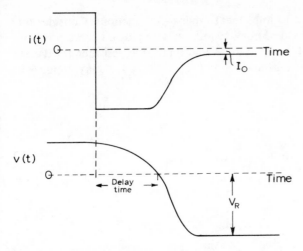

Fig. 15.6 Voltage response to step reversal of current

15.2.3 Logic and Boolean algebra

The simple diode circuit which we considered in Section 15.1 gives a low output when the input is high and vice versa. The basis of practically all switching circuits is the restriction of voltages in the circuit under steady state conditions to two distinct values, HIGH and LOW. Values of voltage between these extremes will only exist in a properly designed circuit during the transitional period when changing between these two states. In order to utilize the results of Boolean algebra which have been developed to analyse logic networks we associate with these two states the values 1 or 0. If we choose to adopt positive logic the higher voltage corresponds to logic 1 and the lower to logic 0; the opposite convention results in negative logic. Unless stated otherwise positive logic will be assumed throughout this book.

If we consider the input variable to the diode switch of Fig. 15.1a as A, which may take either the value 1 or 0, and the output voltage as X we may express the logical function of the circuit by the truth table (Table 15.1) which gives the output X corresponding to an input A.

Table 15.1

A	X
1	1
0	0

This shows that in terms of logic variables, X is always identical with A, i.e. 'X is A'. In Boolean algebra convention this is written

$$X = A$$

Diodes may also be used to implement other logic functions. Consider the circuit of Fig. 15.7. Let us assume positive logic in which logic 1 corresponds to V_{CC} and logic 0 to 0 volts. If both inputs A and B are at logic 1 then the diodes are non-conducting and the output rises to V_{CC}, i.e. $X = 1$. If, say, input A is

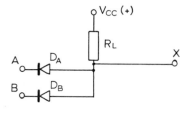

Fig. 15.7 Diode AND gate

at logic 0 then diode D_A conducts and, assuming ideal diodes, the output voltage falls to zero and $X = 0$. A similar situation exists if input B is at 0. This gives the truth table of Table 15.2.

Table 15.2

A	B	X
1	1	1
0	1	0
1	0	0
0	0	0

We may thus see that the output exists (i.e. is logic 1) if both A and B exist and not otherwise. Thus the logic function of this circuit is

$$X = A \quad \text{AND} \quad B$$

or written in Boolean notation

$$X = A.B$$

which is frequently contracted to

$$X = AB$$

Another useful logic function is that which satisfies the relationship

$$X = A \quad \text{OR} \quad B$$

written in Boolean notation

$$X = A + B$$

The truth table for this is given in Table 15.3.

Table 15.3

A	B	X
1	1	1
1	0	1
0	1	1
0	0	0

This logic function may be realized by the circuit of Fig. 15.8 where the input may vary between the two levels of $+V_{CC}$ and zero.

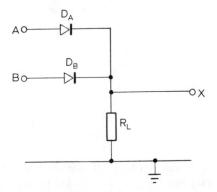

Fig. 15.8 Diode OR gate

15.2.4 Limitations of diodes as logic elements

We have seen in Section 15.2.3 that the circuit of Fig. 15.7 acts as an AND gate. We made the assumption there that all the diodes were ideal. A more accurate analysis would need to take into consideration the effect of the finite forward conduction resistance of the diodes; this is usually accounted for by ascribing a finite value to the break point in the piecewise linear model of the diode. For a silicon diode this may be assumed at 0·6 V.

A modified truth table showing the values of the various voltages rather than logic levels is given in Table 15.4. It may be seen that the output logic levels are not the same as those at the input. The low logic level has risen from 0 to 0·6 V.

Table 15.4

V_A	V_B	V_X
10	10	10
10	0	0·6
0	10	0·6
0	0	0·6

The transfer characteristic relating output voltage V_X to input voltage V_A, when V_B (and any other inputs) are held at high logic level, is shown in Fig. 15.9. From this it is possible to observe the effect of small variations in the magnitude of the input voltages in both the high and low logic states. This is

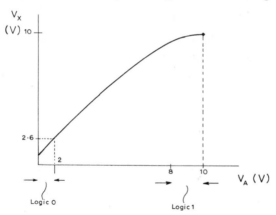

Fig. 15.9 Transfer characteristic for AND gate

termed the noise immunity of the circuit. It is customary to define the two levels of a logic circuit as two bands; thus logic **1** may be defined as all voltages greater than 8 V and logic **0** as all voltages less than 2 V. It may be seen that if the input A has a value of 2 V at the upper limit of logic **0**, the output X will be at about 2·6 V which is outside the bounds of logic **0**. Thus the circuit will fail to hold to the desired specification of logic levels. A deviation in the high logic level is not so serious as it will tend to be reduced at the output. All logic circuits are ultimately intended as units to be interconnected into a large system, the inputs to several logic elements being connected to the output of another. We see therefore that if the 0 logic level rises by 0·6 V through one diode gate it will rise to 1·2 V after a second gate has been cascaded with the first. This trend will continue until it no longer lies within the defined bounds of the given logic level.

Again, if any spurious voltages, possibly due to noise, are induced in the input circuit and superimposed on the input level of **0**, this is likely to cause the output voltage to exceed the bounds of its appropriate logic level. The circuit has no noise immunity. The magnitude of the voltage which needs to be superimposed on the nominal voltage at a given logic state before the circuit changes state is defined as the noise immunity for that logic state of the circuit.

An OR gate may be constructed using diodes and resistors in a similar manner as shown in Fig. 15.8. If we take into consideration the effects of the forward voltage drop of 0·6 V across the diode we shall observe that the voltage table will be as in Table 15.5.

Table 15.5

V_A	V_B	V_X
10	10	9·4
10	0	9·4
0	10	9·4
0	0	0

The above are the only logic gates which may be designed using diodes alone; they are insufficient to implement all combinational logic networks.* To complete the set, a NOT gate is required and this cannot be constructed using diodes alone.

15.3 NOT, NAND, and NOR gates

As noted in Section 15.2.4, in order to form a complete logic set we require an element whose output **X** is related to an input **A** by

> **X** is NOT **A**

written in Boolean algebra as

$$\mathbf{X} = \overline{\mathbf{A}}$$

The truth table, Table 15.6, shows this relationship.

Table 15.6

A	X
1	0
0	1

* Combination logic includes all systems which do not have a memory. Although the time delays inherent to all switching circuits do constitute a form of memory, this is usually an undesired phenomena. Those circuits which are designed with a memory storage are termed *sequential* logic networks.

A gate which realizes this logic function is termed a NOT gate. It is sometimes convenient to cascade an AND gate by a NOT gate and form a composite unit, a NAND gate, represented by the Boolean expression

$$X = \overline{AB}$$

Its truth table is given in Table 15.7.

Table 15.7

A	B	X
1	1	0
0	1	1
1	0	1
0	0	1

Another composite unit, the NOR gate, is obtained by cascading an OR gate with a NOT gate. It is represented by the logic expression

$$X = \overline{A+B}$$

and has the truth table given in Table 15.8.

Table 15.8

A	B	X
1	1	0
0	1	0
1	0	0
0	0	1

Although these last two have been developed as composite units, it is often simpler to construct a NAND or NOR gate basically and build up the others from it.

15.3.1 The transistor as a switch

A NOT gate may be simply constructed using the circuit of Fig. 15.10. The operation of this circuit for slow changes in input voltage from 0 to V_{CC} may be seen by reference to the static characteristic of the transistor in Fig. 15.11. Using the numerical values chosen the transistor has a value of $\beta_F = 50$ and the load resistor $R_L = 2$ kΩ. Assume that v_A may vary from 0 to 10 V; then if $R_B = 100$ kΩ, the base current will be given by

$$i_B = \frac{v_A - V_{BE}}{R_B}$$

and thus varies between 0 and approximately 94 µA (if we assume that $V_{BE} \simeq$ 0·6 V). The operating point of the transistor will thus range from A to B and will remain at all times just within the active region of operation.

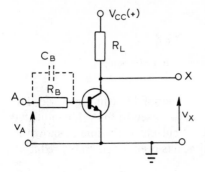

Fig. 15.10 Circuit of inverter or NOT gate

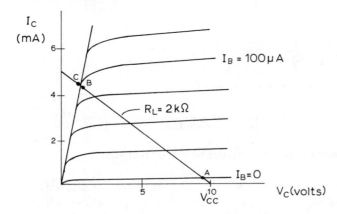

Fig. 15.11 Transistor collector characteristics showing operation of inverter

The speed of transition between these two states may be determined from the charge control model developed in Sections 4.5 and 4.6. The charge shown on Fig. 4.14a needs to be injected into the base and removed from the base by the base current i_B and this is done on a time constant τ_{BF}. The switching currents are shown in Fig. 15.12; the rise and fall time constants are equal and limited by τ_{BF}.

Since the rise time is determined by the speed at which charge may be injected into the base, a faster circuit would be produced if the base current drive were increased by reducing R_B. If R_B were reduced to say 10 kΩ, the available charging current would now be 940 µA and the rise time of the collector current would be 10 times faster. However, the collector current is limited to a maximum of 5 mA by the load resistor of 2 kΩ across the supply voltage of 10 V. Hence the additional base current above that required to maintain the transistor in saturation (approximately 100 µA) will be used to store excess charge in the

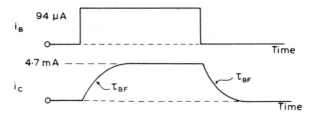

Fig. 15.12 Collector current response of inverter when not bottomed

base. The charge profile in the base is shown by the solid line in Fig. 4.14c. The full collector current will flow when the charge has risen to the value indicated by the broken line. The variation of collector current with time is shown in Fig. 15.13 from which we note that the rise time has been considerably reduced.

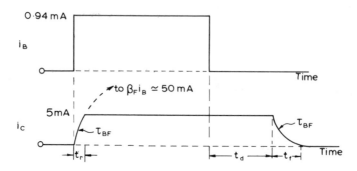

Fig. 15.13 Collector current response of inverter when hard bottomed

In switching off the transistor by reducing the base current to zero we note that there is a time delay, t_d, before any change is noticed in the collector current. This may be seen from Fig. 4.14c since the collector current cannot start to change until the collector–base junction is reverse biased; this will not occur until the charge density in the base at the collector boundary has been reduced to zero by recombination in the base. Thus the delay time is directly related to the amount of overdrive, excess of base current over that required just to bottom* the transistor. Thus a large overdrive will result in a reduction of rise time but also a large increase in delay time. Once the collector base junction has become reverse biased, the collector current falls on a time constant τ_{BF}.

The charge storage delay time may be reduced in a manner similar to that used for the diode, which was discussed in Section 15.1. The base is now switched between a negative and a positive voltage. The switch on rise time is unchanged. However, when switching off the transistor the stored charge may now be removed by a reverse current flowing into the base, in addition to the

* A transistor is bottomed when its collector voltage is limited by the minimum emitter–collector voltage (about 0.2 to 0.3 V).

normal recombination process. The variation of currents and voltages with time is shown in Fig. 15.14.

The above analyses have been based on somewhat idealistic conditions, in which all effects occurring in the depletion layers at collector and emitter have been ignored. This will certainly cause any observed switching waveforms to differ from those deduced above,

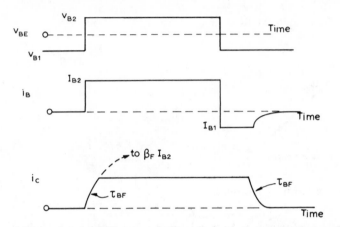

Fig. 15.14 Collector current response of inverter when base is hard overdriven

15.3.2 An inverter or NOT gate

The circuit of Fig. 15.10 may be used directly as a NOT gate since for a high input voltage the output voltage is low and vice versa. The truth table is given in Table 15.6.

If we consider the transistor operating with a supply voltage of 10 V and the two logic levels as before being 0 and 10 V, the voltage table (Table 15.9) may be drawn up taking into account the various typical voltage drops in the transistor.

Table 15.9

V_A	V_X
10	0·3
0	< 10

When the transistor is bottomed the emitter–collector voltage is about 0·3 V. When the base current is zero a small collector saturation current flows causing a small (usually negligible) voltage drop across R_L. The transfer characteristic is plotted in Fig. 15.15 for two values of base series resistor. With $R_B = 100$ kΩ the transistor is only just bottomed when $V_A = 10$ V, and thus the transition from OFF to ON extends over a wide range of input voltage. The noise immunity of such a stage is therefore poor. If R_B is reduced to 20 kΩ the transition between

OFF and ON occurs between 0 and 2 V and the output voltages are stably defined at values of 10 V and 0·3 V.

It has been noted that one of the main limitations to operation of transistors at high speeds is the stored charge in the base. Even if the base is forced negative to turn the transistor off, the speed of removal of the stored base charge is

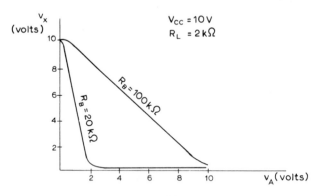

Fig. 15.15 Transfer characteristic of inverter

limited by the magnitude of the series base resistor R_B. If, however, a capacitor C_B is connected across R_B, as shown by broken lines in Fig. 15.10, all this excess charge may, in theory, be removed from the base instantaneously thus eliminating the storage delay. In practice, since the base–emitter voltage is not truly linearly related to charge, this can only give an approximate improvement, valid for a particular voltage.

It is found that the time constant $R_B C_B$ should be of the same order as that of the base charging time constant τ_{BF}, usually about two or three times greater.

A transistor used in this manner is thus a practical implementation of a NOT gate and may therefore be used in conjunction with the diode AND and OR gates to form a complete set of logic elements.

15.4 Diode transistor logic

We have commented earlier that it is often convenient as well as common practice to amalgamate diode AND and OR gates with the inverter to form NAND and NOR gates. An example of a NAND gate, assuming positive logic, is given in Fig. 15.16.

If either input, say A, is at logic **0** (in this case we shall assume 0 volts) then diode D_A will be conducting and P will be clamped at about 0·6 V. Resistors R_2 and R_3 may be so chosen that under this condition the base of the transistor is at a negative voltage; the transistor is therefore cut off and the collector X is at V_{CC} (say +10 V), corresponding to logic **1**.

If both inputs A and B are at logic **1** (say +10 V) then diodes D_A and D_B are both cut off since P is slightly lower than V_{CC}. However, suitable choice of R_1,

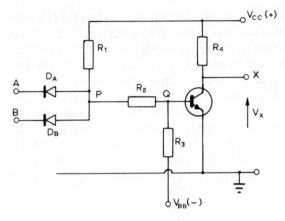

Fig. 15.16 NAND gate using DTL circuit

R_2, and R_3 will ensure that under these conditions the base of the transistor is positive and thus the transistor is conducting. If correctly designed the transistor may be bottomed and hence X will be at about 0·3 V corresponding to logic **0**. The truth table is given by Table 15.10.

Table 15.10

A	B	X
1	1	0
1	0	1
0	1	1
0	0	1

This corresponds to the logic statement

$$\mathbf{X = \overline{AB}}$$

corresponding to a NAND gate. A similar circuit may be constructed to implement the NOR function.

15.4.1 Fan-in and fan-out

The fan-in of a logic network is defined simply as the number of input terminals which are provided to the switch.

Fan-out is a property of the switch which limits the number of similar switches which may be connected in parallel to its output. The fan-out is defined as the number of similar gates which may be connected without impairing operation. The fan-out is often specified in terms of the number of unit loads which may be connected to the circuit; a unit load is a circuit which will model the input of a gate in the state where it takes the heaviest current.

The fan-out of the circuit is then defined as the number of unit loads which may be connected in parallel across the output before the circuit ceases to function correctly. With reference to the circuit of Fig. 15.16 we may see that the maximum input current to, say, terminal A occurs when $V_A = 0$ volts and all other inputs are high. The input current is then approximately equal to V_{CC}/R_1 and this is the unit load current I_u.

Let us assume a fan-out of n; the total collector current flowing will then be $nI_u + V_{CC}/R_4$. Since the base current is sensibly independent of load and determined exclusively by the resistors R_1, R_2, and R_3 we see that the transistor will come out of the bottomed state when the collector current is greater than $\beta_F I_B$. Using the most pessimistic value for β_F we may determine the maximum value of n.

15.5 Diode transistor logic IC form

DTL gates are available in integrated circuit form and this necessitates some modification of the circuit of Fig. 15.16. In particular, resistor R_3 is usually high in discrete transistor form, so that the transistor may be switched between small positive and negative voltages. This is unsuitable for integrated circuit construction. It is also desirable to operate the circuit from a single power supply, and thus we are confronted with the problem of obtaining a voltage drop between P and Q without using a negative supply. One solution to this is to use two forward biased diodes as in Fig. 15.17.

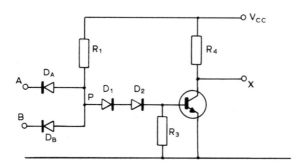

Fig. 15.17 NAND gate with single power supply

When one or more of the inputs is at logic **0** (0 volts) the junction P is at approximately 0·6 V. Since the transistor will start to conduct at a threshold base–emitter voltage of, say, 0·4 V, direct connection between P and the base of the transistor would allow it to conduct and the output X would not be at logic **1** as desired. However, the inclusion of diode D_1 (having a conduction voltage of about 0·6 V) will keep it truly cut off. In fact the output at A will now need to exceed 0·4 V before the transistor will conduct. The noise immunity to logic **0** is thus 0·4 V. Inclusion of a second diode D_2 will require the input to exceed 1·0 V before conduction and hence greater noise immunity is achieved. Sometimes a third series diode is included to give even greater noise immunity.

15.6 Transistor–transistor logic

One of the most common forms of logic circuit in IC construction is TTL. The basic form of a TTL NAND gate is shown in Fig. 15.18. If this is compared with the DTL circuit of Fig. 15.17 it will be noted that the two diodes D_A and D_B have been replaced by the emitter–base diodes of the multiple emitter transistor Tr1. Similarly, the diodes D_1 and D_2 have been replaced by the single base–collector diode of Tr1.

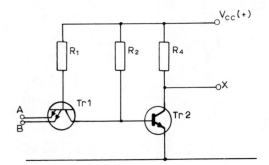

Fig. 15.18 Basic TTL logic NAND gate

The operation of the circuit is thus very similar to the operation of a DTL gate. Consider the circuit of Fig. 15.18 with $V_{CC} = 10$ and two logic levels **0** (= 0 volts) and **1** (= 10 V).

If both A and B are at +10 V, the multiple emitter of Tr1 is cut off. The collector diode of Tr1 and emitter diode of Tr2 (in series with it) are both forward biased and hence base current will be injected into Tr2 which, for a suitable choice of R_4, will be bottomed and X will be at logic **0**.

If either input A or B is at logic **0**, the appropriate emitter–base diode will be forward biased and the base of Tr1 will be at a low voltage. The collector of Tr1 is reverse biased and therefore it will be bottomed at about 0·3 V. This will not give sufficient forward bias on the base of Tr2 to cause appreciable conduction, and thus Tr2 will be effectively cut off and X will be at logic **1**.

The TTL circuit is eminently suitable for construction in IC form. However as it stands the fan-out is low and some form of power drive circuit is required.

15.6.1 Integrated circuit form of TTL NAND gates

Figure 15.19 shows the circuit of an integrated TTL NAND gate incorporating an output stage which allows considerably greater fan-out than the simple circuit. Another disadvantage of the simple circuit of Fig. 15.18 is that when the output transistor changes from ON to OFF any capacitors attached between X and earth require charging through R_4. The main contribution to the capacitance loading at the output is the successive stages of logic which will be connected. Thus, increase in the fan-out will increase the capacitive loading and thus increase the response time of the circuit. An improvement may be obtained by reducing the effective output resistance of the circuit to switch off transients.

(The switch-on transient is not so serious as the stray capacitance discharges to ground through the base-emitter diode which has a very low resistance.)

The first two transistors of the circuit of Fig. 15.19 act in a manner similar to those of Fig. 15.18. Resistor R_2 has been omitted but when Tr1 is switched ON, Tr2 is still kept biased off by the small voltage drop across R_5.

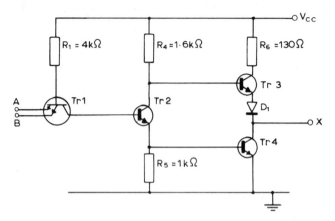

Fig. 15.19 TTL logic NAND gate with output driver circuit

Consider both A and B at logic 1. The emitter–base diode of Tr1 is reverse biased but its collector–base diode is forward biased. Tr2 is now conducting and thus the base of Tr4 is at a small positive voltage causing Tr4 to conduct hard (it is designed to bottom) and point X is at about 0·3 V, logic 0; the collector current for Tr4 will flow through the external load since Tr3 is effectively cut off. This may be confirmed by noting that the base voltage of Tr3 is about 0·9 V (Tr2 is bottomed at 0·3 V and base–emitter voltage of Tr4 is about 0·6 V); there is thus insufficient voltage between the base of Tr3 and X to bias the diode D_1 and transistor Tr3 into appreciable conduction.

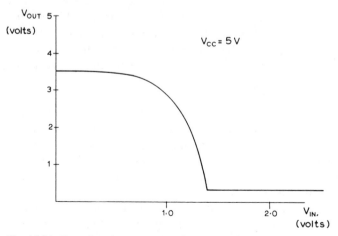

Fig. 15.20 Transfer characteristic of TTL NAND gate

When either A or B is at logic **0**, Tr1 conducts across the forward biased base–emitter junction. The collector junction of Tr1 is now reverse biased and the base of Tr2 is low thus allowing very little conduction through this transistor. Thus, Tr3 is driven hard into conduction and Tr4 is cut off. The advantage of this circuit with regard to fan-out is now obvious. For either transitional change of state one or other of the transistors Tr3 or Tr4 is conducting and hence the stray output capacitance is quickly charged and discharged on changes of state.

The transfer characteristic of the TTL gate of Fig. 15.19 is shown in Fig. 15.20. Assuming that the gate is being operated with a supply voltage of 5 V the two logic levels will be about 3·5 V and 0·3 V. From Fig. 15.20 we may see that the noise immunity of this circuit is about 0·6 V for logic **0** and 2·0 V for logic **1**. These values will be somewhat reduced at low temperatures and slightly increased at elevated temperature.

The manufacturers quoted value of fan-out for this circuit is 10.

15.7 Interconnection of digital gates

Integrated circuit digital gates are currently made in a wide variety of forms, NAND, NOR, AND, OR, INVERTER circuits and single chips containing two or more such circuits. The basic unit from which they are constructed, using TTL design, is the NAND gate. Note that a NAND gate with only one input used and all others connected to logic **1** forms a NOT gate. Similarly a NOR gate with all inputs except one connected to logic **0** is identical to a NOT gate.

It is interesting to note that any logic function may be implemented using only NAND gates (alternatively NOR gates may be used throughout). We shall not attempt to prove this statement in the general case but show how we can implement one specific logic function using NAND gates alone. To assist in this it is desirable to extend our study of the theory of logic networks.

15.8 Boolean logic theorems

All of the following theorems may be proved by drawing up a truth table incorporating every possible combination of input variable

(a) $A + 1 = 1$

The truth table for the function $X = A + 1$ is shown in Table 15.11; this shows

Table 15.11

A	X
1	1
0	1

that $X = 1$ for all values of A.

(b) $\mathbf{A.1} \quad = \mathbf{A}$

(c) $\mathbf{A+0} = \mathbf{A}$

(d) $\mathbf{A.0} \quad = \mathbf{0}$

(e) $\mathbf{A+\overline{A}} = \mathbf{1}$

(f) $\mathbf{A.\overline{A}} \quad = \mathbf{0}$

The commutative rule also holds:

(g) $\mathbf{AB} \quad = \mathbf{BA}$

(h) $\mathbf{A+B} = \mathbf{B+A}$

and the distributive rule:

(i) $\mathbf{A(B+C)} = \mathbf{AB+AC}$

(j) $\mathbf{A+BC} \quad = \mathbf{(A+B)(A+C)}$

A pair of extremely valuable transformations are known as De Morgan's theorems, namely

(k) $\overline{\mathbf{A+B}} = \overline{\mathbf{A}}.\overline{\mathbf{B}}$

(l) $\overline{\mathbf{AB}} \quad = \overline{\mathbf{A}}+\overline{\mathbf{B}}$

The above theorems may be used to reduce complex logical statements to more simple ones before any attempt is made at realization by electronic circuitry.

An example of simplification will illustrate the use of some of the theorems. Consider

$$
\begin{aligned}
\mathbf{X} &= \overline{\mathbf{C}}\,(\overline{\mathbf{A+B}})+\mathbf{C}\,(\overline{\mathbf{A}}+\overline{\mathbf{B}})+\mathbf{A}\,\overline{\mathbf{B}}\,\mathbf{C}+\mathbf{A}\,\mathbf{B}\,\mathbf{C} \\
&= \overline{\mathbf{C}}\,\overline{\mathbf{A}}\,\overline{\mathbf{B}}+\mathbf{C}\,\overline{\mathbf{A}}+\mathbf{C}\,\overline{\mathbf{B}}+\mathbf{A}\,\mathbf{C}\,(\mathbf{B}+\overline{\mathbf{B}}) && \text{using (k) and (i)} \\
&= \overline{\mathbf{C}}\,\overline{\mathbf{A}}\,\overline{\mathbf{B}}+\mathbf{C}\,\overline{\mathbf{A}}+\mathbf{C}\,\overline{\mathbf{B}}+\mathbf{C}\,\mathbf{A} && \text{using (e) and (g)} \\
&= \overline{\mathbf{C}}\,\overline{\mathbf{A}}\,\overline{\mathbf{B}}+\mathbf{C}\,(\mathbf{A}+\overline{\mathbf{A}})+\mathbf{C}\,\overline{\mathbf{B}} && \text{using (i)} \\
&= \overline{\mathbf{C}}\,\overline{\mathbf{A}}\,\overline{\mathbf{B}}+\mathbf{C}+\mathbf{C}\,\overline{\mathbf{B}} && \text{using (e)} \\
&= \overline{\mathbf{C}}\,\overline{\mathbf{A}}\,\overline{\mathbf{B}}+\mathbf{C}+\mathbf{C}\,\overline{\mathbf{B}}\,(\mathbf{1}+\overline{\mathbf{A}}) && \text{using (a)} \\
&= \overline{\mathbf{A}}\,\overline{\mathbf{B}}\,(\mathbf{C}+\overline{\mathbf{C}})+\mathbf{C} && \text{using (i)} \\
&= \overline{\mathbf{A}}\,\overline{\mathbf{B}}+\mathbf{C} && \text{using (e)}
\end{aligned}
$$

15.9 Symbols for logic elements

The symbols which will be used for AND, OR, NAND, and NOR gates throughout this book are shown in Fig. 15.21a to d for three inputs; any number of inputs up to eight is quite common. The symbol for a NOT gate or inverter is shown in Fig. 15.21e; it should be obvious that a NOR gate with all inputs except one set at logic **0** will act as an inverter, as will a NAND gate with all inputs except one set to logic **1**.

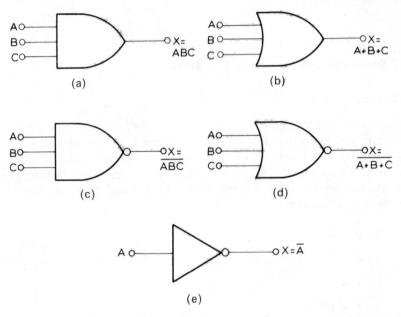

Fig. 15.21 Conventional symbols for logic gates: (a) AND, (b) OR, (c) NAND, (d) NOR, (e) NOT

15.10 Logic network design

Once a logic relationship has been expressed in its simplest form it is relatively straightforward to implement it using logic function blocks. The logic function used as an example in Section 15.8 has a simplified form

$$X = \overline{A}\,\overline{B} + C$$

This may be realized by the network shown in Fig. 15.22. Frequently both the input functions **A, B, C** and their complements $\overline{A}, \overline{B}, \overline{C}$ are available and thus the two input inverters would not be needed.

It may be seen that this implementation requires four gates of three different types. It is often much more convenient to construct a circuit using only one type of gate, either NAND or NOR. This is always possible but will sometimes involve the use of extra gates; the standardization introduced may sometimes offset the disadvantage.

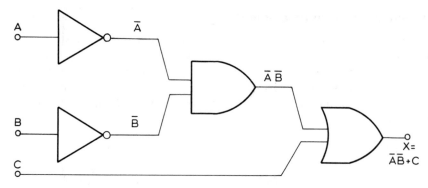

Fig. 15.22 Simple logic network to realize $X = \overline{AB} + C$

15.11 Binary arithmetic

One very common application of logic circuits is in the arithmetic operations on binary numbers. Every digit of a binary number can have two values, **1** or **0**. Thus logic circuits, which operate with only two input voltage levels, are ideal for manipulating binary numbers. As an example of this arithmetic manipulation we may consider the addition of two binary digits **A** and **B**. We may write out all possible combinations of **A** and **B** thus

$$0+0 = 0$$
$$1+0 = 1$$
$$0+1 = 1$$
$$1+1 = 10$$

We note that in the first three sums there is only a sum digit **S**, whereas in the fourth situation a carry digit **C** occurs. We may thus draw up a truth table for this summation in Table 15.12. From the truth table we may construct the

Table 15.12

A	B	S	C
0	0	0	0
1	0	1	0
0	1	1	0
1	1	0	1

appropriate logic statements

$$S = A\overline{B} + \overline{A}B$$
$$C = AB$$

The circuit for the construction of the carry term, **C**, is relatively straight-forward involving an AND gate only. The circuit required for the sum term **S** comprises what is termed an 'exclusive OR' gate which is defined as 'either **A** or **B** but not both'. We may set about designing such a circuit; one very straight-forward realization is shown in Fig. 15.23 in which is shown both the sum and carry circuits. The complete circuit is known as a half adder. In most computations there is a carry digit brought forward from the next lower order summation and this also needs to be added. A circuit which achieves this is termed a full adder.

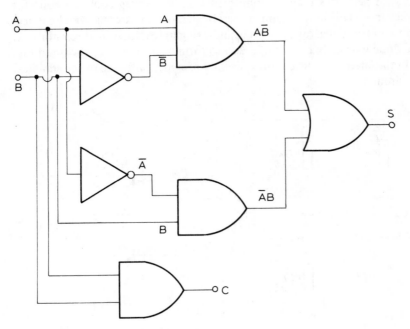

Fig. 15.23 Block diagram of half adder

15.12 Integrated circuit logic

The basic unit from which a logic gate may be constructed is frequently the TTL NAND gate. Other forms of construction such as DTL gates are used but less frequently. In many cases a number of NAND or other gates are incorporated in one package, the limitation of complexity of a circuit being the number of external leads required. One of the most usual forms of package has 14 external leads; two are required for power supplies leaving 12 for logic inputs and output. Thus a single 8 input circuit is possible, dual 5-input, triple 3-input, quadruple 2-input, and hexuple 1-input. Apart from the last, these may be AND, OR, NAND, or NOR gates; the last is only possible in the form of inverters. Such packaging allows a large number of logic units to be accommodated in a small space.

In addition to these basic units, other more complex units are available. As an example, the exclusive OR unit is frequently constructed and encapsulated as a single unit or, more likely, with three others to form a quadruple unit.

15.12.1 Wired connection

It is possible to interconnect two primary logic gates by direct wiring so as to implement a more complex logic function. This is done by using the basic gates without their power driver stages and with the collector of the inverter transistor brought out to an external terminal, no collector load being provided internally. Such a network is known as an open collector circuit.

Two such networks may be connected together with a common collector load resistance. This may be done either by making an external wired connection to two open collector AND gates and using an external load resistor or it may be done internally on a single chip to form a single unit as shown in Fig. 15.24. In the latter case the circuit is usually provided with one of the standard output circuits.

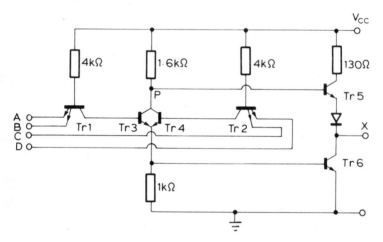

Fig. 15.24 Two open collector NAND gates connected to form an AND-OR-NOT gate

The function of the circuit may be appreciated by studying the behaviour of transistors 1 to 4 in the circuit of Fig. 15.24.

When both inputs A and B are high, transistor Tr3 will conduct and hence P will be low. Similarly when C and D are high Tr4 will conduct and P will be low. Otherwise P will be high since neither Tr3 nor Tr4 is conducting. The operation is thus described by

$$P = \overline{A\,B + C\,D}$$

and the gate is known as an AND-OR-NOT (sometimes AND-OR-INVERT) gate and is available as a standard unit.

Other methods of connecting open collector circuits may be developed which, like the above, may perform more complex functions without the introduction of additional units.

15.13 A fet as a switch

Since the drain characteristics of a fet are very similar to the collector characteristics of a junction transistor, it may also be used as a switch with the gate voltage switched between two values such that at one extreme the transistor bottoms and the drain–source voltage is very low, and at the other extreme the transistor is cut off and the drain current is small.

A typical value of the current flowing when the transistor is off, $I_{D(off)}$, ranges from about 1·0 nA for a junction fet down to 30 pA for a p-channel mosfet.

The drain–source voltage when the transistor is switched on is a fairly strong function of the operating point and it is more customary to quote a value for $R_{DS(on)}$, the resistance between drain and source when the transistor is hard bottomed; this does of course need to be specified at a particular gate–source voltage. A typical value for either a junction fet or a mosfet is from 30 to 500 Ω.

We may compare these values with those for a bipolar transistor. When switched off this will pass a current of the order of the leakage current, about 10 nA, which is larger than that of the mosfet; however, they are both sufficiently small for there to be no strong advantage for the mosfet. When switched on the voltage across a bipolar transistor at, say, 50 mA is about 0·3 V, whereas for a fet it will be of the order of 2 to 3 V. This suggests that bipolar transistors have a considerable advantage.

We must, however, consider other factors. Let us consider, in particular, a p-channel depletion mosfet whose drain characteristics are given in Fig. 2.8. We shall use it in the circuit of Fig. 15.25 as an inverter. The transistor characteristics with load line superimposed are shown in Fig. 15.26. The supply voltage is taken as -20 V and the load resistance as 2·5 kΩ. When the transistor is

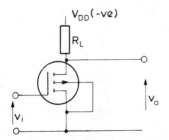

Fig. 15.25 Mosfet as a NOT gate

switched off by application of a gate voltage greater than -5 V, the operating point is at A and the output voltage is at -20 V. When the transistor is switched on, the operating point moves to the vicinity of point B and the output voltage is approximately -3 V; this occurs for a gate voltage less than about -18 V. The two output logic levels of this inverter are -3 V and -20 V corresponding to logic **0** and logic **1**. (When dealing with p-channel fets we shall use negative logic for simplicity of the discussion, although there is no fundamental reason for this choice.) We may now plot a transfer characteristic for this device for input voltages varying from 0 to -20 V. This is shown by the solid line in

Fig. 15.27 from which we can see that for input voltages between about −5 V and −12 V, the circuit is in an active state between the two logic levels of −3 V and −20 V. This is rather a large transitional region, although comparable with that of a bipolar transistor.

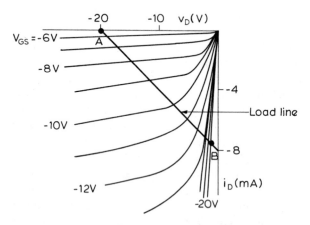

Fig. 15.26 Drain characteristics of mosfet with superimposed load line

If we define the **0** logic level as greater than −5 V and the **1** level as less than −12 V we see that the circuit will perform satisfactorily as an inverter with no degradation of logic level in passing through the stage. Thus a succession of such circuits may be cascaded without cumulative shift in output level.

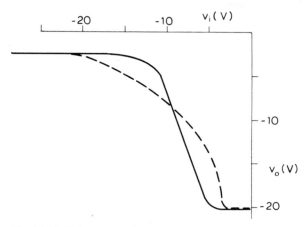

Fig. 15.27 Voltage transfer characteristic of mosfet NOT gate; solid line using resistive load, broken line using active load

The noise immunity is relatively good. Assuming that the nominal logic voltages are −3 V and −20 V we see that the noise immunity to the low input state is 2 V and to the high input state 8 V—quite satisfactory values.

If we attempt to use an enhancement type fet in a switching circuit we shall

see that it is not possible to cascade logic units directly without some intermediate level shifting network. Thus depletion mosfets are almost invariably
used in logic networks.

15.13.1 Fet switches for IC fabrication

In general, it is as simple in integrated circuit construction to fabricate a mosfet
as a linear resistor and hence the tendency has been to replace resistors where
possible by mosfets. We saw in Section 2.3 that at very low drain voltages a
field-effect transistor acted as an efficient resistor. However, in switching circuits the drain current is not restricted to low values and hence this method of
use of a fet is inapplicable. Again in switching circuits we are not primarily
concerned with the linearity of a resistance with current and thus other possibilities are available.

For enhancement devices it is convenient to connect the gate to the drain;
this does not bias the gate with a fixed voltage, since the drain voltage is varying
when used as a two-terminal device. We may see the form of the characteristic
obtained by selecting those points on Fig. 2.8a at which $V_{GS} = V_{DS}$ and plotting
the resulting curve of I_D against V_{DS}. This is shown in Fig. 15.28.

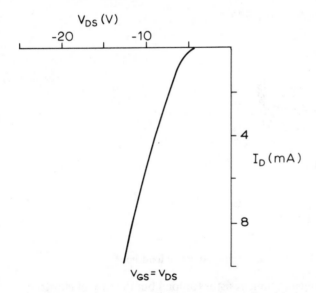

Fig. 15.28 Characteristic of enhancement mosfet with gate connected to drain

The characteristic which we have obtained in Fig. 15.28 represents a nonlinear relationship between current and voltage and thus this two-terminal
device may be used in place of the load resistor in the inverter of Fig. 15.25. The
inverter circuit suitable for IC construction is shown in Fig. 15.29. Transistor
Tr2 has its gate connected to the drain and hence acts as the load resistor to
transistor Tr1. We may plot the load line for this on the characteristics of Tr1
as shown on Fig. 15.30, for a supply voltage of 20 V; the load line is obtained

by plotting I_D against V_{DS1} where $V_{DS1} = V_{DD} - V_{DS2}$ and the relationship between V_{DS2} and I_D is given in Fig. 15.28. The transfer characteristic relating input to output voltage is shown by the broken line in Fig. 15.27. The two logic levels are still about -3 V and -20 V although the transition between these two levels is less clear cut. The noise immunity of this circuit is therefore poorer

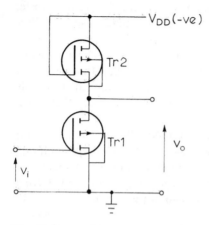

Fig. 15.29 Mosfet NOT gate using active load

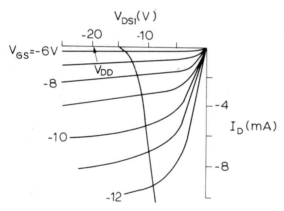

Fig. 15.30 Drain characteristics with superimposed active load line

than that of the circuit using an ohmic resistor for load but the ease of construction in integrated circuit form usually overrides this consideration. When modified for integrated circuit construction, the substrates of both transistors must be earthed and the result is that the source–substrate voltage of Tr2 is now variable. The overall performance of the circuit is little changed as far as static operation is concerned.

15.13.2 Complementary symmetry circuits

Since field-effect transistors of both channel types are available, complementary symmetry circuits may be constructed. These have the considerable advantage that the power dissipated in either the ON or OFF states is negligibly small, but they are not so useful in integrated circuit fabrication as additional manufacturing processes are required to construct transistors of the two types on a single chip. An inverter using a complementary pair of fets is shown in Fig. 15.31. It may be noted that both transistors are enhancement mode types since these permit the direct cascading of stages. The gates of both transistors are driven by the input so that one or other transistor is always cut off for the extreme values of input voltage and no current flows. Only during transition between these two values does appreciable current flow.

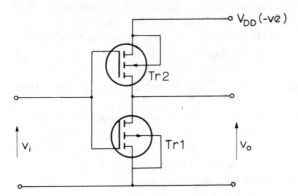

Fig. 15.31 Complementary mosfet NOT gate

Assume the supply voltage $V_{DD} = -20$ V. When the input voltage v_i is negative (say at -20 V) the gate–source voltage of Tr2 is zero and hence it is cut off and only leakage current can flow. The gate–source voltage of Tr1 is -20 V permitting the transistor to conduct fully with a very low voltage drop across it. The output voltage, v_o, will thus be very small, of the order of a few millivolts or less, and the current through the circuit will be the leakage current of the order of a few nanoamperes.

When the input voltage is at its other logic level of zero, the roles of the two transistors are reversed and the output voltage rises to approximately 20 V.

The noise immunity of this circuit is very high, being determined by the excess of the input voltage in either state over the cut-off gate voltage of the appropriate transistor. The transfer characteristic of such a circuit shows that the transition from the high to low state occurs over an extremely small range of input voltage.

15.14 Switching speed of fet circuits

The speed of switching of a fet circuit is determined almost entirely by the time required to establish the corresponding charge in the channel. If we assume that the switch is driven from a low impedance source, R_1, we may determine the time to charge the gate–source capacitance through this source resistance. In general, this time constant will be small compared with others in the external circuit.

The dominant effect in switching a mosfet switch exists at the output where the loading capacitance on the stage may be relatively high, contributed to in part by the input capacitance of subsequent stages; the magnitude of this capacitance is determined principally by the fan-out of the circuit.

Considering first an inverter using a resistive load, the switch-off from low output voltage to high is limited by the stray capacitance C_O at the output charging up to V_{DD} through the load resistor R_L. If we assume C_O to be of fixed magnitude, this will have a time constant of $R_L C_O$ and a rise time of $2 \cdot 2 \ R_L C_O$.

When the transistor is switched on the output capacitance needs to discharge through the transistor, which presents a very variable resistance depending upon the voltage across it. Analysis of this is cumbersome but it may be shown that the fall time is up to 10 times the rise time and its start may well be delayed after the change of gate voltage by a delay time which is of the same order of magnitude as the rise time. Typical values for a p-channel mosfet working into a 10 kΩ load give a rise time of 25 ns, fall time of 240 ns, and delay in fall time of 50 ns.

If a second fet, connected as a nonlinear resistor, is used as a load, as in Fig. 15.29, the analytic problem is very complex. The results show that the rise time using a fet load is increased by a factor of about eight above that using a resistive load working into the same load capacitance. The fall time and delay will be of a similar magnitude to that with a resistive load. In integrated form, as noted in Section 15.13.1, all substrates must be common and this results in a further deterioration in the rise time of the circuit; however in this case it is likely that the output capacitance will be very much lower than that in a discrete circuit.

15.15 Logic gates using fets

We have already studied the use of a field-effect transistor as an inverter or NOT gate. Several fets in conjunction, either of one type or in complementary pairs, may be used to construct any of the basic forms of logic gate, namely AND, OR, NAND, NOR. Since either of the last two comprise a set by themselves, we shall limit our study to these. Furthermore since the majority of such gates are constructed in integrated circuit form we shall restrict ourselves to those circuits which use a fet in place of an ohmic load resistor.

15.15.1 Fet NOR gate

The circuit of a NOR gate for two inputs is shown in Fig. 15.32, using p-channel enhancement type fets. Transistor Tr3 acts as a nonlinear load as discussed in Section 15.13.1. If v_A is negative Tr1 will be conducting hard and, regardless of the potential on the other gate B, current will flow through Tr3 causing v_o to be at a low (magnitude) potential. The same situation will exist if v_B is negative. If both the inputs are at 0 volts, neither of the two input transistors will conduct and the output voltage will approach V_{DD}. If we use negative logic to describe this situation where the negative value of V_{DD} corresponds to logic 1 and zero

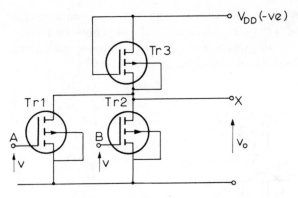

Fig. 15.32 Mosfet NOR gate

volts to logic **0**, we may write the truth table as in Table 15.13. This relationship

Table 15.13

A	B	X
1	0	0
1	1	0
0	1	0
0	0	1

between the input and output variables may be expressed by

$$X = \overline{A+B}$$

which is the Boolean relationship for a NOR gate. The circuit may be readily extended to more than two inputs by connecting additional transistors in parallel with Tr1 and Tr2.

15.15.2 Fet NAND gate

A very similar gate may be constructed to satisfy the NAND function. It utilizes two series transistors (for a two-input gate) with a third transistor acting as a load. If both input voltages are sufficiently negative both transistors will conduct and the output will be low. If either input is zero, conduction will be prevented and the output will rise to V_{DD}. The number of transistors is limited to about four.

15.15.3 Complementary symmetry circuits

The principle of complementary symmetry may be used in the construction of both NAND and NOR gates. Each input requires one pair of transistors. The circuit for a NOR gate is shown in Fig. 15.33. To achieve the NOR function it is necessary for one of the parallel transistors Tr1 and Tr2 to be conducting at the same time as one of the series transistors Tr3 and Tr4 is non-conducting. This can only be achieved by using one pair of transistors driven by each input. A similar circuit may be designed to achieve the NAND function.

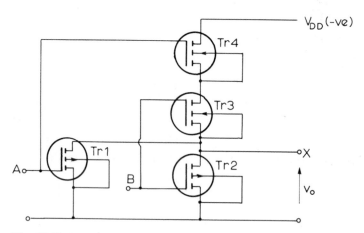

Fig. 15.33 Complementary mosfet NOR gate

These two circuits have the same advantages and drawbacks which were discussed in relation to the inverter circuits in Section 15.13.2.

References

HARRIS, J. N., *et al.*, *Digital Transistor Circuits*, Wiley, 1966, Chapters 1, 2, 4–6.

MILLMAN, J. and TAUB, H., *Pulse, Digital and Switching Waveforms*, McGraw-Hill, 1965, Chapters 9 and 10.

NASHELSKY, L., *Digital Computer Theory*, Wiley, 1966, Chapters 5 and 6.

REEVES, C. M., *An Introduction to Logical Design of Digital Circuits*, Cambridge University Press, 1972, Chapters 3 and 4.

LYNN, D. K., *et al.*, *Analysis and Design of Integrated Circuits*, McGraw-Hill, 1967, Chapters 9 and 10.

HIBBERD, R. G., *Integrated Circuits*, McGraw-Hill, 1969, Chapters 4 and 5.

SCARLETT, J. A., *Transistor-Transistor Logic and its Interconnections*, Van Nostrand Reinhold, 1972, Chapters 3–9.

MORRIS, R. L. and MILLER, J. R., *Designing with TTL Integrated Circuits*, McGraw-Hill, 1971, Chapters 1–6.

CARR, W. N., *MOS/LSI Design and Application*, McGraw-Hill, 1972, Chapter 4.

Problems

15.1 The diode AND gate of Fig. 15.7 works from a supply of 5 V using a load resistor of 1·2 kΩ and diodes which may be assumed to conduct fully for forward voltage greater than 0·6 V. The **0** logic level is defined as any voltage less than 2·5 V and the **1** logic level as voltages greater than 2·5 V. If both A and B are connected together, how many of these gates may be cascaded, if the input level is zero volts, to operate correctly?

15.2 An attempt is made to implement the function

$$X = AB + C$$

by interconnecting the output of an AND gate to one of the inputs of an OR gate as shown in Fig. P15.1. Write the truth table to confirm that this should achieve this. The input logic levels are 5 V and 0 volts, and output logic level **1** between 3·5 and 5 V and level **0** between 0 and 1 V. If $R_1 = 1 \text{ k}\Omega$ and $R_2 = 10 \text{ k}\Omega$ show that the circuit will realize this functional relationship (assume a conducting voltage across the diodes of 0·5 V).

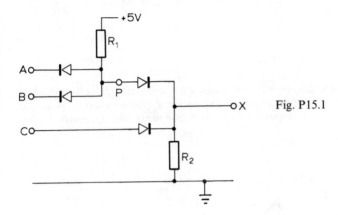

Fig. P15.1

15.3 By setting out a table of voltages for all combinations of input voltage, show that the circuit of Fig. P15.1 will not work if $R_1 = R_2 = 1 \text{ k}\Omega$. In particular determine the voltages at P and X when $v_A = v_B = 5$ V and $v_C = 0$.

15.4 Design a similar logic circuit using diode gates to implement the function

$$X = (A + B)C$$

Which set of input variables is likely to cause malfunction of the circuit? What minimum ratio of load resistors in the AND and OR gates is necessary for acceptable operation?

15.5 The circuit of Fig. 15.10 (without C_B) is used as an inverter. The transistor has a large signal current gain $\beta_F = 45$; $V_{CC} = 10$ V and $R_L = 2\cdot2 \text{ k}\Omega$. Determine the minimum value of R_B in order that the transistor is fully bottomed when $v_A = 10$ V. Make reasonable assumptions for transistor voltages when conducting.

15.6 The circuit of Problem 15.5 has R_B set at 12 kΩ. The input voltage v_A is varied from -10 V to $+10$ V. Estimate the value of v_A when the transistor (a) just bottoms, (b) is cut off. Sketch the transfer characteristic between v_A and v_X.

15.7 What is the effect on the transfer characteristic of increasing R_B to 150 kΩ in the circuit of Problem 15.6.

15.8 Sketch, without dimensions, the charge profile in the base of the transistor of Problem 15.6 when (a) $R_B = 12 \text{ k}\Omega$, (b) $R_B = 150 \text{ k}\Omega$. Sketch the collector current response when the base voltage is suddenly changed from 0 to 10 V and then at some later time is restored to zero.

15.9 The inverter circuit of Fig. P15.2 uses a silicon transistor whose large signal current gain lies in the range 20–100. The logic levels are 0·3 V and 12 V. Can the circuit maintain these output logic

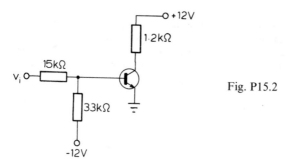

Fig. P15.2

levels for all randomly selected transistors? If not, what will be the output levels for the best and the worst transistors? If the manufacturing probability distribution of transistors having a given value of β_F is given by Fig. P15.3 what fraction of the circuits will fail to meet the specification?

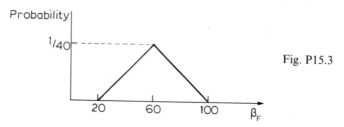

Fig. P15.3

15.10 The circuit of Fig. 15.16 employs a transistor with $\beta_F = 60$; $R_1 = 12$ kΩ, $R_2 = 12$ kΩ, $R_3 = 100$ kΩ, $R_4 = 1·8$ kΩ, and the supply voltages are $+10$ V and -10 V. Determine the voltages at points P, Q, and X for all combinations of $+10$ V and 0 V on the inputs A and B.

15.11 What is the minimum value of β_F for which the circuit of Problem 15.10 will operate?

15.12 What is the minimum value of R_4 in the circuit of Problem 15.10 for which the circuit will still operate?

15.13 What is the current flowing in to terminal A of the circuit of Problem 15.10 when terminal B is at 10 V and terminal A is at (a) 10 V, (b) 0 V? Determine unit load for this circuit.

15.14 What is the fan-out of the circuit of Problem 15.10?

15.15 The circuit of Fig. 15.17 is used as a NAND gate. $V_{CC} = 10$ V, $R_1 = 4$ kΩ, $R_4 = 1·6$ kΩ, $R_3 = 1·0$ kΩ. Logic input levels are 0·3 V and 10 V. Determine the currents and voltages in the circuit for all input voltage combinations.

15.16 Repeat Problem 15.15 for the circuit of Fig. 15.18. Assume $R_2 = 22$ kΩ.

15.17 Determine the voltages throughout the circuit of Fig. 15.19, where $V_{CC} = 10$ V, (a) when $V_A = V_B = 0·3$, (b) when $V_A = V_B = 10$ V. A value of 50 may be assumed for β_F for all transistors.

15.18 Assuming $\beta_F = 50$, what is the maximum current which may be drawn from X in the circuit of Fig. 15.19 when it is at logic 0? Logic 0 extends from 0 to 0·3 V.

15.19 What is the current drawn from source when $V_A = 0$ (V_B may be assumed at 10 V) in the circuit of Fig. 15.19? Estimate the fan-out of the circuit.

15.20 The symbol $\oplus$ is used to represent the exclusive OR function. Verify by drawing up a truth table that

$$(A \oplus B) \oplus C = A \oplus (B \oplus C)$$
$$= A \oplus B \oplus C$$

15.21 A, B, C in Problem 15.20 represent three binary numbers. Show that the sum S of A, B, and C is given by

$$S = A \oplus B \oplus C$$

15.22 Obtain a logic expression for the carry digit Cy in Problem 15.21.

15.23 Implement the logic functions

$$S = A \oplus B \oplus C$$

and $\qquad Cy = AB + BC + CA$

using NAND gates only.

15.24 The inverter of Fig. 15.25 is used with a mosfet whose characteristics are given in Fig. 15.26. Assume $V_{DD} = -20$ V and $R_L = 2.5$ kΩ. Determine the output voltage when $v_i = 0$ and -20 V.

15.25 How many inverters as described in Problem 15.24 may be cascaded if the logic level limits are:

$$\text{logic } 0 \text{ corresponds to} < -16 \text{ V}$$
$$\text{logic } 1 \text{ corresponds to} > -3 \text{ V?}$$

How could the specification of logic bounds be modified to allow more stages to be cascaded?

15.26 Using NAND gates only, realize the following functions of two inputs (a) NOT, (b) NOR, (c) AND, (d) OR, (e) Exclusive OR.

16
Sequential Logic

16.0 Introduction

In Chapter 15 we studied a number of circuits in which transistors are used for switching between two defined logic states. The output of the circuit is, in all cases, determined by the present state of the various inputs.

Another class of circuits are those which may be set into a given state by a certain combination of inputs and which will remain in that state after the removal of the inputs. Such a circuit is considered to have a memory since it is capable of storing a given logic state indefinitely. This chapter will be concerned with such memory circuits; the most important of these is the bistable or flip-flop.

One other characteristic of most memory circuits is that the transition from one state to another, once it has been initiated, is completed by regenerative action within the device. The initial stage of the transition is caused by the trigger voltage at the input; the change in output voltage will be fed back to the input to reinforce the trigger and thus the transition will continue even if the input is removed. This makes for very fast switching speeds; the regeneration is caused by the positive feedback.

We must therefore look at the form of a logic circuit which can remain stable with no externally applied voltage in either the logic 1 state or the logic 0 state. This will form the basic unit of a memory circuit; we may then consider methods of switching this circuit between its two stable states. Such circuits are the basic units from which digital computers are constructed and the science of their interconnection is known as sequential logic.

16.1 Fundamental form of bistable circuit

If we use the logic blocks which were introduced in Section 15.9 we will find that the circuit shown in Fig. 16.1 of two inverters cross connected will satisfy our requirements for a bistable circuit (or flip-flop). We may confirm this by assuming a value for Q and showing that the states of other points in the circuit are logically consistent with this arbitrary choice. If $\mathbf{Q} = \mathbf{1}$, then $\mathbf{B} = \mathbf{1}$ and $\mathbf{A} = \mathbf{0}$; this forces $\mathbf{Q}$ to be $\mathbf{1}$ which is consistent with our original chosen value. Similarly a choice of $\mathbf{Q} = \mathbf{0}$ will also be a stable state which does not violate any logic requirements of the circuit. In addition it may be observed that in all cases the output labelled $\overline{Q}$ is the complement of Q; when $\mathbf{Q} = \mathbf{1}, \overline{\mathbf{Q}} = \mathbf{0}$ and vice versa. The circuit may thus be used to give two complementary outputs.

Although this gives a circuit which has two stable states and therefore satisfies our criterion for a bistable logic block, we must expand the circuit slightly to

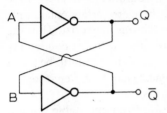

Fig. 16.1 Interconnection of two NOT gates to form a basic bistable circuit

allow us to set it into some assigned state and to change that state as required; in other words we must provide some input terminals to the unit. We shall postpone consideration of this to Section 16.2.

16.1.1 Bistable circuits using bipolar transistors

We may implement this circuit using discrete transistors by substituting, for each of the two inverters, one of the transistor circuits given in Fig. 15.10. With the addition of the two resistors R_3 and R'_3 needed to ensure that the transistor bases are below zero volts when cut off, we obtain the circuit of Fig. 16.2.

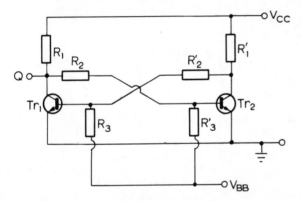

Fig. 16.2 Bistable circuit using bipolar transistors

There are several ways in which we can look at the operation of this circuit. For simplicity of initial study we may assume the circuit to be symmetrical, and as a consequence we may reasonably assume that both transistors are in the same condition, namely at some point in their active region.

Since all circuits are subject to slight fluctuations of voltage due to noise, or other sources, let us consider what happens if the base of transistor Tr1 rises slightly in potential. This will cause the collector current of Tr1 to increase and thus its collector voltage will fall. This will be conveyed to the base of Tr2 via the potential divider R_2, R_3' so that its potential also falls, resulting in a decrease in the collector current of Tr2 and a consequent rise in its collector potential. This rise in potential is conveyed through R_2' and R_3 back to the base of Tr1.

Thus an initial assumed rise in base potential of Tr1 will result in a further rise in its potential and this cumulative action will continue until Tr1 is fully bottomed and Tr2 is cut off. As we shall see below there are certain constraints on the values of the resistors in order that these conditions may exist. Thus the only stable conditions that the circuit can adopt are either with Tr1 conducting and Tr2 cut off or vice versa. If we consider the output of the circuit to be taken from Q, the collector of Tr1, and using positive logic, Q can be either at logic **1** or logic **0**.

Before proceeding to a discussion of the transition from one stable state to another it is profitable to determine the conditions necessary to obtain one of these stable states. Let us assume that we need to design the circuit such that Tr1 is conducting and bottomed and Tr2 is cut off. We may draw a simple equivalent model for this on the assumption that when bottomed both the collector–emitter voltage and base–emitter voltage are zero. When cut off we assume that both base and collector currents are zero. Thus the simple model of Fig. 16.3 applies to this condition.

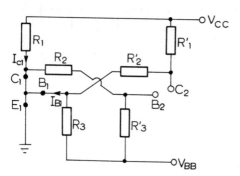

Fig. 16.3 Circuit model of bistable circuit with Tr1 conducting

For Tr1 to be bottomed,

$$\beta_F I_{B1} \geqslant I_{C1} \tag{16.1}$$

Now
$$I_{B1} = \frac{V_{CC}}{R_1' + R_2'} + \frac{V_{BB}}{R_3}$$

(V_{BB} is numerically negative for an n-p-n transistor)

and
$$I_{C1} \simeq \frac{V_{CC}}{R_1}$$

This neglects the current flowing through R_2 and R_3; this is justified as these resistors are usually large compared with R_1. Hence this requirement gives the condition

$$\frac{V_{CC}}{R_1' + R_2'} + \frac{V_{BB}}{R_3} \geqslant \frac{V_{CC}}{\beta_F R_1} \tag{16.2}$$

For Tr2 to be cut off

$$V_{B2} < 0 \tag{16.3}$$

From Fig. 16.3 we see that,

$$V_{B2} = \frac{R_2}{R_2 + R_3'} V_{BB}$$

and since V_{BB} is negative condition 16.3 is always satisfied. It is a relatively straightforward matter to modify the above inequalities to include the effects of finite voltages which exist when a transistor is bottomed.

To summarize the above, we see that one transistor is always hard ON and the other OFF. If we consider the collector of Tr1 as the output corresponding to Q in Fig. 16.1, it will be at logic 1. It is also apparent that the output from the collector of Tr2 must be $\overline{Q}$, the complement of Q.

16.1.2 Bistable circuit using field-effect transistors

The basic bistable circuit using field-effect transistors may be constructed by using two inverter circuits such as that shown in Fig. 15.29. These may be interconnected in a manner similar to the bipolar transistor circuit. We shall study flip-flop circuits using mosfets in greater detail when we have considered the problems of triggering the circuit between its two stable states.

16.2 R-S flip-flop

We now need to investigate the manner in which the state of the circuit may be changed between the two stable states. The simplest circuit whose state can be changed by an external stimulus is the R-S flip-flop. This is constructed from the circuit of Fig. 16.1 by replacing the two inverters by NAND gates and adding two further inverters as shown in Fig. 16.4. We may now write the truth table for this system as shown in Table 16.1, where $Q_{n+\delta}$ and $\overline{Q}_{n+\delta}$ are the values of the output variable in the stable state immediately after application of the two inputs; Q_{n+1} and $\overline{Q}_{n+1}$ are the values of the output variable when the inputs have been removed and both R and S have reverted to the 0 state. We may observe from this that when both inputs are at zero the output variables are identical with their previous values. When $R = 1$ and $S = 0$, the output variables are set to the same values as the input, namely $Q = 1, \overline{Q} = 0$. Similarly when $R = 0$ and $S = 1$, the output is identical with the input at $Q = 0$, $\overline{Q} = 1$ regardless of its original state. When $S = R = 1$ exists at the input terminals the output state of $Q = \overline{Q} = 1$ is stable. However as soon as the input is removed and S and R both revert to 0, this is no longer a permitted state

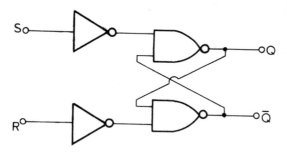

Fig. 16.4 R-S flip-flop using NAND and NOT gates

and the output of the circuit must revert to either the $1, 0$ or $0, 1$ condition; the choice between these two depends upon lack of symmetry in the circuit and is thus random between different circuits. It is thus obvious that the input state $R = S = 1$ must not be allowed to exist at the input to an R-S flip-flop. This may be relatively easy to achieve in principle by always deriving the R input as the complement of the S input. If there were no time delays in the circuit we

Table 16.1

S	R	Q_n	$\overline{Q}_n$	$Q_{n+\delta}$	$\overline{Q}_{n+\delta}$	Q_{n+1}	$\overline{Q}_{n+1}$
0	0	1	0	1	0	1	0
0	0	0	1	0	1	0	1
0	1	1	0	0	1	0	1
0	1	0	1	0	1	0	1
1	0	1	0	1	0	1	0
1	0	0	1	1	0	1	0
1	1	1	0	1	1	Indeterminate	
1	1	0	1	1	1	Indeterminate	

could always be certain that the two inputs never overlapped. However, small delays in the inverter and associated circuits may cause a slight overlap of the two inputs at the time of transition between the two states and thus allow the R-S flip-flop to be set into its indeterminate state. Care must be taken to ensure that this does not occur.

The truth table (Table 16.1) may be rewritten in the more compact form of Table 16.2 which shows the above relationships more clearly.

To summarize, an R-S flip-flop will set the output to the same condition as the input (so long as the two inputs are complementary) and it will remain in

Table 16.2

S	R	Q_{n+1}	$\overline{Q}_{n+1}$
0	0	Q_n	$\overline{Q}_n$
0	1	0	1
1	0	1	0
1	1	Indeterminate	

that state when the input is removed. It will thus act as a store of the input data.

A block diagram showing an alternative form of realization of an R-S flip-flop is shown in Fig. 16.5 using two NOR gates. This may be shown to have an identical truth table to that given in Table 16.1.

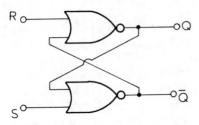

Fig. 16.5 R-S flip-flop using NOR gates

The R-S flip-flop is the basic unit from which a large number of other memory units are derived. The conventional symbol used for it is shown in Fig. 16.6.

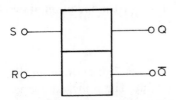

Fig. 16.6 Conventional symbol for R-S flip-flop

16.2.1 R-S flip-flop using bipolar transistors

One form of implementation using discrete transistors is shown in Fig. 16.7. Transistors Tr1 and Tr2 together with load resistor R_1 and input resistor R_2 acts as one NOR gate, the other pair of transistors as the second NOR gate. The circuit thus performs the function of an R-S flip-flop.

16.2.2 Transition between stable states

We have seen how it is possible to change the state of a flip-flop by applying an appropriate voltage at the input. We now need to study the manner in which the circuit switches between states in order to determine the speed of operation and the way in which transition may be initiated.

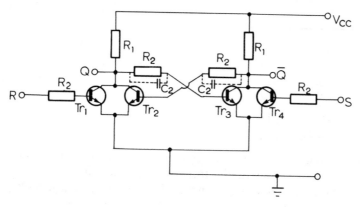

Fig. 16.7 R-S flip-flop using bipolar transistors

Let us assume that the circuit of Fig. 16.2 is in the state in which Tr1 is conducting (ON) and Tr2 is OFF. Let us apply a small negative voltage pulse to the base of Tr1 causing it to cut off. Then the collector of Tr1 will rise, carrying the base of Tr2 with it, and thereby causing Tr2 to conduct. The collector potential of Tr2 will therefore fall and consequently the base of Tr1 will also fall further and the circuit will have changed over into its second stable state. At this point the initial negative pulse may cease since the circuit is now in a stable state. We can thus see that it is possible to trigger the circuit from one state to the other by application of negative pulses (logic 0) to the two bases alternately; alternatively the negative pulses may be connected to the collectors, with the opposite collectors connected to the supply (logic 1).

This is precisely the input conditions which result in correct operation of the R-S flip-flop as may be seen from Table 16.2.

16.2.3 Switching speed

The speed at which the circuit will change state is dependent entirely on the rate of removal or supply of charge around the circuit. The principal charge storage location in the network is in the base of the transistor; we saw in Section 15.3.1 that the delay and rise time of a transition from cut off to bottoming is dependent on the time taken to supply the charge in the base; the reverse transition is usually longer if the transistor is heavily bottomed since considerable excess charge needs to be removed. It has also been shown in Section 15.3.2 that the speed of switching a transistor may be improved by connecting a capacitance in parallel with the base series resistor. The same technique may be utilized here by connecting a capacitor C_2 across each of the coupling resistors R_2 as shown in Fig. 16.7. The value of C_2 is best determined experimentally.

16.3 Clocked R-S flip-flop

The R-S flip-flop is the basic unit in numerous digital circuits. It is frequently desirable to set up the input data on a large number of such circuits without initiating an immediate change of state in the circuit. When all the inputs have been set, a clock pulse is fed to all the binaries which permits the individual circuits to respond to the inputs and take up the appropriate output state. A clocked R-S flip-flop is easily developed from the circuit of Fig. 16.5 by adding a second input to each of the inverters, thus transforming them into NAND gates. This second input is connected to the clock pulse generator, **Ck**, which produces a train of pulses at a fixed frequency, thereby causing all changes of state to occur at regular intervals. This circuit is shown in Fig. 16.8a together with the conventional symbol in Fig. 16.8b.

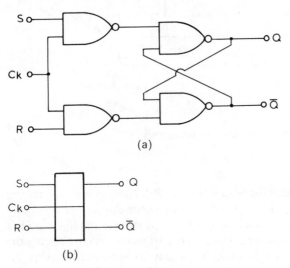

(a)

(b)

Fig. 16.8 (a) Clocked R-S flip-flop using four NAND gates, (b) Conventional symbol for clocked R-S flip-flop

It may be seen that no change of state can occur when **Ck** = **0**; only when **Ck** = **1** will the input data be transferred into the flip-flop and stored there. The truth table of a clocked R-S circuit will be identical with that given in Table 16.2 if it is assumed that the table is only valid when the clock pulse exists and thereafter the outputs remain unchanged.

16.3.1 R-S flip-flop using mosfets

As an example of a clocked R-S flip-flop we shall consider one using mosfets. We shall use p-channel transistors throughout as this will enable us to use similar transistors with gate and drain connected to simulate the load resistors (see Section 15.13.1). A circuit of such an R-S flip-flop is shown in Fig. 16.9. Since we are using p-channel transistors the analysis will be conducted assuming positive logic.

Transistors Tr1 and Tr2 form a NOR gate with Tr3 acting as a load resistor. Similarly Tr4, 5, and 6 form a second NOR gate. As discussed in Section 16.2 the cross connection of two NOR gates will form an R-S flip-flop. Transistors Tr1 and Tr7 in series form an AND gate, as do Tr4 and Tr7 ensuring that any signals on S or R do not operate the circuit unless the clock pulse, **Ck**, is present in addition.

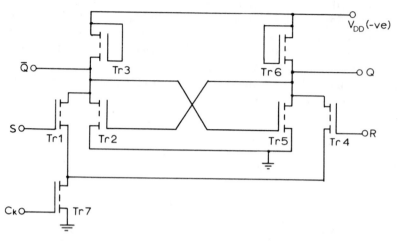

Fig. 16.9 Clocked R-S flip-flop using field-effect transistors

16.4 J-K flip-flop

We have seen in Table 16.2 that the output from an R-S flip-flop is indeterminate when both **R** and **S** = **1**, a situation which may occur during the transition between two states. The use of clocked R-S flip-flops reduces the likelihood of this condition existing to occasions when such a transition occurs during the period of the clock pulse. This indeterminacy may be removed by using the present output of the flip-flop to gate the input and thus to inhibit the existence of this input state which generates the indeterminacy. This produces the J-K flip-flop whose block diagram and circuit symbol are shown in Fig. 16.10a and b.

If we ignore the P and C input initially we shall be able to draw up a truth table for the stable state of the circuit. This is given in Table 16.3. We may see that it is identical to the R-S flip-flop except when both inputs are **1**. Let us assume that the present output state is with $Q_n = 1$ and $\overline{Q}_n = 0$. Then NAND gate ① will be held with its output at **1** regardless of the input, because of the feedback connection between the output $\overline{Q}_n$ and the input of gate ①. NAND gate ② will be enabled (it will respond to other inputs) since the feedback from Q_n to one of the inputs is at logic **1**. Thus if both $J = K = 1$ the outputs of NAND gates ① and ② at points A and B will be logic **1** and **0** respectively and these will now be the inputs to the gates ③ and ④, setting it into the state $Q_{n+1} = 0$, $\overline{Q}_{n+1} = 1$, the only state which is permitted by the logic units. This state is stable and will therefore remain after cessation of the clock pulse. Thus the state $J = K = 1$ forces the output to change its state.

Although we have resolved the difficulty which existed due to the indeterminacy of the output, we have introduced a problem known as 'race-around' condition. In the transition between states which occurs when $J = K = 1$, there will be a finite time delay inherent in the circuit. As a result of the feedback from the output of gates ③ and ④ to the input of gates ① and ② the fed back input will change from its initial value of, let us say, $Q_n = 1$, $\overline{Q}_n = 0$, to its new value after a time delay δt of $Q_{n+\delta} = 0$, $\overline{Q}_{n+\delta} = 1$ and this will immediately cause a further change of output to $Q_{n+2\delta} = 1$, $\overline{Q}_{n+2\delta} = 0$ and so on.

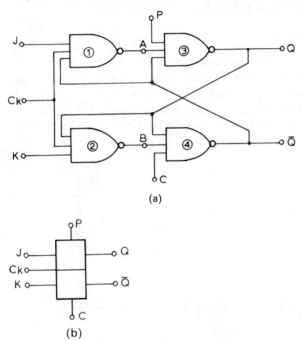

Fig. 16.10 (a) J-K flip-flop using four NAND gates, (b) Conventional symbol for J-K flip-flop

This oscillation between the two possible output states will continue so long as the clock pulse $\mathbf{Ck} = 1$. We see therefore that in order for there to be only a single change of state when $\mathbf{J} = \mathbf{K} = 1$, the clock pulse duration must be less than the delay time δt in the flip-flop. In integrated circuit construction this constraint is very difficult to implement since the internal delays in the circuit are usually much shorter than the clock pulse; in this case it is customary to introduce an additional delay in the feedback connection.

Table 16.3

J	K	Q_{n+1}	$\overline{Q}_{n+1}$
0	0	Q_n	$\overline{Q}_n$
0	1	0	1
1	0	1	0
1	1	$\overline{Q}_n$	Q_n

The inputs Preset, P, and Clear, C, are used to preset the flip-flop into any desired state. If both **P** and **C** are at logic **0** the operation of the circuit is completely inhibited, it will not respond to any input. If both **P** and **C** are set at logic **1** the circuit will operate in the manner outlined above. If **C = 0, P = 1**, while the clock pulse is absent, the output will be set with **Q = 0, Q̄ = 1** and into the opposite state when **C = 1, P = 0**.

16.4.1 Master-slave J-K flip-flop

A more sophisticated approach to the resolution of the race-around problems of a J-K flip-flop mentioned in Section 16.4 is to use the master–slave principle. The circuit consists essentially of two R-S type flip-flops connected in cascade with feedback from output to input. As opposed to the simple J-K flip-flop, in the master–slave the feedback is taken across both the R-S stages. This is shown in Fig. 16.11. During the clock pulse the input data are used to set the state of

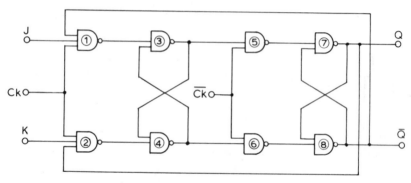

Fig. 16.11 Master–slave J-K flip-flop

the first flip-flop involving gates ① to ④. The second flip-flop does not change its state as it has no clock pulse applied. On cessation of the clock pulse the first flip-flop remains in its set state and this becomes the input to the second flip-flop comprising gates ⑤ to ⑧. This second flip-flop responds and transfers the input data to the output **Q** and **Q̄** during the period between clock pulses when **C̄k = 1**. There is thus no possibility of a race-around condition since there is no loop transmission around the network, only one half of the circuit operating during any part of the clock cycle.

The circuit may be implemented using either bipolar transistors or field-effect transistors. They are both readily adaptable to construction in integrated circuit form.

A master–slave R-S flip-flop may be constructed in a similar manner to the master–slave J-K flip-flop by connecting two R-S flip-flops in cascade and clocking the first with **Ck** and the second with **C̄k**. This allows data to be temporarily stored in the first flip-flop during the interval between pulses. The advantage of this occurs when it is desired to 'read in' data to an R-S flip-flop at the same time as the existing data are being 'read out'.

16.5 T-type flip-flop

A J-K flip-flop may be simply converted into a circuit which may be triggered from a single pulse, usually known as a T type circuit, by connecting the J and K inputs together to form the trigger T input. Such a circuit is also usually clocked from a master clock circuit so that all transitions occur synchronously. As a consequence the trigger pulse applied to T must be of long duration compared with that of the clock pulse. The T-type flip-flop is not normally available as a commercial integrated circuit as it may easily be constructed by connecting together the J and K inputs of a J-K flip-flop.

16.6 D-type flip-flop

Yet another form of flip-flop occasionally encountered is the D type. This enables an R-S flip-flop to be used with a single pulse input. This is applied directly to the S input and through an inverter to the R input. Thus the state when $R = S = 1$ is completely precluded from occurring.

16.7 Shift register

One important application, amongst many, of the flip-flop is as an element in a shift register. The function of a shift register is to store a binary number immediately prior to its use by some arithmetic unit in a computer. This binary number is often available as a time sequence of binary digits (usually abbreviated to *bits*). Each bit may have a value of 1 or 0 and each successive bit represents the magnitude of the next higher power of 2. Thus the binary number 1001_2 (the subscript indicating that the number base is 2) represents $(1 \times 2^3) + (0 \times 2^2) + (0 \times 2^1) + (1 \times 2^0) = 9_{10}$. A sequence of bits such as this is usually known as a binary word.

A shift register capable of storing four bits utilizing four master–slave R-S flip-flops is shown in Fig. 16.12. The register is cleared (all binaries set into the

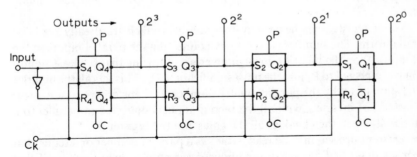

Fig. 16.12 Shift register using R-S flip-flops

zero state) by application of a **0** to all the C inputs which are connected to a common line. Both P and C of all the binaries are then held at **1** to 'enable' them to operate as flip-flops.

The serial binary number is now applied to the input and on the occurrence of the clock pulses, which must be synchronized to the input binary number, the first digit (which is usually the least significant digit) is transferred to the 'master' section of the first R-S flip-flop. On the cessation of the clock pulse this digit is then transferred to the slave section of this flip-flop. On the second clock pulse the second binary digit is stored in the master section of the first flip-flop and the first digit is transferred from the slave of the first flip-flop to the master section of the second. When this clock pulse ceases both bits are transferred to the slave section of their respective flip-flops.

This continues with the bits shifting through the cascade of flip-flops until the complete sequence has been stored (in our case 4 bits); the clock pulses are then interrupted and the binary number remains permanently stored in the group of flip-flops. Although master–slave R-S flip-flops have been used in this illustration, J-K flip-flops are satisfactory alternatives; in addition, the master–slave technique is not essential since the inherent delays in the circuit are sufficient to prevent the occurrence of race conditions.

We may now 'read' this number by noting the state of the various outputs of the individual flip-flops. If the input were the number 1001_2 the outputs would read $Q_4 = 1, Q_3 = 0, Q_2 = 0, Q_1 = 1$. These states could be indicated, for example, by small lamps connected across the output of each flip-flop. It may be noted that the binary number which was initially available as a time sequence has now been converted into one which is available at a single instant of time but available at four points. We may thus take an output from this register by means of four wires over a much shorter time interval; the bits may be read in parallel. When such a shift register is used in a computer, the bits are almost always handled in parallel since this can permit a great reduction in processing time, particularly with long words.

Alternatively, it is possible to continue to operate with the word in sequential form and it may be read out from the shift register as a time sequence by setting the input to zero and applying a sequence of clock pulses. This will feed in a series of 0 bits at the input, and the output, read from Q_1 will give the original number in its correct sequence.

The same register may be used to store numbers which are already available in parallel form. The number is read in by connecting the parallel inputs to the preset inputs P on the binaries through inverters, ensuring that the least significant bit is connected to P_1 and the most significant to P_4. This sets all the binaries into the state given by the parallel binary number. It is customary to use NAND gates instead of inverters with all the second inputs connected in parallel to a WRITE signal which then enables all the gates and the register is set.

The register once set, may be read, either as a parallel number or a sequential number, as described above. In the above description it may be noticed that when the output is read in parallel the word is still left set up in the shift register and may only be removed by operation of the clear pulse. When the word is read sequentially it is automatically destroyed. The data may be retained in the register by connecting the output, from Q_1, through a gate back to the input. Thus as the read pulses applied to Ck shift the word through the register and out along the output line, they are at the same time being read back into the input and thus at the end of the read sequence of clock pulses the original word is once again stored in the register.

16.7.1 Pseudo-random binary number generator

Another interesting application of a shift register is to generate a randomly ordered sequence of numbers. As no truly random process is involved, this sequence can only approximate to a strict random sequence since it will repeat itself once the cycle has been completed. However if the cycle of numbers is long enough the difference between this sequence and a truly random sequence will not usually be of significance. We may illustrate the operation of the circuit by considering a 4-bit number, which is capable of representing 16 different integers. A pseudo-random number generator will produce a sequence of 15 numbers apparently randomly arranged. A circuit which will achieve this is shown in Fig. 16.13.

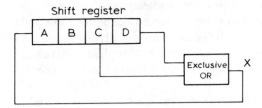

Fig. 16.13 Pseudo-random binary number generator

The input to the exclusive OR gate is taken from the outputs of flip-flops C and D of the shift register. The truth table for this network is shown in Table 16.4.

Table 16.4

A	B	C	D	X	Decimal number
1	0	0	0	0	8
0	1	0	0	0	4
0	0	1	0	1	2
1	0	0	1	1	9
1	1	0	0	0	12
0	1	1	0	1	6
1	0	1	1	0	11
0	1	0	1	1	5
1	0	1	0	1	10
1	1	0	1	1	13
1	1	1	0	1	14
1	1	1	1	0	15
0	1	1	1	0	7
0	0	1	1	0	3
0	0	0	1	1	1
1	0	0	0	recycle	

Table 16.4 shows that, after 15 different arrangements of the state of the shift register, the circuit recycles. The numbers in the right hand column give the decimal equivalent to the binary number stored in the shift register at each clock pulse, assuming that **A** corresponds to the most significant digit; these numbers may be seen to be passably random (for a shift register containing many more bits this is much more convincing).

The circuit shown in Fig. 16.13 is one of several possibilities, all using an exclusive OR gate fed from two or more of the shift register outputs. It may be demonstrated by constructing a truth table that not all combinations of input to the exclusive OR gate will result in a random sequence of 15 numbers; in some cases the sequence may recycle before all permutations have been generated. The design of such circuits using long word shift registers is very complex.

16.8 Asynchronous binary counters

Another common application of flip-flops is the construction of counters. The simplest of these uses a succession of flip-flops to record the number of events in binary form. The events which are to be counted are usually translated into a sequence of pulses. In general the sequence of pulses is applied to the clock-input terminal of one of the standard binary circuits discussed earlier. Alternatively a T type circuit may be used.

One form of such a 4-bit counter is shown in Fig. 16.14. Since both J and K inputs of all the flip-flops are held at logic **1** they will change state only when a

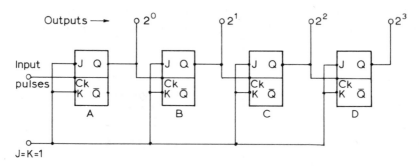

Fig. 16.14 Asynchronous binary counter

pulse is received on the clock input. If we assume that all the circuits are of the master–slave type, we see that the master flip-flop changes state on the rising transition of the input pulse and the slave section causes the output **Q** to change state on the following falling edge of the pulse. Each binary will therefore change state when the output **Q** of the preceding binary changes from **1** to **0**.

We may therefore draw up a truth table for the state of the counter after each successive input pulse; this is given in Table 16.5.

If we consider Q_A output to correspond to the least significant bit, 2^0, and Q_D to the most significant, 2^3, the four outputs from the binaries will record the number of input pulses in the binary code.

Table 16.5

Number of input pulses	Q_A	Q_B	Q_C	Q_D
0	0	0	0	0
1	1	0	0	0
2	0	1	0	0
3	1	1	0	0
4	0	0	1	0
5	1	0	1	0
6	0	1	1	0
7	1	1	1	0
8	0	0	0	1
9	1	0	0	1
10	0	1	0	1
11	1	1	0	1
12	0	0	1	1
13	1	0	1	1
14	0	1	1	1
15	1	1	1	1
16	0	0	0	0

We may see that the binaries in this counter chain do not all change state at the same instant in time and hence it is termed an asynchronous counter; the pulses from the input cause changes of state which ripple along the counter from the input end and this gives rise to its alternative name of a ripple counter.

One of the drawbacks in all ripple counters is the total time taken for all the binaries to take up their steady values after receipt of a pulse. On many occasions only one or two binaries need to change state but as may be seen from Table 16.5, in the worst case, for example on the eighth pulse, all the binaries need to change state.

These transitions will occur in a steady progression through the counter, a finite time being required by each binary to respond to its own input. This total of the transition times of all the binaries is termed the 'carry-time'. In a counter consisting of a large number of binaries, the carry-time may occasionally exceed the time interval between pulses. Although this, of itself, will not limit the frequency of pulses which may be counted, it will be apparent that the counter will never be in a stable state between inputs; when the early stages of the counter are responding to an input pulse, the later stages of the counter will still be reacting to the ripple set up by the previous pulse. Thus it is not possible to read the output of such a counter in the interval between pulses and the only way in which an output may be obtained is to stop the count.

16.8.1 Synchronous binary counter

The carry time may be greatly reduced if we can ensure that all circuits respond at the instant when the pulses occur. This will result in a synchronous counter.

We may observe from Table 16.5 that binary A changes state on each input.

Thus this condition will be satisfied if:

$$\mathbf{J_A} = \mathbf{K_A} = \mathbf{1}$$

Again binary B changes state whenever $\mathbf{Q_A} = \mathbf{1}$. We may therefore apply the input pulses directly to the clock input of binary B and ensure that it is only able to operate when $\mathbf{Q_A} = \mathbf{1}$. We may do this by setting

$$\mathbf{J_B} = \mathbf{K_B} = \mathbf{Q_A} \tag{16.4}$$

from which we see that when $\mathbf{Q_A}$ is zero the input pulses do not cause any transition in binary B.

Similarly from Table 16.5 we see that the operation of flip-flop C must not occur unless both $\mathbf{Q_B}$ and $\mathbf{Q_A}$ have the value **1**. Hence

$$\mathbf{J_C} = \mathbf{K_C} = \mathbf{Q_A Q_B} \tag{16.5}$$

Finally we deduce similarly that

$$\mathbf{J_D} = \mathbf{K_D} = \mathbf{Q_A Q_B Q_C} \tag{16.6}$$

If we introduce logic gates to effect these constraints, it will be apparent that all transitions, which are necessary, will take place on the occurrence of each input pulse and in all cases only one binary circuit is involved and thus the carry time may be reduced to the transition time of one binary only together with the transmission time of the logic gates.

A practical implementation of this is shown in Fig. 16.15 in which 2-input AND gates only are used. Many modifications of this circuit may be developed, which may have advantages in reduction of carry time but which almost invariably increase the complexity of the circuit.

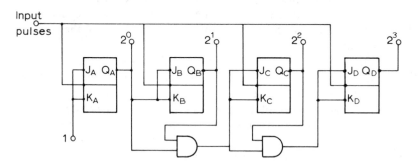

Fig. 16.15 Synchronous binary counter

16.9 Synchronous decade counter

Since the decimal scale is more familiar than the binary scale it is often desirable to present the numerical count in decade form. This involves designing a counter which will recycle after a count of ten instead of after some power of 2 which is the natural cyclic period of a binary counter. To achieve this we require a 4-bit binary counter since this is the smallest binary counter which can count up to

10. The only problem now is to design a logic network which will force the circuit to recycle after 10 pulses.

Reference to Table 16.5 shows that at the ninth pulse, reading the table from right to left, we have the binary number 1001 and on receipt of the 10th pulse this becomes 1010. In the decade counter the reading on the 10th pulse should be 0000. Hence we need a circuit which will reset all the binaries to zero when $Q_A = 0$, $Q_B = 1$, $Q_C = 0$, $Q_D = 1$; and a logic network is needed producing an output X whose value is

$$X = \overline{Q}_A Q_B \overline{Q}_C Q_D \qquad (16.7)$$

We may then connect the complement of **X**, namely $\overline{X}$, to the clear (C) inputs of all the binaries of the binary counter shown in Fig. 16.14.

Before designing this we may note that the logic expression 16.7 is more complex than we need. Since we are never going to allow the states shown by rows 11 to 15 to exist in the decimal counter we may make use of this fact to simplify equation 16.7. We find that the logic expression $X = Q_B Q_D$ will suffice. This may be confirmed by noting that the only rows which satisfy this condition are 10, 11, 14, and 15 of which the only one of interest to us is row 10. Using this simplified logic expression we may construct the decade counter shown in Fig. 16.16.

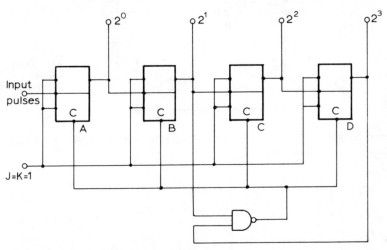

Fig. 16.16 Asynchronous decade counter

There are, of course, a wide variety of alternative ways in which we can force a 4-bit binary counter to recycle after 10 operations, by omitting any six rows of Table 16.5.

This counter will give a transition from **1** to **0** on the output of binary D after every ten input pulses and this may be used as the input to a similar decade counter to count the next higher power of 10. The final readout of the number will be in decade form but each digit of the decade will be expressed in binary form. This is termed a binary coded decimal number (BCD). If this is needed to give a visual readout in decade form we need to design a decoder which will have 4 inputs from the binaries A to D and will give 10 outputs corresponding to the 10 digits of a decimal display device for each digit of the decimal number. Such devices are generally termed decoders.

16.9.1 Decoders

We shall consider the general problem of design of decoders with reference to the decoding of a BCD number. We may write the truth table for the output in Table 16.6 in which each row represents the output required to indicate one specific digit. There are thus 10 output lines Q_0 to Q_9 corresponding to the rows of the table. Each column represents the output from one of the binaries of a decade counter.

Table 16.6

A	B	C	D	Decimal Number	Output
0	0	0	0	0	Q_0
1	0	0	0	1	Q_1
0	1	0	0	2	Q_2
1	1	0	0	3	Q_3
0	0	1	0	4	Q_4
1	0	1	0	5	Q_5
0	1	1	0	6	Q_6
1	1	1	0	7	Q_7
0	0	0	1	8	Q_8
1	0	0	1	9	Q_9

From this we may write logic expressions relating the output variables, **Q**, with the input variables **A, B, C, D**

$$Q_0 = \overline{A}\,\overline{B}\,\overline{C}\,\overline{D}$$

$$Q_1 = A\,\overline{B}\,\overline{C}\,\overline{D}$$

$$Q_2 = \overline{A}\,B\,\overline{C}$$

$$Q_3 = A\,B\,\overline{C}$$

$$Q_4 = \overline{A}\,\overline{B}\,C$$

$$Q_5 = A\,\overline{B}\,C$$

$$Q_6 = \overline{A}\,B\,C$$

$$Q_7 = A\,B\,C$$

$$Q_8 = \overline{A}\,D$$

$$Q_9 = A\,D$$

The logic functions given above are simpler than those which might be determined directly from Table 16.6 as use has been made of the fact that certain combinations of variables (those corresponding to the numbers 11–15) cannot occur in a decimal counter. It is now a straightforward task to implement the above logic functions using standard combinational logic gates.

The complexity of some of these functions has lead to the development of certain codes which permit the use of simpler circuitry to decode into, say, decimal form. A large variety of decoders may be constructed according to the output requirement and the code used for the original binary number.

References

MILLMAN, J. and TAUB, H., *Pulse, Digital and Switching Waveforms*, McGraw-Hill, 1965, Chapters 10, 18, and 19.

BABB, D. S., *Pulse Circuits*, Prentice Hall, 1964, Chapters 11 and 12.

HIBBERD, R. G., *Integrated Circuits*, McGraw-Hill, 1969, Chapter 7.

SCARLETT, J. A., *Transistor-Transistor Logic and its Interconnections*, Van Nostrand Reinhold, 1972, Chapters 11–13.

MORRIS, R. L. and MILLER, J. R., *Designing with TTL Integrated Circuits*, McGraw-Hill, 1971, Chapter 7, 8, 10, and 11.

CARR, W. N., *MOS/LSI Design and Application*, McGraw-Hill, 1972, Chapters 4 and 5.

Problems

16.1 The circuit of Fig. 16.2 has $R_1 = R'_1 = 4 \cdot 7$ kΩ, $R_2 = R'_2 = 22$ kΩ, $R_3 = R'_3 = 22$ kΩ, $V_{CC} = +10$ V, $V_{BB} = -5$ V. The transistors have a large signal current gain β_F lying in the range 15 to 60. Specification requires an output voltage, switching from $0 \cdot 3$ V to 9 V. Will the circuit work for all transistors? If not, what selection must be made amongst the transistors?

16.2 In the circuit of Problem 16.1, transistors have been selected with $\beta_F = 40$. The output voltage at the collector of Tr2 must switch between $0 \cdot 3$ V and 10 V. What is the minimum resistance which may be shunted across R'_1?

16.3 The circuit of Problem 16.1 is constructed using transistors with $\beta_F = 15$. What is the range of output voltage available at the collector of Tr2?

16.4 What will be the stable voltages in the circuit of Problem 16.1 if the transistors have a β_F of 8?

16.5 Considering the circuit of Fig. 16.2 as a feedback amplifier, determine the loop gain. What must be the minimum value of β_F for the circuit to be unstable?

16.6 The binary circuit of Fig. P16.1 has two stable states. In each stable state one transistor is bottomed and one is cut off. Determine the minimum value of β_F for the circuit to fulfil its specification. What are the two output voltage levels at the collector of Tr2 in the two stable states?

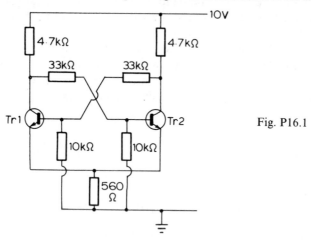

Fig. P16.1

16.7 The circuit of Fig. 16.9 uses fets whose characteristics are identical with those in Fig. 15.26. The two input levels are 0 and -20 V. Determine the voltage at Q for the four possible input combinations, immediately after a clock pulse.

16.8 It is proposed to construct a pseudo-random sequence of 15 digits by modifying the circuit of Fig. 16.13 such that the input to the exclusive OR gate is obtained from the output of the binaries B and D. Draw up a truth table and investigate this.

16.9 Can you make a pseudo-random sequence by replacing input C to the exclusive OR gate by input A in the circuit of Problem 16.8?

16.10 An asynchronous (ripple) decade counter is to be designed. The circuit will consist of the 4-bit binary counter shown in Fig. 16.14 with certain modifications. When the counter has counted up to 9 all the flip-flops will be cleared to zero and a carry pulse sent on to the next decade. Draw up a truth table for this counter based on Table 16.5. Write down the Boolean expression for the clear pulse after a count of 9. (Ensure that you obtain the minimum form of this expression by noting those states which cannot exist.) Implement this logic expression using NAND gates.

16.11 A similar decade counter to that of Problem 16.10 is constructed from the truth table of Table 16.5 by jumping from line 7 to line 14 by omitting lines 8 to 13. Write the truth table for this counter. Write the Boolean expressions for the feedback combinational logic and implement it using NAND gates.

16.12 Write logic expression from the truth table of Problem 16.11 which will decode the BCD (binary coded decimal) number set in the counter, as pulses along 10 separate wires. Simplify as far as possible. Implement some of these using NAND gates only.

16.13 By drawing a truth table show that the circuit of Fig. P16.2 is a scale of three counter.

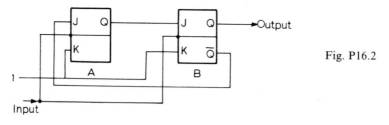

Fig. P16.2

16.14 Show that the circuit of Fig. P16.3 is a scale of five counter. The J-K flip-flops are master–slave circuits.

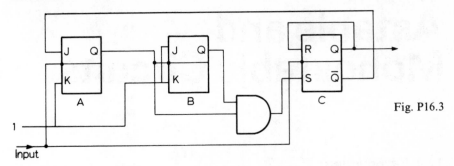

1

Input

Fig. P16.3

16.15 Design an asynchronous counter to divide by seven.

16.16 A digital display is constructed of seven light emitting diodes, a to g, as shown in Fig. P16.4. Selected elements are illuminated to display the digits 0 to 9. Such a display is used to give a visual

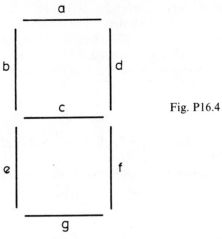

Fig. P16.4

read-out of the state of one decade of a binary coded decimal counter whose truth table is given in Table 16.6. Show that the logic relationships for each of these elements are

$$a = B + D + A\,C + \overline{A}\,\overline{C}$$
$$b = D + \overline{A}\,(\overline{B} + C) + \overline{B}\,C$$
$$c = D + \overline{A}\,B + \overline{B}\,C + B\,\overline{C}$$
$$d = \overline{C} + D + A\,B + \overline{A}\,\overline{B}$$
$$e = D + \overline{A}\,(B + \overline{C})$$
$$f = A + \overline{B} + C + D$$
$$g = D + \overline{A}\,(B + \overline{C}) + B\,\overline{C} + A\,\overline{B}\,C$$

17
Astable and Monostable Circuits

17.0 Introduction

Closely associated with the bistable circuits studied in Chapter 16 are two other groups of circuits. The monostable circuits (sometimes known as monostable multivibrators) are identical with the bistable circuits except that one of the states is stable for only a limited period of time; at the end of this interval the circuit reverts to its one fully stable state and remains there. The temporarily stable condition is termed a metastable state. In the astable circuits both states are metastable and the condition of this circuit is one of continual oscillation between these two states; this circuit is known as an astable multivibrator or often simply as the multivibrator.

17.1 Monostable multivibrator

If one of the coupling resistors, say R_2, in the bistable flip-flop circuit of Fig. 16.2 is replaced by a capacitance, the circuit may be shown to have only one stable state. The circuit of one such monostable multivibrator (sometimes known as a one-shot multivibrator) is given in Fig. 17.1.

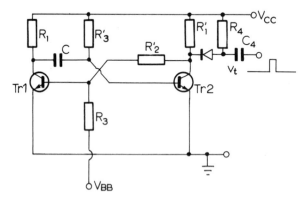

Fig. 17.1 Monostable multivibrator using bipolar transistors

From the circuit it is obvious that normally transistor Tr2 will be bottomed since a large base current will flow as its base is connected directly to the positive rail through R_3'. Thus the stable state of this circuit is with Tr2 ON and

Tr1 OFF. In this stable state the collector of Tr2 is at a low voltage and hence the base of Tr1 is negative. Transistor 1 is non-conducting and hence the collector of Tr1 is at V_{CC}. As the base of Tr2 is close to zero (at approximately 0·6 V), the capacitor C is charged up to V_{CC} with its left side positive. This situation is shown by the initial values in Fig. 17.2.

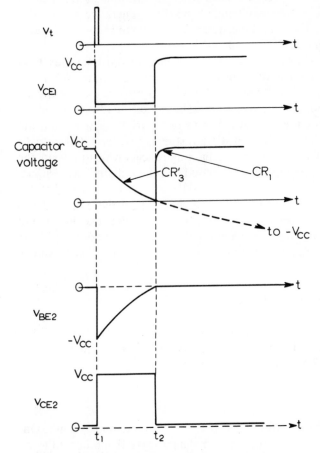

Fig. 17.2 Waveforms of monostable multivibrator

Suppose now that a positive triggering pulse is applied at the collector of Tr2 sufficient to drive the base of Tr1 positive and hence into its conducting region. The collector potential of Tr1 will fall and, since the capacitor voltage cannot change instantaneously, the base potential of Tr2 will fall by the same amount, cutting Tr2 off and allowing its collector potential to rise further. It is seen that this transition is regenerative as positive feedback exists around the circuit and thus the change of state will be very fast. At the end of the transition Tr2 will become fully cut off and Tr1 will be bottomed.

Since the collector of Tr1 is now bottomed at approximately 0·2 V and the capacitor voltage is unchanged, the base of Tr2 will be very close to $-V_{CC}$.

There is thus a voltage of approximately $2V_{CC}$ across R'_3 and current will now flow through R'_3 and the bottomed transistor Tr1 into the capacitor C, discharging it and attempting to recharge it with opposite polarity towards V_{CC}; this will take place with a time constant CR'_3. During this period the base voltage of Tr2 will be rising similarly. However, when the capacitor is fully discharged, the base voltage of Tr2 will have reached zero and it will start to conduct again. Its collector potential will fall and the regenerative process will again commence and continue until the system is once again in its initial stable state with Tr2 bottomed and Tr1 cut off.

These transitions are shown in Fig. 17.2. One final readjustment needs to take place before the circuit is fully restored to its original state. When the second switching occurs at time t_2 the capacitor C has approximately zero volts on it, whereas we may note that in the initial stable condition it was charged to approximately V_{CC}. Thus the capacitor has to recharge through R_1 and the base–emitter resistance of the bottomed transistor Tr2 to its initial voltage, V_{CC}, with time constant $R_1 C$. It is this recharging process which prevents the circuit being triggered again until a finite period has elapsed after the termination of the preceding cycle.

It may be seen that the circuit is capable of generating a pulse of predetermined duration when it is triggered by an external short pulse. The duration of the pulse is determined by the time taken for the capacitor to discharge from V_{CC} to zero when the total applied voltage is $-2V_{CC}$. This may be easily shown to be given by

$$t_2 - t_1 = CR'_3 \ln 2 = 0.69 CR'_3 \tag{17.1}$$

A slight modification of this circuit may be made in which the upper end of resistor R'_3 is disconnected from the positive supply rail and connected to an independent supply V'. The pulse duration is now given by

$$t_2 - t_1 = CR'_3 \ln [(V_{CC} + V')/V'] \tag{17.2}$$

It may be seen that by variation of V' we may vary the pulse length. One application of this would be in the generation of a pulse width modulated wave. The modulation signal would be connected in series with V' and the circuit triggered at the carrier frequency; the width of each pulse would be a function of the signal amplitude. In this rather crude form the relationship between pulse width and signal amplitude would be linear for only a very small amplitude of the modulating voltage.

Another application of the circuit is to regenerate pulses. After passage through a transmission system a pulse shape frequently becomes very distorted. The 'poor' pulse may be used to trigger this circuit and generate a 'clean' pulse at the output.

Again since the pulse duration can be accurately defined, the circuit may be used to delay pulses by an accurately timed interval. The initial pulse triggers the multivibrator which is designed to give a pulse equal to the desired delay.

The trailing edge of this pulse may now be used to regenerate a delayed pulse, the time delay being precisely determined by the parameters of the monostable circuit.

The trigger pulse is shown in Fig. 17.1 connected to the collector of Tr2 through a diode, resistor, R_4, and capacitor, C_4. The diode isolates the trigger source from the multivibrator as soon as transition has been initiated. The resistor–capacitor combination differentiates the input with respect to time, thus forming a very sharp triggering pulse.

17.2 Astable multivibrator

An extremely common circuit developed from the monostable multivibrator is the free-running or astable multivibrator. This is obtained by replacing the second coupling resistor, R'_2, in Fig. 17.1 by a capacitor C' as shown in Fig. 17.3.

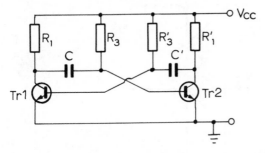

Fig. 17.3 Astable multivibrator using bipolar transistors

We noted in Section 17.1 that in the monostable multivibrator there was only one stable state with Tr2 ON; we should therefore surmise that if both coupling elements are capacitive there will be no permanently stable state, the circuit switching from one stable condition to the other depending upon the charging times of the two timing capacitors C and C'. The waveforms of this circuit are shown in Fig. 17.4; it is evident that no triggering pulse is necessary since this circuit is entirely free running.

A similar analysis to that which was outlined for the pulse duration in Section 17.1 will show that the period of each pulse in a multivibrator is given by

$$t_1 \simeq 0\cdot7R_3C$$
$$t_2 \simeq 0\cdot7R'_3C'$$

where t_1 and t_2 are defined in Fig. 17.4. The circuit may therefore be used to give an output pulse from the collector of Tr2 having a duration t_1 and repetition period $t_1 + t_2$.

The main application of this circuit is in the generation of square pulses of controllable duration and repetition rate. The frequency stability of the circuit is high, being dependent only on resistors and capacitors; the gain of the transistors is relatively unimportant so long as they remain sufficiently high for regenerative switching to occur.

The upper frequency at which a multivibrator may operate is determined mainly by the speed at which the capacitors may be recharged to their original

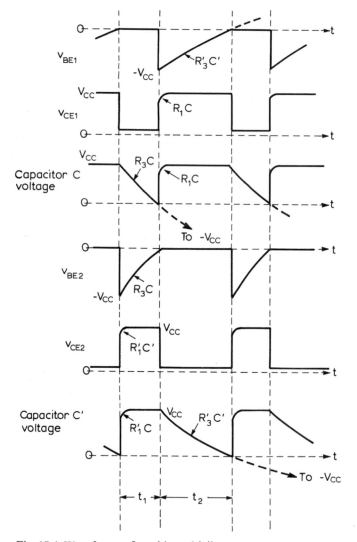

Fig. 17.4 Waveforms of astable multivibrator

voltage. In Section 17.1 we noted that an astable multivibrator could not be retriggered until the capacitor had recharged to V_{CC}. The same restriction applies here with regard to both C and C', hence for high frequency operation R_1 and R'_1 must be small.

In order to obtain very low repetition frequencies, it is necessary to make either C or R_3, or both, very large. An upper limit is imposed on the value of capacitance since large valued capacitors usually have a relatively low leakage resistance across them. The value of R_3 in the circuit is limited by the consideration that in the metastable state when Tr2 is conducting the base current must be sufficiently large to keep Tr2 in a bottomed condition. This has been discussed in Section 16.1.1 in relation to a bistable circuit.

When Tr2 is bottomed the collector current is approximately V_{CC}/R'_1 and the base current V_{CC}/R_3. Thus for Tr2 to be bottomed

$$V_{CC}/R'_1 < \beta_F V_{CC}/R_3$$

Thus $$R_3 < \beta_F R'_1 \tag{17.3}$$

This imposes an upper bound on the value of R_3 and hence on the lowest frequency of operation.

Very low frequency multivibrators may be constructed·using field-effect transistors as it is possible to use values of R_3 of 10 MΩ and above since the bottoming of fets is determined by gate voltage, not current. This permits the use of smaller valued capacitors having higher leakage resistance.

17.3 Schmitt trigger circuit

Another very useful circuit closely allied to the multivibrator is the Schmitt trigger circuit. It is used to generate pulses with steep sides from a more slowly varying waveform and is closely allied to the comparator discussed in Section 13.10. It uses the regenerative principle to cause a fast transition from one stable state to the other when the applied voltage exceeds a certain threshold value, and to revert to its first state when the input falls below another threshold level. The circuit of a Schmitt trigger is shown in Fig. 17.5.

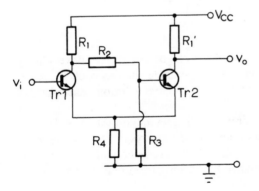

Fig. 17.5 Schmitt trigger circuit using bipolar transistors

Assume initially that the input voltage, v_i, is zero. This will cause Tr1 to be cut off and its collector voltage to equal V_{CC}. As a result of the potential divider R_2, R_3 the base potential of Tr2 will be positive and hence Tr2 will be conducting; consequently its emitter current, flowing through R_4, will cause the common emitter potential v_E to be positive, of value V_E', ensuring that Tr1 is well cut off. The choice of resistors R_2 and R_3 is such that in this state Tr2 will be bottomed.

If v_i is now increased Tr1 will commence to conduct when $v_i > V_E'$. Its collector potential will fall reducing the base voltage of Tr2 and hence reducing the current and causing the emitter potential to fall. This further increases the base–emitter voltage of Tr1 and we have a regenerative transition to a second stable state with Tr1 ON and Tr2 OFF in which the common emitter potential is now V_E'' ($V_E'' < V_E'$). The circuit will remain in this state so long as $v_i > V_E''$. If v_i is reduced below V_E'' transistor Tr1 will cut off and the regenerative transition will restore the original state.

The circuit will thus switch between two stable states depending on the magnitude of the input voltage. The waveforms at various points in the circuit are shown in Fig. 17.6. The output from the circuit is normally taken from the

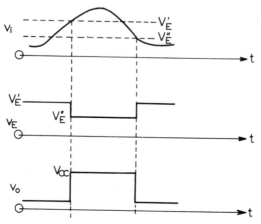

Fig. 17.6 Switching waveforms of Schmitt trigger circuit

llector of Tr2. This has the advantage that it can be designed to have a fairly low output impedance and also that, as it is outside the feedback loop, the effect of stray capacitance at this point will not have a serious effect on the switching speed.

One of the shortcomings of this circuit is the inevitable hysteresis between the input and output; the switching ON and OFF do not occur at the same input voltage. To set against this we require that, for fast regenerative switching, the loop gain during switching, namely when both transistors are in an active state, should be equal to or greater than unity.

It may be shown theoretically that, if the loop gain is made exactly equal to unity, the hysteresis is eliminated. Such a critical adjustment is not practicable and, in order to allow for variations in parameters due to ageing, temperature changes, etc., a loop gain greater than unity is used with a consequential small amount of hysteresis. In addition, reduction of the loop gain to unity will result in very poor regeneration during transition between stages.

17.4 Operational amplifier multivibrators

The circuits which we have been considering in this chapter may be viewed as two stage amplifiers with positive feedback overall. The astable multivibrator is effectively a two stage amplifier with feedback through a capacitance. The bistable multivibrator has a.c. coupling between the two stages of the amplifier and also in the feedback loop. The Schmitt trigger circuit has a directly coupled feedback connection in which the two states are determined by the magnitude of the input voltage.

It is possible to design similar circuits using operational amplifiers by starting from a circuit which has been designed with two stable states in which the choice of state is determined by the magnitude of an applied voltage. Modification may then be made to this circuit to render one or both of the states as metastable producing either a monostable or astable (free-running) multivibrator.

17.5 Operational amplifier Schmitt trigger

In Section 13.10 we showed how a high gain operational amplifier could be used to give an output voltage approximately equal to the positive supply voltage of the operational amplifier for positive excursions of the input voltage greater than the threshold voltage, and similarly an output equal to approximately the negative supply voltage for negative inputs. For the small range of inputs in the 'dead' zone between the two threshold voltages, the output lies between these two limits, as the comparator behaves as a linear amplifier in this range. If we take such a comparator and increase the gain by means of positive feedback, the dead zone may be reduced until it becomes zero when the amplifier gain is infinite, corresponding to a loop gain of unity. This is identical to the marginal design of a Schmitt trigger mentioned in Section 17.3 where the loop gain was exactly unity. Further increase in positive feedback will produce a more conventional Schmitt trigger circuit in which there is appreciable regeneration associated with a certain amount of hysteresis between the two switching levels.

The circuit of a Schmitt trigger using an operational amplifier is shown in Fig. 17.7. We may directly study the manner in which this circuit operates. For simplicity we shall assume that in the active region of operation the voltage between the two inputs of the amplifier is effectively zero and that $R_3 = 0$. Consider the input v_1 to be large and negative. The output voltage v_0 will now

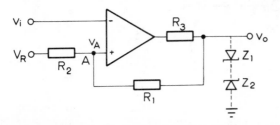

Fig. 17.7 Schmitt trigger circuit using operational amplifiers

be saturated at approximately V^+, (approximately 3 V for the 710 amplifier). The voltage, V'_A, appearing at the non-inverting input A of the amplifier is

$$V'_A = (R_1 V_R + R_2 V^+)/(R_1 + R_2) \tag{17.4}$$

So long as v_I is less than this value the output will remain at V^+ and the circuit is stable. As soon as v_I exceeds V'_A by a few millivolts the amplifier will come into its active region of operation, v_O will start to fall and as a result v_A, the voltage at A will fall also. This increases the differential voltage applied between the two inputs of the amplifier and thus accelerates the fall in v_O. This is thus a regenerative transition and continues until the output of the amplifier saturates in the negative sense at V^- (about -0.5 V for the 710 amplifier). The value of voltage at A now becomes V''_A given by

$$V''_A = (R_1 V_R + R_2 V^-)/(R_1 + R_2) \tag{17.5}$$

and the output remains clamped at V^- so long as $v_I > V''_A$. The circuit thus acts exactly as a Schmitt trigger; the two switching threshold levels being determinable by choice of R_1 and R_2 and the reference voltage. The two output levels are determined by the clamping levels of the comparator. They may be more precisely determined by connecting two back-to-back Zener diodes, Z_1 and Z_2, between output and earth and a series resistor R_3 in the output, as shown by the broken lines in Fig. 17.7. The upper clamping level is then equal to the breakdown voltage of Z_2 namely V_{Z2}, and the lower level to $-V_{Z1}$.

17.6 Astable multivibrator

The Schmitt trigger circuit may easily be converted to an astable multivibrator. The reference voltage level is set at zero for convenience and a resistor capacitor circuit, R_4, C, is connected to give negative feedback as shown in Fig. 17.8. Assume, as previously, that at some instant of time v_B is at a low potential. Then v_O will be high at V_{Z2}. Capacitor C will now commence to charge up

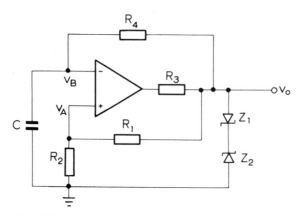

Fig. 17.8 Astable multivibrator using operational amplifiers

towards V_{Z2} through R_4 and the voltage v_B will rise exponentially with time constant CR_4 until it reaches a value, V_A', given by equation 17.4. The Schmitt trigger will now switch over and v_O will fall to $-V_{Z1}$ and capacitor C will now discharge through R_4 towards $-V_{Z1}$ with the same time constant until it has fallen to the value, V_A'', given by equation 17.5. The waveform across the capacitor v_B and the output waveform under steady state conditions are shown in Fig. 17.9. For this circuit, since $V_R = 0$, the values of V_A' and V_A'' are given by:

$$V_A' = R_2 V_{Z2}/(R_1 + R_2) \tag{17.6}$$

$$V_A'' = R_2 V_{Z1}/(R_1 + R_2) \tag{17.7}$$

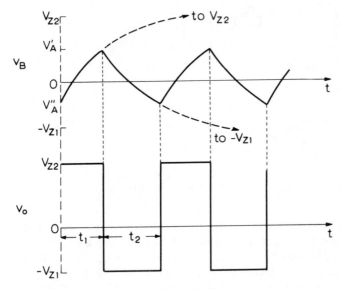

Fig. 17.9 Waveforms of astable multivibrator

The output voltage is seen to be a square wave in which the relative duration of each half cycle is determined by the values of V_{Z1} and V_{Z2}. The period, t_1, of the positive excursion of v_O is obtained by determining the time taken for the capacitor C to charge up to a voltage $V_A' - V_A''$ through a resistor R_4 from a supply $V_{Z2} - V_A''$. We obtain

$$t_1 = R_4 C \ln \left[(V_{Z2} - V_A'')/(V_{Z2} - V_A') \right]$$

$$= R_4 C \ln \frac{(R_1 + 2R_2)V_{Z2}}{(R_1 + R_2)V_{Z2} - R_2 V_{Z1}} \tag{17.8}$$

The period t_2 may be similarly determined as

$$t_2 = R_4 C \ln \frac{(R_1 + 2R_2)V_{Z1}}{(R_1 + R_2)V_{Z1} - R_2 V_{Z2}} \tag{17.9}$$

17.7 Monostable multivibrator

A monostable circuit may be easily developed from the circuit of Fig. 17.8 by clamping the voltage, v_B, across the capacitor at one or other of its extreme values. This may be done by using a diode across C as shown in Fig. 17.10 to hold the capacitor voltage at zero volts or below. This ensures that there is always one stable state since the voltage v_A cannot follow the output voltage when v_O becomes positive.

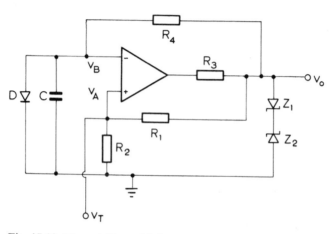

Fig. 17.10 Monostable multivibrator using operational amplifiers

The stable state of the circuit exists when v_B is approximately zero (more accurately at about $+0.6$ V), v_O is at V_{Z2} and v_A at its stable value V'_A given by equation 17.6. The circuit may be triggered out of this stable state by applying a negative pulse v_T to the non-inverting input as shown in Fig. 17.10; this is usually done through a diode RC circuit as discussed in Section 17.1. This negative trigger must be sufficient to cause v_A to fall below v_B; it must therefore have a magnitude greater than $V'_A - 0.6$. The input differential voltage has now been reversed in polarity and so v_O will commence to fall, and v_A will follow it; thus the differential input voltage will increase and a regenerative transition will occur, causing the output to fall to its low value $- V_{Z1}$, and v_A to take up the value V''_A.

This condition cannot last since the capacitor will now charge up negatively from $- V_{Z1}$ through R_4. When the negative voltage across C, namely v_B, is equal to, or just exceeds V''_A, the input to the comparator will reverse and v_O will rise instantaneously to V_{Z2}. The capacitor will now attempt to reverse its polarity by charging up towards V_{Z2} through R_4 but this will be terminated when $v_B \simeq 0.6$ V and the diode D clamps the input voltage. The circuit has now reverted to its original state in which it is completely stable. The voltages at the inverting input v_B and at the output for one cycle of operations are shown in Fig. 17.11.

The duration, t_0, of the pulse may be determined in a manner similar to that for the astable circuit leading to

$$t_0 = R_4 C \ln \left[(R_1 + R_2)/R_1 \right]$$

From Fig. 17.11 it may be seen that a finite time exists after the end of one pulse before the circuit has returned to its stable condition and is ready to receive a further trigger pulse.

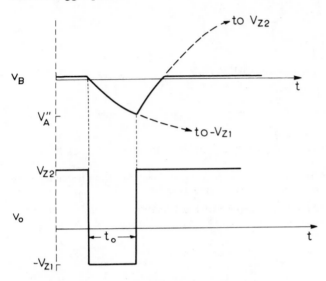

Fig. 17.11 Waveforms of monostable multivibrator

References

MILLMAN, J. and TAUB, H., *Pulse, Digital and Switching Waveforms*, McGraw-Hill, 1965, Chapter 11.
BABB, D. S., *Pulse Circuits*, Prentice Hall, 1964, Chapter 11.

Problems

17.1 The circuit of Fig. 17.1, omitting R_4, C_4, and the diode, is used to generate short pulses. The circuit values are: $R_1 = R_1' = 2 \text{ k}\Omega$, $R_2' = 50 \text{ k}\Omega$, $R_3 = R_3' = 100 \text{ k}\Omega$, $C = 1.0 \text{ nF}$, $V_{CC} = 20 \text{ V}$; V_{BB} is set at -5 V. Determine the potentials at the bases and collectors of both transistors and also the base and collector currents. What is the minimum value of β_F if Tr2 is to be bottomed?

17.2 The voltage V_{BB} in the circuit of Problem 17.1 is now raised momentarily to $+5$ V and then returned to -5 V. Sketch and dimension the waveforms of the voltages at the collector of Tr1, and the base and collector of Tr2. What is the duration of the pulse at the collector of Tr2?

17.3 The circuit of Fig. P17.1 is designed to have two stable states. The transistors have minimum values of β_F of 20. Determine values for R_2 and R_3. Sketch the transfer characteristics relating v_o with v_i.

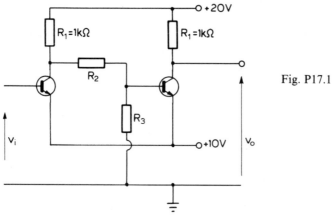

Fig. P17.1

17.4 The circuit of Fig. 17.5 has $R_1 = 5$ kΩ, $R_1' = 1$ kΩ, $R_2 = 10$ kΩ, $R_3 = 120$ kΩ, $R_4 = 3$ kΩ, $V_{CC} = 15$ V.

Sketch the transfer characteristic of the network and sketch and dimension the output wave when a 50 Hz sinewave of r.m.s. value 10 V is applied to the input.

17.5 The circuit of Fig. P17.2 is used as a clipper. Sketch and dimension the transfer characteristic relating output and input voltages when the input voltage ranges from -10 V to $+10$ V. The transistors have a current gain of 60.

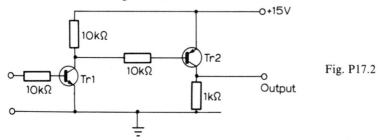

Fig. P17.2

17.6 A resistor of 20 kΩ is connected between the output and the base of Tr1 in the circuit of Fig. P17.2. Sketch the transfer characteristic of the modified circuit.

17.7 Sketch the waveforms and determine the repetition frequency of the circuit of Fig. P17.3 and also the duration of the output pulses.

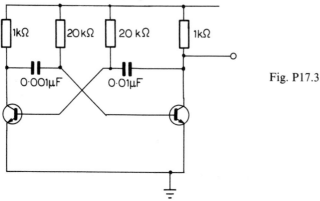

Fig. P17.3

Appendix A: Two-port Parameters

A.1 Introduction

The majority of systems used in the transmission of information, whether in communication, control, or computer applications, involve devices and networks with one pair of input terminals, the input port, and one pair of output terminals, the output port. Such networks are termed two-port networks, since voltages may be applied between the two terminals of each port, and the current flowing in to one terminal of a port is identical with the current flowing out from the other terminal of the same port.

In many situations it will be noticed that one of the input terminals is identical with one of the output terminals and hence the two-port network reduces to a three-terminal network. A bipolar transistor and, in most situations, a field-effect transistor are examples of devices which may be represented as three terminal networks.

A.2 Networks represented by two-port parameters

The network of Fig. A.1 shows an arbitrary two-port network having ports 11′ and 22′. Four external parameters, V_1, I_1, V_2, I_2 may be specified for this network. By definition the current leaving terminal 1′ is I_1 and that leaving 2′ is I_2, furthermore we are not concerned with the voltage between 1′ and 2′.

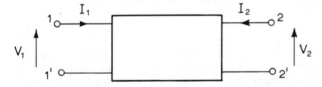

Fig. A.1 Two-port network specified by voltages and currents at the two ports

These four variables may be interrelated by two equations connecting two dependent variables with two independent variables. We may choose the two independent variables from the given four variables in six ways and hence there are six possible sets of equations defining the properties of the network. We shall mainly be interested in only two of these sets; the first shows the dependence of the two currents on the voltages; the second shows how V_1 and I_2 are related to I_1 and V_2. Although the two-port representation of a network is valid for any voltages and currents, we shall restrict our study to the relationship between sinusoidal phasors applied at the two ports.

Considering the first mode of representation we may write

$$I_1 = y_{11}V_1 + y_{12}V_2$$
$$I_2 = y_{21}V_1 + y_{22}V_2$$

(A.1)

and the parameters $y_{11}, y_{12}, y_{21}, y_{22}$ completely specify the network. Since the voltages and currents are phasors, these four y parameters will be admittances and will usually be complex functions of frequency. Thus, for example, $y_{11} = g_{11} + jb_{11}$ where g_{11} is the conductance and b_{11} the susceptance. At low frequencies all the parameters are usually real.

If we short circuit the output terminals ($V_2 = 0$) and measure the input admittance, I_1/V_1, we see from equation A.1 that this is equal to y_{11}; similarly the forward transfer admittance, y_{21}, is the ratio I_2/V_1 with the same termination. The other two parameters may be determined similarly by shorting the input and making measurements on the output.

The four parameters, when applied to transistors, are frequently named differently using suffixes i, r, f, o to represent input, reverse, forward, and output; thus $y_{11} = y_i, y_{12} = y_r, y_{21} = y_f, y_{22} = y_o$.

The other common method of specification is by the hybrid parameters (strictly this term could be applied to two representations but by common usage it is restricted to the ones now being considered). The defining equations are

$$\left. \begin{array}{l} V_1 = h_{11}I_1 + h_{12}V_2 \\ I_2 = h_{21}I_1 + h_{22}V_2 \end{array} \right\} \tag{A.2}$$

It is fairly obvious that no new data have been introduced by this rearrangement of the equations and hence we may obtain certain interrelations between the h and y parameters. For example

$$y_{11} = 1/h_{11}$$
$$y_{12} = -h_{12}/h_{11}$$
$$y_{21} = h_{21}/h_{11}$$
$$y_{22} = (h_{11}h_{22} - h_{12}h_{21})/h_{11}$$

It should be noted that the h parameters are not dimensionally identical; h_{11} is an impedance, h_{12} and h_{21} are dimensionless, and h_{22} is an admittance. In general the h parameters, like the y parameters, are complex functions of frequency.

The hybrid parameters are capable of measurement by short circuiting the output terminals to obtain h_{11} and h_{21}, or open circuiting the input terminals to determine h_{12} and h_{22}.

Although the above discussion has concentrated on the application of two-port parameters to transistors, they may also be used to model an amplifier or other system having an input and an output port, so long as the restrictions imposed on a two-port network are satisfied. The choice of parameter set used will depend mainly on the properties of the network. Some networks do not possess certain of the parameter sets; for example an idealized field-effect transistor with zero conductance between gate and source has an h parameter set which is undefined.

Yet another set of parameters which are sometimes used to describe networks are the transmission parameters. They may be used in the description of devices also but this is less common. They are defined by the equations

$$V_1 = a_{11}V_2 + a_{12}(-I_2)$$
$$I_1 = a_{21}V_2 + a_{22}(-I_2)$$

and thus relate the current and voltage at the input with that at the output of the network. The negative sign associated with I_2 appears as a result of the sign convention which was adopted when this representation of networks was first adopted. The a parameters may be directly determined in terms of any of the other parameters. For example, in terms of the y parameters:

$$a_{11} = -y_{22}/y_{21}$$
$$a_{12} = -1/y_{21}$$
$$a_{21} = -(y_{11}y_{22}-y_{12}y_{21})/y_{21}$$
$$a_{22} = -y_{11}/y_{21}$$

These parameters are principally of value when cascading networks as the output port of one network is the input port of the succeeding network.

A.3 Three-terminal networks

When one of the input terminals is common with one of the output terminals, the network reduces to the three-terminal network of Fig. A.2. This can be described in exactly the same way as the two-port network by means of the admittance or hybrid parameters, if one pair of terminals is identified as the input, say 1 and 3, and a second pair, 2 and 3,

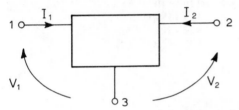

Fig. A.2 Three-terminal network as a special case of two-port network

as the output. This could, for example, describe a transistor in common emitter configuration, namely with the emitter terminal common between input and output. In this case terminal 3 corresponds with emitter, 1 with base, and 2 with collector, and the circuit equation may be written

$$\left.\begin{array}{l} I_b = y_{ie}V_{be}+y_{re}V_{ce} \\ I_c = y_{fe}V_{be}+y_{oe}V_{ce} \end{array}\right\} \tag{A.3}$$

Where the second suffix on the y parameters indicates the terminal chosen as common between input and output.

Similarly in common base configuration with the input to the emitter and the output from the collector

$$\left.\begin{array}{l} I_e = y_{ib}V_{eb}+y_{rb}V_{cb} \\ I_c = y_{fb}V_{eb}+y_{ob}V_{cb} \end{array}\right\} \tag{A.4}$$

It is possible from equation A.4 and A.5 to derive one set of parameters from the other by noting that

$$I_e + I_b + I_c = 0$$

and
$$V_{eb} + V_{bc} + V_{ce} = 0$$

Thus
$$y_{ie} = y_{ib} + y_{fb} + y_{rb} + y_{ob}$$
$$y_{re} = -(y_{rb}+y_{ob})$$
$$y_{fe} = -(y_{fb}+y_{ob})$$
$$y_{oe} = y_{ob}$$

Similar relationships may be obtained for a transistor in common collector configuration.

Appendix B: Physical Constants

Electronic charge (e)	$1·602 \times 10^{-19}$ J
Electron mass (m)	$9·109 \times 10^{-31}$ kg
Permeability of free space (μ_0)	$4\pi \times 10^{-7}$ H/m
Permittivity of free space (ε_0)	$8·86 \times 10^{-12}$ F/m
Boltzmann's constant (k)	$1·380 \times 10^{-23}$ J/K
Resistivity of copper	$1·72 \times 10^{-8}$ Ωm
Mobility of electrons in silicon	$0·16$ m^2/Vs
Mobility of holes in silicon	$0·04$ m^2/Vs
Diffusion constant of electrons in silicon	4100 m^2/s
Diffusion constant of holes in silicon	1030 m^2/s

Appendix C: Symbols

C.1 Voltage and current conventions

The convention for voltages and currents is to use lower case letters for instantaneous values and capital letters to represent average, r.m.s., peak, or steady values. Suffixes are written using capital letters for total quantities and lower case letters for small fluctuations. These variations are shown in Fig. C.1. Double capital letters indicate the magnitude of the appropriate supply battery.

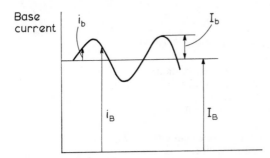

Fig. C.1 Explanation of symbols used for current and voltage

The convention used for impedances, admittances, resistances, etc., is to use capital letters to represent the value of a physical resistor or other impedance; lower case letters are restricted to the value of the ratio of an incremental voltage to an incremental current. In some cases this may represent a physical resistor but in other cases this may represent a model of some physical process.

C.2 List of principal symbols

Significance of additional suffixes is given in Section C.3. A bar over a symbol indicates that the mean value is to be taken. An asterisk beside a symbol denotes the complex conjugate of that quantity.

a	Elements of A matrix with appropriate suffixes
A	Forward transmission in feedback system
A	Cross-sectional area of transistor base or of fet channel
A_A	Available gain
A_P	Power gain
A_T	Transducer gain
B	Bandwidth

B	Feedback transmission
C	Capacitance
C_{ak}	Anode–cathode capacitance
C_b	Base capacitance
C_{ds}	Drain–source capacitance
C_{gd}	Gate–drain capacitance
C_{gk}	Grid–cathode capacitance
C_{gs}	Gate–source capacitance
C_{jc}	Collector junction capacitance
C_{je}	Emitter junction capacitance
C_{ox}	Capacitance across oxide layer of mosfet
C_π	Capacitance between base and emitter in hybrid π model
C_μ	Capacitance between base and collector in hybrid π model
D_L	Viscous friction constant
D_n	Diffusion constant of electrons
D_p	Diffusion constant of holes
e	Electronic charge
$\mathscr{E}$	Electric field strength
f	Frequency
F	Noise figure
f_T	Frequency at which common emitter current gain falls to unity
f_β	Corner frequency of common emitter current gain
g	Conductance
$G(\ )$	Network transfer function
G_D	Total drain conductance of fet
g_d	Small signal conductance of diode
g_{do}	Low frequency small signal conductance of diode
g_m	Transfer conductance of transistor, fet, or valve
g_π	Input conductance of hybrid π model
h	Elements of h-matrix with appropriate suffixes
i, I	Current
I_{CBO}	Collector–base current with emitter open circuit
I_{CBS}	Collector–base current with emitter shorted to base
I_{CEO}	Collector–emitter current with base open circuit
I_{DSS}	Drain current at zero gate–source voltage
I_{EBO}	Emitter–base current with collector open
I_{EBS}	Emitter–base current with collector shorted to base
$Im(\ )$	Imaginary part of (...)
j	Complex operator
j, J	Current density
J_L	Load inertia
k	Boltzmann's constant
k_a	Amplifier constant
k_d	Differential mode gain
k_f	Motor field constant
k_i	Current gain
k_{io}	Mid-frequency current gain
k_m	Motor constant
k_p	Potentiometer constant
k_v	Voltage gain
k_{v0}	Mid-frequency voltage gain
L	Inductance
l	Channel length in fet
L_n, L_p	Diffusion length for electrons, holes
M	Mutual inductance

n	Electron density
n	Transformer ratio; gear ratio
N_A	Density of acceptor atoms
N_D	Density of donor atoms
p	Pole of network function
P	Power
P_{ia}	Available input power
P_{oa}	Available output power
q, Q	Charge
Q_A	Accumulation charge
q_{BF}, q_{BR}	Forward and reverse base stored charge
Q_D	Depletion charge
Q_G	Gate charge
Q_I	Inversion charge
q_{JC}	Collector junction charge
q_{JE}	Emitter junction charge
R, r	Resistance
R_a	Armature resistance
Re( )	Real part of (...)
r_o	Output resistance in hybrid π model
r_x	Base resistance in hybrid π model
r_π	Base–emitter resistance in hybrid π model
s	Complex frequency variable
t	Time
T	Temperature (absolute)
$T( )$	Transfer function
t_d	Delay time
t_f	Fall time
t_r	Rise time
u_n	Velocity of electrons
V, v	Voltage
V_{th}	Threshold voltage
V_z	Zener diode voltage
V_O	Break point voltage of diode
W	Base width of transistor
x	Linear dimension
y	Admittance parameters of network
z	Zeros of network function
Z, z	Impedance
α	Common base current gain
β	Constant of fet
β	Common-emitter small-signal current gain
β_F	Large variable common-emitter current gain
β_0	Low-frequency common-emitter small-signal current gain
ζ	Damping factor of second order system
θ	General input variable to feedback system
Θ, θ	Angular position in control system
μ_n	Mobility of electrons
μ_p	Mobility of holes
σ	Conductivity
τ_{BF}	Charge control parameter
τ_F	Charge control parameter
τ_m	Torque of motor
τ_o	Output torque
ω	Angular frequency

ω_T Unity current gain, common-emitter frequency
ω_β Corner frequency of common-emitter current gain
Ω_o, ω_o Output angular velocity in control system

C.3 List of principal suffixes

Suffixes used on current, voltage, resistance, capacitance, impedance, etc.

A, a   Anode of thyristor or thermionic valve
B, b   Base of bipolar transistor
C, c   Collector of bipolar transistor
D, d   Drain of fet
E, e   Emitter of bipolar transistor
E, e   Error (voltage or current) in feedback system
F, f   Forward
F, f   Feedback (voltage or current)
G, g   Gate of fet or thyristor; grid of thermionic valve
I, i   Input
K, k   Cathode of thyristor or thermionic valve
L, l   Load
N, n   Noise
n   Electrons
O, o   Output
p   Holes
R, r   Reverse
S, s   Source

Answers to Selected Problems

1.1 (a) -57 mV; (b) 17 mV.
1.2 (a) 100 nA, 9·97 V, 27 mV; (b) 150 nA, 150 V, 50 V.
1.3 D_1, 0·55 V; D_2, 0·7 V.
1.4 2·5 mA.
1.7 610 mA; 113 mA; 14·8 mA.
1.8 (a) 1·0 A; (b) 1·16 A.
1.10 From 0·34 to 0·935.
1.11 (a) 69·0 MHz; (b) 923 MHz.
1.12 840 kHz to 1·09 MHz; BW from 59 to 93 kHz.
1.14 $R_A = 127$ kΩ; $R_B = 12·7$ kΩ—depending on linear approximation.
1.15 $3·6 \times 10^{-3}$ m^2/Vs; 54 μm/s.
1.16 (a) 0·48 mS/m; $0·45 \times 10^{12}$; (c) 12·8 S/m.
1.17 16·8 V, 12·9 μs.

2.7 (a) $V_{GS} = -9$ V, ± 3 V, 2·2; (b) $V_{GS} = -9·7$ V, $\pm 2·3$ V.
2.8 (a) -6 V, -12 V, $-1·8$ mA; (b) -15 V, $-0·6$ V, $-2·94$ mA.
2.9 (a) 0·98 to 0·62; (b) 0·98 to 0·48.

3.1 $g_m = 4·2$ mS; $V_{GS} = -1·0$ V.
3.2 $g_m = 2$ mS; $g_{ds} = 80$ μS.
3.3 (a) $g_m = 4$ mS, $C_{gs} = 9·4$ pF; $C_{gd} = 1·6$ pF.
     (b) $g_m = 12$ mS, $C_{gs} = 9·4$ pF.
     (c) $g_{ds} = 32$ μS, $C_{ds} = 1·2$ pF.

4.1 1·09 ns.
4.2 52·3 ns.
4.3 $I_C = I_B = -735$ nA.
4.4 5·43 pC; 104 μA; 5·1 mA.
4.8 (a) 13·2 nA; (b) 53 nA; (c) 40 nA; (d) 10 nA; (e) 40 nA.
4.9 40 nA.
4.10 (a) 0·75 mA, 7·5 μA, 43·4 mV.
      (b) 3·75 mA, 300 μA, 219 mV ($v_{CB} = 316$ mV).
4.11 142 mV.
4.12 67, 22.
4.13 -45 pC, 150 μA.
4.14 -1460 pC.

5.1 $r_o = 5$ kΩ approx., $g_m = 240$ mS, $r_\pi = 625$ Ω, $r_x = 200$ Ω approx.
5.2 $g_m = 200$ mS, $r_\pi = 300$ Ω, $C_\pi = 1000$ pF.

5.3 $g_m = 80$ mS, $r_\pi = 2.5$ kΩ, $C_\pi = 51$ pF.
5.4 $g_m = 50$ mS, $r_\pi = 2.4$ kΩ, $C_\pi = 32$ pF.
5.5 $g_m = 400$ mS, $r_\pi = 350$ Ω, $r_x = 100$ Ω, $r_o = 20$ kΩ, $C_\mu = 1$ pF, $C_\pi = 2000$ pF.
5.7 $y_{ie} = 1.67 + j169$, $y_{re} = -j2.5$, $y_{fe} = 200 - j2.5$, $y_{oe} = 0.02 + j2.5$, mS.

6.3 Min $R_S = 400$ Ω; $\Delta V_o = 0.9\%$.
6.7 (a) 3.33 mW, 450 µW.
    (b) 5.16 mW, 240 µW.
6.8 21 Ω, 4.8 mW.
6.9 41.7 m^2; 780 in series, 535 in parallel.

7.1 663 pF, 11.1 nF.
7.2 15.7 kΩ.
7.3 16.2 Ω.

8.1 7.5 mA, 6.75 V.
8.2 ±3.75 V.
8.3 ±0.75 V; ±1.95 V.
8.4 A possible solution is $R_1 = 12.6$ kΩ, $R_2 = 7.0$ kΩ, $R_E = 595$ Ω, $R_L = 730$ Ω.
8.5 ±3.28 V; ±3.14 V.
8.6 A possible solution is $R_S = 2.8$ kΩ, $R_1 = 470$ kΩ, $R_2 = 120$ kΩ, $R_L = 7.2$ kΩ.
8.7 A possible solution is $R_S = 286$ Ω, $R_1 = 10$ MΩ, $R_2 = 3$ MΩ.
8.8 $R_E = 1.18$ kΩ, $R_1 = 32.5$ kΩ.
8.9 A possible solution is $R_E = 500$ Ω, $R_1 = 10.4$ kΩ, $R_2 = 15$ kΩ.
8.10 $R_S = 196$ Ω.
8.11 (a) 12 nA, (b) 67 nA; (a) 0.45 V, (b) zero biased.
8.12 $y_{ic} = 1.5$, $y_{rc} = -1.45$, $y_{fc} = -101.5$, $y_{oc} = 101.5$ mS.
     $y_{ib} = 101.5$, $y_{rb} = -0.05$, $y_{fb} = -100.1$, $y_{ob} = 0.1$ mS.

9.1 77; 539 Hz.
9.2 5.8 µF; −25; no; reduce R_L to 176 Ω; $k_{vo} = 11.5$.
9.3 $R_L = 106$ Ω, $C_C = 1.45$ µF.
9.4 0.3 pF.
9.6 −121; 2.1 kΩ; 590 Hz.
9.7 $-3.6 + j25$.
9.8 0.80 µH; $\begin{bmatrix} 5+j6 & 0 \\ 50-j66 & 2 \end{bmatrix}$
9.9 $1.19 + j10.8$ mS; 545.
9.10 $t_r = 1.5$ ns, $C_C = 0.23$ µF.

10.1 2.2×10^{-16} A^2.
10.2 3.2×10^{-18} A^2.
10.3 145 K.
10.4 (a) 1.98×10^{-17}; (b) 1.78×10^{-16}; (c) 6.1×10^{-16} V^2.
10.5 (a) 8×10^{-16} V^2; (b) 3.8×10^{-16} V^2; (c) 16×10^{-18} V^2; 1.82×10^{-11} V^2.
10.6 6.48×10^{-10} V^2.
10.7 1.38×10^{-13} W; 1.33; 99 K.
10.8 $\overline{V}_n^2 = 86.6 \times 10^{-20}$ B, $\overline{I}_n^2 = 2.65 \times 10^{-23}$ B.
10.9 180 Ω, 1.6.
10.10 1.02, 5.5 kΩ.
10.11 1.54.

11.1 36 dB, 49 ns.
11.2 1.37 mV.
11.3 23 kΩ.

11.4 $I_{B1} = 0.66\ \mu A$; $I_{C1} = 98.7\ \mu A$; $I_{B2} = 99.4\ \mu A$; $I_{C2} = 14.9\ mA$.
11.5 $g_{m1} = 4\ mS$; $r_{\pi 1} = 50\ k\Omega$; $g_{m2} = 600\ mS$; $r_{\pi 2} = 330\ \Omega$.
11.7   0.991.

12.1   1/100.
12.2   0.04%.
12.3   19.2 MΩ; −1560.
12.4   −500; 0.4%; 1.9×10^{-4}.
12.5   −333, −286.
12.6 $A = 3\ M\Omega$, $B = 2.49\ \mu S$.
12.7 $A' = 354\ k\Omega$, $r'_{in} = 177\ \Omega$.
12.8 $k_v = -301$, $r_{in} = 1.18\ k\Omega$.
12.9 $k_v = 477$, $r_{in} = 515\ \Omega$.
12.10 $k_v = 3.92$, $r_{in} = 420\ k\Omega$.
12.11 $k_v = 281$.
12.12 99×10^{-6}, 8 mV.
12.13   10 V, 0.3 mV.
12.14 $B = 0.75 \times 10^{-4}$; $A' = 8000$, BW = 15 MHz.
12.15   750 Hz; 1; 15 MHz; 750 Hz.
12.16   1 MHz and 0.03 Hz; 75 dB and 6 dB; 89 dB; 45°.
12.17   35.7 dB, 1.5 kΩ, 9.1 kΩ, 1 MHz.
12.18 $6.25\ \angle\ -135°$.
12.20   (a) Inverting, 29; (b) Inverting, 2.6; (c) Non-inverting, 3.
12.21   4; 119 Hz.

13.1 $I_{C1} = 9\ mA$; $R_2 = 1.56\ k\Omega$, $R_3 = R_4 = 4.37\ k\Omega$.
13.2   −30 μA.
13.3   −295.
13.4   15 kΩ.
13.5 $I_{E1} = 3.2\ mA$; $I_{E3} = 300\ \mu A$; $V_{bias} \simeq 3.0\ V$.
13.6   Approx. ±3 V.
13.7   At A 75 Ω; at B 10 kΩ.
13.8   −57.
13.9   31 mV.
13.10   At A, 7.33 V and 5.47 V; at B 7.93 V and 6.07 V; ±2.07 V.
13.11   0.96.

14.1   85 V; 0.68 V.
14.2 $5.5/(4.72 + j\omega)$.
14.3   210 ms.
14.4   Sudden fall to 90.4 V.
14.5   100.68 V.
14.6   1.41 ms.
14.7   Sudden fall to 90.4 V, rising to 99.93 V on time constant 14 ms.
14.8 $A_0 = 148$, $B = 1$.
14.12 $A = 1/4$, $B = 0.8$.

15.1   4.
15.3 $v_P = 2.8\ V$, $v_x = 2.2\ V$.
15.4 $v_A = v_B = 0$, $v_C = 5$; 10.
15.5   97 kΩ.
15.6   1.8 V, approx. 0.
15.7   When $v_A = 10\ V$, $v_x = 3.73\ V$.
15.9   No; 0.3 and 12, 2.9 and 12; 1%.
15.11   520 Ω.

15.13 (a) 0; (b) 690 μA.

15.14 18.

15.15 If either V_A or $V_B = 0.3$ V, $V_P = 0.9$ V, $V_X = 10$ V.
   If both V_A and $V_B = 10$ V, $V_P = 1.8$ V, $V_X = 0.3$ V.

15.18 182 mA.

15.19 2·35 mA, 29.

15.22 $C = AB + BC + CA$.

15.24 -20 V; -2 V.

15.25 Even one circuit is unsatisfactory; logic 1 to correspond to greater than -5 V.

16.1 Reject all transistors with $\beta_F < 20$.

16.2 4·7 kΩ.

16.3 Transistors will never be bottomed; 3·0 V to 10 V.

16.4 Transistors will neither be bottomed nor cut off.

16.5 $\beta_F > 5$.

16.6 $\beta_F > 21$; 10 V; 1·3 V.

16.8 No.

16.9 Yes.

16.10 $Cy = \overline{BD}$.

16.11 $P_B = P_C = \overline{\overline{B}\overline{D}} = B + \overline{D}$.

17.1 $v_{B1} = -1.5$ V, $v_{C1} = 20$ V, $v_{B2} = 0.6$ V, $v_{C2} = 0.3$ V, $i_{B1} = 0$, $i_{C1} = 0$, $i_{B2} = 194$ μA, $i_{C2} = 9.85$ mA; $\beta_F > 50$.

17.2 69 μs.

17.3 $R_2 = 15$ kΩ, $R_3 = 100$ kΩ.

17.4 *On* when $v_i > 13$ V, *off* when $v_i < 6$ V.

Index